Probiotics

Probiotics

The scientific basis

Roy Fuller

Springer-Science+Business Media, B.V.

First edition 1992

Originally published by Chapman & Hall in 1992

Typeset in 10/12 Melior by Falcon Typographic Art Ltd, Edinburgh

ISBN 978-0-412-40850-2 ISBN 978-94-011-2364-8 (eBook)
DOI 10.1007/978-94-011-2364-8

A catalogue record for this book is available from the British Library

Library of Congress Cataloging-in-Publication data available

Contents

List of contributors

Alvarez, S. Centro de Referencia para Lactobacilos (CERELA), Chacabuco 145, 4000 Tucumán, Argentina.

Barrow, P. A. AFRC Institute for Animal Health, Houghton Laboratory, Huntingdon, Cambridgeshire, PE17 2DA, UK.

Berg, R. D. Department of Microbiology and Immunology, Louisiana State University Medical School, Louisiana State University Medical Center-Shreveport, Shreveport, Louisiana 71130, USA.

Conway, P. Department of General and Marine Microbiology, Gothenburg University, Carl Scottsbergs g. 22, S–413 19, Gothenburg, Sweden.

Freter, R. Department of Microbiology and Immunology, The University of Michigan, Ann Arbor, MI 48109–0620, USA.

Fuller, R. Intestinal Microecology Consultant, Russet House, Ryeish Green, Three Mile Cross, Reading RG7 1ES, UK.

Goldin, B. R. Department of Community Health, Tufts University School of Medicine, Boston, MA 02111, USA.

Gorbach, S. L. Department of Community Health, Tufts University School of Medicine, Boston, MA, 02111, USA.

Havenaar, R. Department of Biotechnology, Institute for Biotechnology and Chemistry, TNO–Food and Nutrition, PO Box 360, 3700 AJ Zeist, The Netherlands.

Hentges, D. J. Department of Microbiology, Texas Tech University, Health Sciences Center, Lubbock, Texas 79430, USA.

Huis In 't Veld, J. H. J. Department of Biotechnology, Institute for Biotechnology and Chemistry, TNO–Food and Nutrition, PO Box 360, 3700 AJ Zeist, The Netherlands.

Jonsson, E. Department of Food Science, Swedish University of Agricultural Sciences, PO Box 7051, S-750 01 Uppsala, Sweden.

Newbold, C. J. Rowett Research Institute, Bucksburn, Aberdeen AB2 9SB, UK.

Perdigón, G. Centro de Referencia para Lactobacilos (CERELA), Chacabuco 145, 4000 Tucumán, Argentina.

Raibaud, P. Institut National de la Recherche Agronomique, Laboratoire d'Ecologie et de Physiologie du Système Digestif, Centre de Recherches de Jouy, Jouy–en–Josas 78352 Cedex France.

Rowland, I. R. BIBRA, Woodmansterne Road, Carshalton, Surrey SM5 4DS, UK.

Tannock, G. W. Department of Microbiology, University of Otago, Dunedin, New Zealand.

Ten Brink, B. Department of Biotechnology, Institute for Biotechnology and Chemistry, TNO–Food and Nutrition, PO Box 360, 3700 AJ Zeist, The Netherlands.

Wallace, R. J. Rowett Research Institute, Bucksburn, Aberdeen AB2 9SB, UK.

Preface

In recent years the gastrointestinal microflora has featured strongly in scientific, veterinary and medical research. As a result it has become obvious that the gut microflora is an essential component of the healthy animal. Not only is it involved in digestion of food, it is essential for the optimal resistance to disease.

The first part of this book records the research that has been done on the factors affecting colonization of the gut and the effect that the flora has on the host animal. The second part discusses the way in which this basic knowledge affects the choice of organism being used as a probiotic. The evidence for the involvement of the gut microflora in the health and well-being of the animal is incontrovertible, but the development of probiotics has been largely empirical, failing to capitalize on the relevant research data. The bringing together of the basic information on gut microecology and the development of probiotic preparations is long overdue. It is hoped that this exercise will result in a more scientific approach to probiotic development and the emergence of new and improved preparations for animals and man.

The authors involved are all experts in their field and I am greatly indebted to them for their contributions to the book.

R. Fuller

Abbreviations used for generic names

A.	*Aspergillus*
B.	*Bacillus*
Bact.	*Bacteroides*
Bif.	*Bifidobacterium*
C.	*Clostridium*
Cam.	*Campylobacter*
Can.	*Candida*
Cor.	*Corynebacterium*
E.	*Escherichia*
Eb.	*Enterobacter*
Ent.	*Enterococcus*
F.	*Fusobacterium*
Fib.	*Fibrobacter*
K.	*Klebsiella*
L.	*Lactobacillus*
Lact.	*Lactococcus*
N.	*Neocallimastix*
P.	*Propionibacterium*
Pr.	*Proteus*
Ps.	*Pseudomonas*
R.	*Ruminococcus*
S.	*Streptococcus*
Sac.	*Saccharomyces*
Sal.	*Salmonella*
Ser.	*Serratia*
Sh.	*Shigella*
Staph.	*Staphylococcus*
V.	*Vibrio*

Generic names have been spelt out in full the first time they are used in each chapter. Thereafter, within the chapter, they have been abbreviated as above.

Chapter One

History and development of probiotics

ROY FULLER

1.1 INTRODUCTION

The word 'probiotic' is derived from the Greek meaning 'for life' and has had several different meanings over the years. It was first used by Lilley and Stillwell in 1965 to describe substances secreted by one microorganism which stimulated the growth of another. It thus meant the exact opposite of 'antibiotic' and its etymological pedigree was beyond reproach. However, its use in this form did not persist and it was subsequently used by Sperti (1971) to describe tissue extracts which stimulated microbial growth. It was not until 1974 that Parker used it in the context in which we shall use it in this book. His definition was 'Organisms and substances which contribute to intestinal microbial balance'. This definition related probiotic use to the intestinal microflora but the inclusion of 'substances' gave it a wide connotation which would include antibiotics. In an attempt to improve the definition, Fuller (1989) redefined probiotics as 'A live microbial feed supplement which beneficially affects the host animal by improving its intestinal microbial balance'. This revised definition stressed the need for a probiotic to be viable.

1.2 HISTORY

Although the word 'probiotic' relating to feed supplements only dates from 1974, the history of live microbial feed supplements goes back thousands of years. Probably the first foods that contained living microorganisms were the fermented milks that are recorded in the Old Testament (Genesis 18 : 8). There is also evidence from wall paintings dating back to 2500 B.C. that the Sumarians were in the habit of inoculating milk to induce fermentation (see Kroger *et al.*, 1989). While the health benefits for the individual can only be inferred, the effect on prevention of spoilage would undoubtedly have a beneficial effect on the health of the community.

The consumption of fermented milks in many different forms has continued until the present day. The beneficial effects of yoghurt were put on a scientific basis at the beginning of the century. Elie Metchnikoff, working at the Pasteur Institute in Paris, played a key role in the process. He had long regarded the microflora of the lower gut as having an adverse effect on the health of the human adult. So convinced was he of this that he had advocated surgical removal of the colon. However, he was converted to a less invasive therapy by the finding that Bulgarian peasants, who ingested large amounts of soured milks, also lived to a ripe old age. He was in no doubt that the two observations were related.

It should be emphasized that Metchnikoff was concerned with sour milk rather than what we now call yoghurt, but subsequently when pure cultures became available he advocated the use of milk fermented with a single strain of lactobacillus. The early work was done with a strain called the 'Bulgarian bacillus'. This is almost certainly identical with the organism that was subsequently named *Bacillus bulgaricus* and later became *Lactobacillus bulgaricus*. The lactobacillus that is responsible for the fermentation of yoghurt is now called *L. delbrueckii* subsp. *bulgaricus* and acts in concert with *Streptococcus salivarius* subsp. *thermophilus* to produce the yoghurt we know today. It is impossible to know with any certainty which species Metchnikoff and his contempories were studying but it is likely that unintentional mixtures of lactobacilli were sometimes used (D.J. Bibel, pers. comm.). Based on this and other people's findings with regard to the health benefits of fermented milks, Metchnikoff wrote a book which, in the original French edition published in 1907, was entitled *Essais Optimistes*. In the book he discussed the philosophy, literature, religion, folklore and science of ageing. Only a small part of this discourse contained his views on the lower gut flora and the beneficial effects that fermented milk might have on it. At the end of this section of the book, in the English edition, he concludes:

> If it be true that our precocious and unhappy old age is due to poisoning of the tissues (the greater part of the poisoning coming from the large intestine inhabited by numberless microbes), it is clear that agents which arrest intestinal putrefaction must at the same time postpone and ameliorate old age. This theoretical view is confirmed by the collection of facts regarding races which live chiefly on soured milk, and amongst which great ages are common. However, in a question so important, the theory must be tested by direct observations. For this purpose the numerous infirmaries for old people should be

> taken advantage of, and systematic investigations should be made on the relation of intestinal microbes to precocious old age, and on the influence of diets which prevent intestinal putrefaction in prolonging life and maintaining the forces of the body. It can only be in the future, near or remote, that we shall obtain exact information upon what is one of the chief problems of humanity.

In spite of this guarded statement, he is always quoted as having established a relationship between consumption of fermented milks and long life. This reputation was seemingly endorsed by the English translation of his book which, much to Metchnikoff's annoyance, was given the title *The Prolongation of Life* with 'Optimistic studies' relegated to subtitle status. Whatever his intentions, Metchnikoff's work can be regarded as the birth of probiotics, i.e. microbes ingested with the aim of promoting good health. The habit was given added support by the publication in 1911 of a book by Louden Douglas called *The Bacillus of Long Life*. In it the author reiterated the connection between fermented milks and longevity. He also summarized what was known, at that time, of the bacteriology of fermented milks.

The First World War and Metchnikoff's death in 1916 saw a decline in interest, but there was a resurgence after the war and in the 1920s attention switched to *L. acidophilus* as the dietary supplement. Rettger and his colleagues at Yale had shown that the 'Bulgarian bacillus' could not survive in the human gut and they used, instead, intestinal isolates of *L. acidophilus* (Rettger and Cheplin, 1921a). Later Kopeloff (1926) also became an advocate of this approach. Clinical trials gave encouraging results, especially in return to normal of patients with chronic constipation (Rettger *et al.*, 1935).

The Second World War now intervened and it was not until the post-war era that interest in the gut flora was revived. This was due to two factors: (a) the discovery of the antibiotic growth promotion of animals, which stimulated research into its mode of action and consequently gave rise to attempts to define the composition of the gut microflora; and (b) the improved techniques for rearing germ-free animals. Germ-free animals had been known for many years; Pasteur first suggested their role in research and by 1895 Nuttal and Thierfelder had produced guinea-pigs reared in complete absence of microorganisms. However, it was not until the 1950s that the technique became readily available for research groups throughout the world – particularly in the USA, Sweden, Japan and the UK. The use of this technique confirmed Metchnikoff's basic hypothesis that the gut microflora was having an adverse effect on the host animal and that, in the case of rats reared germ-free, the lifespan was slightly increased. However, it should be remembered that the gut microflora can also make an essential positive contribution to the nutrition of the host, e.g. vitamin production and fibre digestion.

There also grew up during this period a realization that the gut microflora was involved in protection of the host animal against disease. These studies reinforced the view that not all bacteria were having adverse effects on the host and that there was in the gut a population of bacteria that were necessary for the continuing health and well-being of the animal. As Metchnikoff said:

> A reader who has little knowledge of such matters may be surprised by my recommendation to absorb large quantities of microbes, as a general belief is that microbes are harmful. This belief is erroneous. There are many useful microbes, amongst which the lactic bacilli have an honourable place.

Some of the crucial work done in the development of this concept was done by Bohnhoff's group and by Freter in the 1950s. They showed that by administering antibiotics to experimental animals *per os* they could render mice more susceptible to infection with *Salmonella typhimurium, Shigella flexneri* and *Vibrio cholerae* (Bohnhoff *et al.*, 1954; Freter, 1955, 1956). This effect was later to have important implications in agriculture when it was shown that growth-promoting antibiotics in the diet of chickens increased their susceptibility to salmonella colonization of the gut, and also in human medicine when it was found that antibiotic treatment could induce diarrhoeal conditions such as pseudomembranous colitis caused by *Clostridium difficile*. The restoration of the gut microflora by the administration of faecal suspensions has been used experimentally to treat both these conditions successfully.

One of the most convincing demonstrations of the role of the gut microflora in resistance to disease was provided by Carter and Collins in 1978. They showed that the germ-free guinea-pig was killed by 10 cells of *Sal. enteritidis* but it required 10^9 cells to kill a conventional animal with a complete gut microflora.

There is, thus, no doubt that animals have in their intestine a population of microorganisms that protects them against disease. If that is the case, why do we need probiotics? Under normal conditions there would be no need for probiotics; in the wild, the young animal rapidly acquires a protective flora from its mother and the environment. However, modern methods of perinatal care tend to limit the contact with the mother and provide unnatural foods and unnatural environmental conditions. The result is that the gut microflora is deficient in some of the normal components that are responsible for resistance to disease. Even the flora of the adult can be affected by diet, antibacterial drugs and stress (see Tannock, 1983).

The use of probiotic supplements seeks to repair these deficiencies. It is, therefore, not creating anything that would not be present under natural conditions but is merely restoring the flora to its full protective capacity.

The development of this approach has been in part stimulated by the public's misgivings about the side-effects that often follow the use of antibiotics as therapeutic agents and growth promoters. In 1969 the Swann Committee limited the use of antibiotics as growth promoters to those antibiotics that were not being used clinically. More recently the anti-additive lobby has objected to the use of antibiotics even on this scale. Some supermarkets are already selling 'antibiotic-free' meat, and in Sweden antibiotics can no longer be used as growth promoters. There is, therefore, a growing demand for an effective alternative to the antibiotic growth promoters and probiotics could fill the gap.

1.3 COMPOSITION OF PROBIOTIC PREPARATIONS

The original work done by Metchnikoff and his colleagues using the 'Bulgarian bacillus' was almost certainly done with an organism closely related to the lactobacillus starter of yoghurt (*L. delbreuckii* subsp. *bulgaricus*) and to this day lactobacilli have remained the most commonly used probiotic organisms. The use of *L. acidophilus* was stimulated by the desire to ensure that the organism used would survive in the gut. *L. acidophilus* was used because it was thought to be the dominant lactobacillus in the intestine. Later work showed this not be true and a wide range of lactobacilli have subsequently been used. Currently available probiotic preparations contain *L. delbreuckii* subsp. *bulgaricus*, *L. acidophilus*, *L. casei*, *L. fermentum*. *L. plantarum*, *L. brevis*, *L. cellobiosus*, *L. lactis* and *L. reuteri*.

The use of bifidobacteria stemmed from the work of Tissier (1905) who showed that these organisms were the dominant bacteria in the gut of breast-fed infants and implied that they were not to be found in the formula-fed infant. In spite of a great deal of evidence to show that bifidobacteria do occur in formula-fed babies (Hall *et al.*, 1990; Benno and Mitsuoka, 1986) the belief in a relationship between bifidobacteria and the superior disease resistance of breast-fed babies has persisted. While there is not a direct correlation between numbers of bifidobacteria and resistance, the finding that breast-fed and formula-fed babies have different species of bifidobacteria may be of significance (Neut *et al.*, 1980). The bifidobacteria currently being used as probiotics are; *Bifidobacterium adolescentis*, *Bif. animalis*, *Bif. bifidum*, *Bif. infantis*, *Bif. longum* and *Bif. thermophilum*.

The first use of streptococci as probiotics was in the form of soured milk and yoghurt. The yoghurt starter *S. salivarius* subsp. *thermophilus* is still a common probiotic organism. The move towards intestinal isolates resulted in the use of *Enterococcus faecium*. The two most commonly used strains (M74 and SF68) are both human isolates but are used mainly in animal preparations. Other species of streptococci used as probiotics are *S. lactis* (renamed *Lactococcus lactis* subsp. *lactis*), *S. cremoris* (*Lact. lactis* subsp. *cremoris*), *S. diacetilactis* (*Lact. lactis* subsp. *lactis*) and *S. intermedius* (*S. anginosus*). Probiotics also contain bacteria belonging to the genera *Leuconostoc*, *Pediococcus*, *Propionibacterium* and *Bacillus*. Yeasts (*Saccharomyces cerevisiae* and *Candida pintolopesii*) and moulds (*Aspergillus niger* and *A. oryzae*) are also used but mainly in animal products. The currently available probiotics have been extensively reviewed by Lloyd-Evans (1989).

Probiotics may contain one or several (up to nine) strains of microorganisms and may be presented to the animal in the form of powders (loose or in capsules), tablets, granules or pastes. They may be administered by direct insertion into the mouth or by inclusion in the food or water. Experiments have also been done with the administration to newly hatched chicks by spraying into the surrounding atmosphere.

In spite of the careful selection of strains, it seems unlikely that it would be possible to establish permanently the probiotic organism in the intestinal tract and multiple dosing is essential if the full probiotic effect is to be obtained.

An effective probiotic is required to operate under a variety of different environmental conditions and to survive in many different forms. It should, therefore have the following characteristics:

1. It should be capable of being prepared as a viable product on an industrial scale.
2. It should remain stable and viable for long periods under storage and field conditions.
3. It should have the ability to survive (not necessarily grow) in the intestine.
4. It must produce a beneficial effect in the host animal.

Given this sort of preparation what sort of effects can we expect to obtain? The beneficial claims made for probiotic supplementation are numerous and include:

1. *Improved growth rate of farm animals*. This is generally regarded as being due to suppression of a subclinical infection with a growth-depressing microorganism; such a mechanism has been described for antibiotic growth promotion. For example, in the chicken there is evidence that *Ent. hirae* is the organism responsible for restricting its growth (Fuller *et al*., 1979).

2. *Improved utilization of food.* This may be achieved by increased efficiency of existing digestive processes or by promoting the digestion of previously indigestible substances. For example *Ent. faecium* supplementation of chickens can allow them to digest cellulose (Kumprecht *et al.*, 1984).
3. *Improved milk production by dairy cows.* This is an effect obtained particularly with fungal supplements such as *Sac. cerevisiae* or *A. oryzae* (see Williams and Newbold, 1990). The effect may be manifested in increased yield and in increased fat content. This may be a consequence of the observed effect on rumen metabolism.
4. *Increased egg production.* There have been reports of increases in the numbers of eggs produced and individual egg weights, but these are seldom significant.
5. *Improved health.* This includes resistance to infectious diseases either by direct antagonism or by stimulation of immunity. Claims have also been made in relation to heart disease and cancer.

The evidence for these claims is not always good and inevitably the evidence that does exist is often variable and inconsistent (see Chapter 14).

The underlying basis for most of the effects claimed for probiotics is an effect on the gut microflora – either its composition or its metabolic activity. Fundamental to the understanding of the probiotic effect is the knowledge of how the specific microorganisms used can affect other microorganisms such as those which comprise the indigenous gut microflora or invading pathogens.

As well as presenting the basic knowledge of the factors affecting gut colonization and the way in which bacteria in the gut interact, the book will seek to show how this information can be used to select probiotic strains and explain probiotic effects.

The object of the exercise is to produce a solid scientific basis for the understanding of the probiotic concept and in so doing suggest ways in which probiotic preparations may be improved and provide a foundation on which further relevant research in this rapidly growing area may be built.

REFERENCES

Benno, Y. and Mitsuoka, T. (1986) Development of the intestinal microflora in humans and animals. *Bifidobacteria Microflora*, **5**, 13–25.

Bohnhoff, N., Drake, B. L. and Muller, C. P. (1954) Effect of streptomycin on susceptibility of the intestinal tract to experimental salmonella infection. *Proc. Soc. Exp. Biol. Med.*, **86**, 132–7.

Collins, F.M. and Carter, P.B. (1978) Growth of salmonellae in orally infected germfree mice. *Infect. Immun.*, **21**, 41–7.

Douglas, L.M. (1911) *The Bacillus of Long Life*. Putnam, New York.
Freter, R. (1955) The fatal enteric cholera infection in the guinea pig achieved by inhibition of normal enteric flora. *J. Infection Dis.*, **97**, 57–65.
Freter, R. (1956) Experimental enteric shigella and vibrio infection in mice and guinea pigs. *J. Exp. Med.*, **104**, 411–18
Fuller, R. (1989) Probiotics in man and animals. *J. Appl. Bacteriol.*, **66**, 365–78.
Fuller, R., Coates, M.E. and Harrison, G.F. (1979) The influence of specific bacteria and a filterable agent on the growth of gnotobiotic chicks. *J. Appl. Bacteriol.*, **46**, 335–42.
Hall, M.A., Cole, C.B., Smith, S.L. *et al*, (1990) Factors influencing the presence of faecal lactobacilli in early infancy. *Arch. Dis. Childhood.*, **65**, 185–8.
Kopeloff, N. (1926) *Lactobacillus acidophilus*, Williams & Williams, Baltimore.
Kroger, M., Kurmann, J.A. and Rasic, J.L. (1989) Fermented milks – past and present. *Food Technol.*, **43**, 92–9.
Kumprecht, I., Gasnarek, Z., Zobac, P. *et al.* (1984) The effect of a single and continuous administration of *Streptococcus faecium* M-74 germs on the growth of broilers and on metabolic processes in broiler. *Zivoc. Vyr.*, **29**, 949–55.
Lilly, D.M. and Stillwell, R.H. (1965) Probiotics: growth promoting factors produced by microorganisms. *Science*, **147** 747–8.
Lloyd-Evans, L.P.M. (1989) *Probiotics*, PJB Publications Ltd, Richmond.
Metchnikoff, E. (1907) *The Prolongation of Life*, Heinemann, London.
Neut, C., Romond, C. and Beerens, H.A. (1980) A contribution to the study of the distribution of *Bifidobacterium* species in the faecal flora of breast-fed and bottle-fed babies. *Reprod. Nutr. Dev.*, **20** 1679–84.
Nuttal, G.H.F. and Thierfelder, H. (1895) Thierisches Leben ohne Bakterien im Verdauungskanal *Z. Physiol, Chem.*, **21**, 109–21.
Parker, R.B. (1974) Probiotics, the other half of the antibiotic story. *Anim. Nutr. Health.*, **29**, 4–8.
Rettger, L.F. and Cheplin, H.A. (1921) *A Treatise on the Transformation of the Intestinal Flora with Special Reference to the Implantation of* Bacillus acidophilus, Yale University Press, New Haven, Connecticut.
Rettger, L.F., Levy, M.N., Weinstein, L. and Weiss, J.E. (1935) Lactobacillus acidophilus *and its Therapeutic Application*, Yale University Press, New Haven, Connecticut.
Sperti, G.S. (1971) *Probiotics*, Avi Publishing Co., West Point, Connecticut.
Tannock, G.W. (1983) Effect of dietary and environmental stress on the gastrointestinal microbiota, in *Human Intestinal Microflora in Health and Disease*, (ed D.J. Hentges) Academic Press, New York, pp. 517–39.
Tissier, H. (1905) Cited by Mitsuoka, T. (1984), Taxonomy and ecology of bifidobacteria. *Bifidobacteria Microflora*, **3**, 11–28.
Williams, P.E.V. and Newbold, C.J. (1990) Rumen probiosis: The effects of novel microorganisms on rumen fermentation and ruminant productivity, in *Recent Advances in Animal Nutrition* (eds W. Haresign and D.J.A. Cole), London, Butterworth, pp. 211–27.

Chapter Two

Bacterial interactions in the gut

PIERRE RAIBAUD

2.1 INTRODUCTION

Microbial interactions represent the main force which contributes to the homeostasis of the bacterial flora in the gut. This flora forms an ecosystem with its host, comprising: (a) biotic components, e.g. indigenous and transient microbes, and gastrointestinal epithelial cells which delimit the biotope; (b) abiotic components of dietary origin, namely those that have not been digested during their course through the small intestine: and (c) endogenous components, coming from saliva, gastric, pancreatic, hepatic and intestinal secretions or excretions, including enzymes, hormones, mucus, bile salts, urea, immunoglobulins, peptides and probably several other unknown components. All these components interact and the result of such interactions is compatible with the healthy survival of the host. When gastrointestinal disorders arise, the ecosystem becomes destabilized. This emphasizes the importance of maintaining microbial interactions in a way that maintains the stability of the ecosystem and optimal health for the host.

Given the range of biotic and abiotic components of the ecosystem, it is not surprising that bacterial interactions reach a very high degree of complexity. Mechanisms involved in these interactions seem to be mostly multifactorial, rendering their study much more difficult. In this chapter, we shall try to give a critical overview of the main bacterial interactions occurring in the gut.

2.2 METHODS FOR STUDYING BACTERIAL INTERACTIONS

A method that is still frequently used is the study of interactions between one or more bacterial strains in a broth culture. The results obtained are then extrapolated directly to the animal gut. Using this sort of information it has been concluded that lactobacilli act within the gut by destroying enterobacteria, due to their ability to produce D, L or DL lactic acid from fermentable sugars. This is perfectly true when lactic acid accumulates in a culture tube, but gives rise to erroneous conclusions in the gut for two reasons. Firstly, sugars that are present in commercial diets – and which can be fermented by lactobacilli – reach the lower part of the gut at a very low concentration, because they are digested in the upper part of the gut and absorbed. Secondly, even when lactic acid is produced in the caecum or colon, it is readily absorbed across the intestinal mucosa, so that it does not accumulate to the same extent in the gut as it does in a culture tube.

Continuous flow culture (CFC), as described by Freter *et al.* (1983) and Veilleux and Rowlands (1981), has also been used. This method is much more valuable, because the transit of abiotic components can be made comparable to that which occurs in the gut. However, the media used in such CFC do not fully simulate the intestinal environment. In addition, the continuous or discontinuous rhythm of the intake of endogenous components in the gut, which probably play a striking role on the functions of the intestinal flora, cannot be easily reproduced. Neverthless, this method has been proved to be valuable for studying some bacterial interactions.

A third method is the use of gnotobiotic animals. Various germ-free (axenic) animals are now available. The less expensive are mice, then rats. Piglets, chicken, Japanese quail, lambs, dogs are more expensive; for example, it costs US$1000 for one piglet or one dog. These germ-free animals are reared in plastic, Trexler-type isolators. The use of a rapid transfer system instead of the classic transfer through a lock is quicker but nevertheless effective.

When germ-free animals are inoculated with one or more known bacterial strains, they are called gnotobiotic animals. Bacterial interactions can be studied either by sampling faeces from live animals or by sampling the different regions of the intestine after the sacrifice of animals. The effects of the dietary regimen or of the host on these bacterial interactions can also be assessed by using various groups of gnotobiotic animals harbouring the same bacterial strains, or by using different hosts harbouring the same bacteria and receiving the same diets. Physiological parameters can also be modified in gnotobiotic animals by using surgical procedures, e.g. ligation of the pancreatic and hepatic ducts, caecectomy, construction of a blind loop. However, it has to be kept in mind that such gnotobiotic animals are experimental models, allowing the analysis of the mechanisms of bacterial interactions with a known but limited number of biotic components. The findings obtained with these animals have to be validated using conventional animals in order to be sure that such mechanisms can also act in animals harbouring a more complex flora. Gnotobiotic animals can be used most effectively by dosing germ-free animals with the flora of a conventional animal and rearing them in an isolator. The bacterial interactions in the donor and in the recipient animal can then be compared. If these interactions are similar in recipient and donor, attempts can be made to determine the minimum number of strains necessary for the observed effect using gnotobiotic animals. If successful, one can assume that the mechanisms involved in the gnotobiotic recipient represent those involved in the conventional donor. Germ-free animals can be dosed with the intestinal flora belonging either to the same species or to another species. Such polyassociated models have been used by several authors in the form of (a) polyassociated mice harbouring a porcine flora (Ducluzeau *et al.*, 1978a), a human flora (Raibaud *et al.*, 1980; Hazenberg *et al.*, 1981), or a hamster flora (Su *et al.*, 1987); (b) polyassociated rats harbouring a human flora (Mallett *et al.*, 1987) or a chicken flora (Nugon-Baudon *et al.*, 1988); and (c) polyassociated chickens harbouring a rat flora (Nugon-Baudon *et al.*, 1988). These animals give a better – or, in the case of human beings, a unique – opportunity to study bacterial interactions in an *in vivo* ecosystem. However, the results obtained with the polyassociated models have to be submitted to a critical evaluation, because the endogenous components and the physiological parameters can differ from one animal species to another.

2.3 MAIN TYPES OF BACTERIAL INTERACTIONS IN THE GUT

In the gut, the nature of the bacterial interactions can be antagonistic or synergistic. They can affect either the population level of a given strain or the metabolic activity of that strain. In addition, genetic transfers can occur between strains within the gut. The host and the diet can modulate the expression of the bacterial interactions.

2.3.1 Bacterial interactions affecting population levels of various bacterial strains in the gut

When a bacterial strain is ingested by a conventional host, it can either become established at a high or low population level or be eliminated. Figure 2.1 illustrates a useful procedure by which these outcomes can be distinguished. Cells of the bacterial strains to be inoculated to a conventional host are mixed with spores of a thermophilic *Bacillus* which pass through the gut without losing viability and which fail to germinate and multiply (Contrepois and Gouet, 1969). The mixed inoculum is given *per os* to the host, and spores of transit marker and cells of the strain under study are counted. The former are selectively counted in the faeces at their optimum growth temperature (60°C), which precludes the growth of all other intestinal bacteria. The latter have to be counted on selective media. Comparison of both counts allows a clear distinction between the establishment of the inoculated strain and its elimination. When no bacterial interaction has occurred, the strain remains at a stable level for the whole of the experimental period. When the strain is submitted to bacterial interactions, it can either be eliminated or remain at a low population level. When bacterial interactions lead to an elimination of the strain, cells usually enter into bacteriostasis, and are eliminated at the same rate as the transit marker or a little more quickly. Occasionally cells are submitted to bactericidal interactions, characterized by the fact that only very few bacterial cells can be counted in the first hours post-inoculation. This emphasizes the diversity of the expression of bacterial interactions.

2.3.2 Bacterial antagonisms detected in conventional animals

Ducluzeau *et al.* (1970) have pointed out that several bacterial strains given *per os* to conventional mice were eliminated as rapidly as the transit marker or even more rapidly. Few strains remained in the subdominant flora. These authors have called this phenomenon the 'barrier effect', which had already been termed 'bacterial antagonism' by Freter (1956), 'bacterial interference' by Dubos (1963), 'colonization

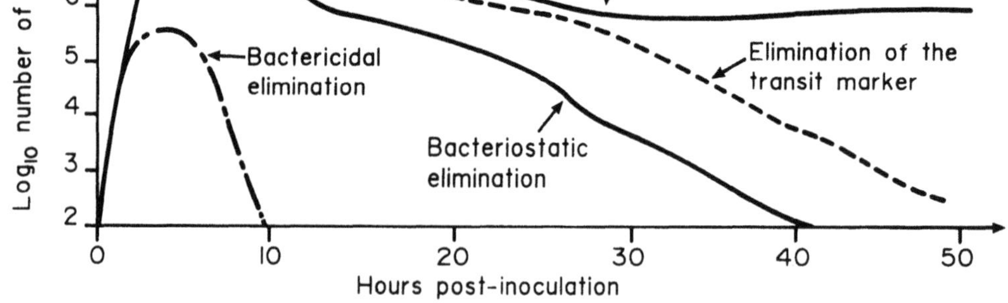

Figure 2.1 Procedure to follow the outcome of a bacterial strain once orally administered to germ-free, gnotobiotic, polyassociated or conventional hosts. Spores of a thermophilic *Bacillus* are used as transit marker. They are admixed with the bacterial inoculum.

resistance' by van der Waaij *et al.* (1971) and 'competitive exclusion' by Lloyd *et al.* (1977). These interactions represent the main function devoted to the indigenous predominant bacteria. They protect the host against the proliferation of alimentary bacteria and potentially pathogenic bacteria, which can produce toxins when they grow in the hindgut, as is the case with many strains of toxigenic *Clostridium*, enterobacteria or *Campylobacter*. Some toxigenic strains can escape the barrier effect, when they adhere to the mucosal cell wall of the human stomach (e.g. *Helicobacter*), or of the small intestine (e.g. enterobacteria).

Bacterial antagonisms become established gradually from birth until the steady state of the gut flora is achieved. The first bacterial species that colonize the gut belong to a small number of species. Then, new species enter the ecosystem, and can repress the earlier colonizers. At the steady state several species and biotypes are present in the predominant flora and exert strong antagonisms against the exogenous bacteria.

Barrier effects are extremely efficient for preventing intestinal infections but their expression can be modified by external factors. Antibiotic therapy can destroy them, leading to dramatic intestinal disorders. Several – but not all – antibiotics have been involved in the onset of diseases such as pseudomembranous colitis or diarrhoea, due to the growth of *Clostridium difficile* in the hindgut as a consequence of the destruction of the barrier effect. Diet can also lead to an elimination of indigenous predominant bacteria. Thus, dietary lactose (48% in the diet) given to conventional rats leads to the elimination of bacteria responsible for the production of hyodeoxycholic acid, a bacterial metabolite of β-muricholic acid (Andrieux *et al.*, 1989).

Barrier effects are also responsible for maintaining some bacterial strains at a low population level. When the homeostasis of the gut is achieved, a rather small number of bacterial species are predominant, reaching population levels of between 5×10^8 g^{-1} to 1×10^{11} g^{-1} of faeces or large intestine content, whereas others reach a population level below 5×10^8 g^{-1}. Often, there are large fluctuations in the subdominant population levels between individuals and in the same individual during a life time. When the population level of a given strain is below 10^7 g^{-1}, this strain does not play any role in the ecosystem due to the continuous renewal of the intestinal content, providing it remains at all times below 10^7 g^{-1}. For instance, populations of less than 10^7 g^{-1} of potentially toxigenic strains such as *Clostridium* or enterobacteria are well tolerated by the host. This situation characterizes a healthy carriage. However, when environmental factors (diet, stress, etc.) disturb the ecosystem, these strains can develop and express their toxigenicity. Strains present in numbers comprising between 10^7 and 5×10^8 g^{-1}

may or may not express their toxigenicity, depending on their enzymatic activity.

2.3.3 Bacterial antagonisms detected in gnotobiotic animals

Several experiments have been performed using gnotobiotic animals harbouring one or more bacterial strains. The aim of such studies was to obtain information on the mechanism of bacterial antagonisms in various environmental conditions.

From *in vitro* experiments, it has been claimed that bacterial antagonisms can be due to the production, by the inhibitory strain, of antibacterial substances acting on the target strain. An early work of Ducluzeau and Raibaud (1968) has shown that a strain of *Enterococcus faecalis* and a strain of *Lactobacillus acidophilus* coexisted without any interaction in the gut of gnotobiotic mice, although the *Enterococcus* lysed the *Lactobacillus* in an agar culture medium. Duval-Iflah *et al.* (1981) showed that a colicinogenic strain of *Escherichia coli* was inhibited by a non-colicinogenic strain. Ikari *et al.* (1969) found that the antagonism that can be demonstrated between colicinogenic and colicin-sensitive strains was not expressed in the gut of gnotobiotic mice. This means that a general assessment cannot be drawn from *in vitro* experiments. Moreover, bacteriocins have never been detected in the intestinal contents of conventional or gnotobiotic animals. It is of interest to note that not one bacterial strain producing antibiotic has been isolated from the predominant intestinal flora. Nevertheless, Ducluzeau *et al.* (1976, 1978b) showed that a strain of *Bacillus licheniformis* was able to produce a bacitracin-like diffusible antibiotic which totally prevented the growth of bacitracin-sensitive strains, such as *Clostridium perfringens*, in gnotobiotic mice (Table 2.1). However, the antibiotic production disappeared when *C. perfringens* was established first or when other strains, such as a *Lactobacillus* strain, were added to the ecosystem, allowing *C. perfringens* to become established. An additional result was that thermoresistant spores of *B. licheniformis* were no longer produced when *C. perfringens* became established, suggesting that the mechanism of interaction could be the inhibition of *B. licheniformis* sporulation resulting from the growth of both *C. perfringens* or the *Lactobacillus*.

This result illustrates the fact that bacterial antagonisms could appear in gnotobiotic animals harbouring a simplified flora, whereas they do not work in a more complex ecosystem.

As previously mentioned, the role of lactic acid-producing bacteria in the *in vivo* bacterial antagonisms has often been claimed. However, experimental models in general fail to confirm this statement. Ducluzeau and Raibaud (1973) showed that a *L. murinus* strain did not exert any antagonism against a *Shigella flexneri* strain or *E. coli* in adult gnotobiotic mice. Suckling mice, delivered from mothers harbouring *L. murinus* and enterobacteria, were sacrificed at different ages before weaning and compared with suckling mice of the same age delivered from mothers harbouring either *L. murinus* or enterobacteria. Enterobacteria were detected at a slightly lower concentration in the stomach of diassociated one-week-old mice than in that of monoassociated mice. No significant difference was found in the hindgut of mice diassociated with *E. coli*, and only a slight difference was observed in mice diassociated with *Sh. flexneri*. No interaction could be detected at two weeks of age, when young mice began to eat solid food. Using gnotobiotic adult

Table 2.1 Antagonism between a *Bacillus licheniformis* strain producing a bacitracin-like antibiotic and a *Clostridium perfringens* strain in the gut of gnotobiotic mice. Effect of a *Lactobacillus murinus* strain. (After Ducluzeau *et al.*, 1976, 1978b.)

Order of strain inoculation	Inhibition of *C. perfringens* in the gut	Sporulation of *B. licheniformis* in the gut
B. licheniformis then *C. perfringens*	+	+
C. perfringens then *B. licheniformis*	–	–
B. licheniformis then *L. murinus* then *C. perfringens*	–	–

mice associated with three strains of *Lactobacillus* representative of the species present in conventional mice, Itoh and Freter (1989) have shown that counts of a strain of *E. coli* in the stomach were 50 to 100 times lower in the triassociated mice than in mice monoassociated with *E. coli*, whereas counts in the hindgut were less than 10 times lower in the former mice than in the latter. According to these authors, the lactobacilli affected the population of *E. coli* only in the stomach and, possibly, in the small intestine, but not in the hindgut, although the pH of the stomach was not different in the different groups of gnotobiotic mice. In another experiment (Ducluzeau *et al.*, 1971) axenic adult mice were diassociated with a strain of *L. murinus* and a strain of *Bacteroides putredinis*. Both strains coexisted without interaction. A 20% solution of lactose was then given as drinking water. It was observed that the *Bact. putredinis* count dropped to below 10^2 g^{-1} of faeces, whereas lactose had no effect on *Bact. putredinis* in mice monoassociated with this strain (Table 2.2). The lactic acid concentration was increased seven-fold in the caecal content, although the pH was not modified following the lactose intake. The diassociated mice were further associated with a strain of *E. coli* (Table 2.2). In the absence of lactose, *Bact. putredinis* was suppressed, whereas there was no interaction between *E. coli* and *L. murinus*. When lactose was given to the mice, no interaction occurred between *L. murinus* and *E. coli*, whereas *Bact. putredinis* was strongly suppressed. Thus, one can assume that lactose played a role in the interaction between *L. murinus* and *Bact. putredinis*, even though the pH was not decreased, whereas *E. coli* was not affected in the same conditions. Interaction between *Lactobacillus* and *E. coli* was also investigated by Hudault *et al.* (1976). A strain of *L. casei*, from a pharmaceutical preparation, and a human strain of *E. coli* were given to a germ-free human baby and to gnotobiotic mice. *L. casei* was established first in both hosts and reached a high population level in their gut, then it was rapidly eliminated in both cases when *E. coli* became established. The faecal pH of the baby monoassociated with *L. casei* fell from 6.8 to 5.0. Nevertheless, *E. coli* became established within 48 hours post-inoculation. All these results are in agreement with those obtained by Cole *et al.* (1984) using conventional rats and Ratcliffe *et al.* (1986) using conventional piglets. They observed that coliform counts were not significantly decreased in the hindgut of neonatal conventional rats fed a rat strain of *Lactobacillus* or in that of early weaned piglets fed yoghurt or milk fermented with a porcine strain of *Lactobacillus*. However, these authors observed a significant but transient decrease in coliform counts in the stomach and, in some cases, in the upper part of the small intestine which was correlated with a decrease in the pH of the stomach content. Ratcliffe *et al.* (1986) concluded from their results that the decrease in coliform counts in the upper gut was due to the low pH produced by the ingested lactic acid. Nevertheless, the results obtained both with conventional and gnotobiotic animals are not always in agreement with the statement that lactobacilli compete with coliforms in the hindgut. This means that other mechanisms may be involved in the production of low counts of coliform in the hindgut of adult men and animals.

Ducluzeau (1967) and Ducluzeau and Raibaud (1974) have also shown interactions between strains of *E. coli* and *Staphylococcus pyogenes* or *Sh. flexneri* in diassociated mice, the mechanism of which has not been elucidated. It could be assumed that it plays a role in human newborns or suckling piglets, where *E. coli* is very often the predominant bacterial species, whereas *Staph. pyogenes* is absent or at a very low level.

Bacterial interactions between strains belonging to the same species can also be demonstrated. Duval-Iflah *et al.* (1981) have observed several antagonisms between isogenic strains of *E. coli* associated with gnotobiotic mice and differing only by their plasmid content or by a chromosomal mutation. In general, the plasmid-free strain inhibited the establishment of plasmid-bearing strains. In addition, these authors have demonstrated that the outcome of the interactions depended on the order of inoculation of the strains in gnotobiotic mice. The plasmid-free strain no longer exerted a barrier effect against the derivative plasmid-bearing strain when the former was inoculated one week after the latter. They have also demonstrated that maintaining the plasmid-free strain in monoassociation with gnotobiotic mice resulted in the 'adaptation' of the bacteria to their host, since the 'adapted' strain

Table 2.2 Effect of dietary lactose on the interaction between a strain of *Lactobacillus murinus*, a strain of *Bacteroides putredinis* and a strain of *Escherichia coli* in the gut of gnotobiotic mice. (After Ducluzeau *et al.*, 1971.)

Bacterial strains	Diet	
	Without lactose	With lactose
L. murinus alone	2×10^9	1×10^{10}
B. putredinis alone	8×10^8	1×10^9
E. coli alone	8×10^9	8×10^9
L. murinus	2×10^9	1×10^{10}
+*B. putredinis*	8×10^8	$< 10^2$
L. murinus	2×10^9	1×10^{10}
+*B. putredinis*	2×10^5	1×10^3
+*E. coli*	8×10^9	8×10^9

Figures are mean counts per gram of faeces.

exerted the same barrier effect regardless of whether it was introduced before or after its plasmid-bearing derivative. This ecological advantage disappeared when the adapted strain was cultured in broth. Ultrastructural differences in cell morphology were observed between the adapted and non-adapted strain. Can such interactions occur in conventional conditions? A positive answer was given by Duval-Iflah *et al.* (1982). Twenty-two healthy human newborns were inoculated *per os* with the plasmid-free strain of *E. coli* used in the previous experiments within 4 hours of delivery. The strain became established at a high level in 19 out of 22 of the 3–6-day-old inoculated neonates. Counts of coliforms bearing antibiotic resistance were low in all neonates. In non-inoculated neonates, this situation was spontaneously observed in 71%, whereas 29% harboured antibiotic-resistant coliforms in the predominant faecal flora. This suggests that intraspecific bacterial interactions could play a role in the balance between antibiotic-susceptible and antibiotic-resistant enterobacteria.

Intraspecific interactions have also been observed in gnotobiotic mice between transconjugant *E. coli* strains (i.e. strains harbouring characters of both parental strains) and their parental strains by Duval-Iflah *et al.* (1981) using two plasmid-bearing strains and by Ducluzeau and Galinha (1967) using Hfr and F-strains. Sarra *et al.* (1989) have also demonstrated intraspecific interactions between two isogenic strains of *L. reuteri*. Corthier and Muller (1988) have succeeded in isolating a non-toxigenic clone from a highly toxigenic clone of *C. difficile*. The non-toxigenic clone exerted a strong antagonism against the toxigenic one in gnotobiotic mice. Unfortunately the mechanisms of all these interactions observed in gnotobiotic animals have not been elucidated.

Several other studies have attempted to elucidate the mechanism of the microbial interactions in the gut. For instance, Ducluzeau *et al.* (1974) examined whether facultative anaerobes can antagonize strict anaerobes in the gut of conventional suckling mice, where the facultative anaerobes were predominant. For that purpose, germ-free mice were associated with six strictly anaerobic strains belonging to the predominant flora of conventional mice. These strains colonized the gut of adult mice, with no interactions. Apart from one strain, present in low count within the first few days of life, none of the other five strains could be found in the gut of their suckling mice before the tenth day of life, whereas a concentration as high as 10^8 g^{-1} facultative anaerobes can be observed in conventional suckling mice of the same age. This suggests that the kinetics of establishment of the anaerobic strains in the young mice depended upon abiotic compounds and not upon bacterial interaction due to facultative anaerobes.

2.3.4 Attempts to elucidate the mechanisms of bacterial antagonisms using polyassociated and gnotobiotic models

Few attempts have been made to elucidate the mechanism of bacterial antagonisms occurring in the gut. Freter *et al.* (1983) have demonstrated for the first time that bacterial antagonisms regulating the faecal *E. coli* count in conventional mice could be reproduced in gnotobiotic mice harbouring a collection of 95 strict anaerobes taken from the predominant gut flora of conventional mice. Moreover, they have demonstrated that such antagonisms could be reproduced in CFC. However, attempts to produce a simpler inhibitory flora in gnotobiotic mice were unsuccessful. Yurdusev *et al.* (1986), working on the antagonism exerted by the faecal flora of conventional piglets against *C. perfringens* have succeeded in isolating three strains which together exerted the same antagonism as did the whole faecal flora from a conventional piglet when inoculated into germ-free mice. Further (Yurdusev *et al.*, 1989) they have described an antagonistic effect exerted by only two strains (*Bact. thetaiotaomicron* and *Fusobacterium necrogenes*) against *C. perfringens*. They also have succeeded in reproducing the phenomenon by using concentrated faecal suspensions incubated *in vitro*, but the activity was lost in CFC. They concluded that the inhibitory strains produced unknown substances having a bacteriostatic effect against the target strain. These substances did not accumulate in the gut and were only produced when the two strains were together associated with the germ-free mice. Attempts to characterize these substances are still under investigation. By submitting concentrated faecal supernatants to HPLC separation, two peaks have been found which exert an *in vitro* inhibitory effect on the target strain. They were only found in faeces of mice harbouring the two inhibitory strains. One can assume that they are continuously produced *in vivo*, probably from endogenous compounds, and enter the cells of the target strain, leading to its elimination. Obviously this experimental model has to be validated, by looking for such substances in the faeces of the conventional host.

Another important finding of the previous work was that the expression of the antagonism was host and diet dependent. Table 2.3 shows that the antagonism exerted by the two inhibitory strains against the same target strain of *C. perfringens* was significantly less efficient in diassociated rats than in diassociated mice. It was also less efficient in diassociated mice fed a milk diet than in those fed a commercial diet, although there were no differences in the counts of the two inhibitory strains in rats or mice. Similar results were obtained in mice triassociated with the previous two strains and a *Clostridium* strain (Yurdusev *et al.*, 1986). In that case the commercial diet was either irradiated or autoclaved. Counts of *C. perfringens* were significantly higher in the gut of mice fed the autoclaved diet than in the gut of those fed the irradiated diet, showing that the antagonism was less efficient when mice were fed an autoclaved diet. It is most likely that the autoclaved diet carried an inhibitory substance which counteracted the antagonism. However, the actual mechanism by which the antagonism was modulated has not been elucidated. It is interesting to note that such an antagonism seems to be very specific. Indeed, in the experiment of Yurdusev *et al.* (1986) the barrier effect exerted against four strains of *C. perfringens* in gnotobiotic mice led to the elimination of only two strains. The other two strains became established for the duration of the experiment at a population level of about 10^5 g^{-1} of faeces.

Other attempts have been made to simplify barrier effects exerted by the whole flora of a given host associated with germ-free animals. For instance, the antagonistic effect exerted against *Salmonella typhimurium* by the chicken flora has been reproduced in germ-free mice as well as in germ-free chickens associated with the whole flora of conventional chickens by Hudault *et al.* (1985). However, various combinations of known strains were unable to reproduce the antagonism both in gnotobiotic mice and chickens. It is likely that the efficient strains have not been cultured on the media used in these experiments. Impey *et al.* (1982) showed that colonization of the caeca of newly hatched conventional chickens by *Sal. typhimurium*

Table 2.3 Effects of host and diet on the expression of the antagonism exerted by two strains (*Bacteroides thetaiotaomicron* and *Fusobacterium necrogenes*) against a strain of *Clostridium perfringens*. (After Yurdusev et *al.*, 1989.)

Host	Diet	Mean counts per gram of faeces		
		B. thetaiotaomicron	*F. necrogenes*	*C. perfringens*
Mice	Commercial	4×10^{10}	9×10^{9}	$< 10^{2}$
Mice	Milk replacer	4×10^{10}	8×10^{9}	2×10^{8}
Rats	Commercial	3×10^{10}	2×10^{9}	4×10^{6}

Counts were made 14 days post-inoculation of *C. perfringens*. Those of the inhibitory strains were not significantly different, whereas those of *C. perfringens* are significantly different ($p < 0.001$).

was prevented by oral administration of a mixture of 48 bacterial strains. However, it has not been determined whether the mixture was also efficient in gnotobiotic conditions. A commercial preparation has been successfully used to protect newly hatched chicks against *Salmonella* colonization (Schneitz *et al.*, 1990). However, the bacterial composition of the preparation has not been described.

2.3.5 Bacterial synergism

Bacterial interactions affecting the population levels can also involve synergistic forces. The bacterial antagonism exerted by *Bact. thetaiotaomicron* and *F. necrogenes* against *C. perfringens* (Yurdusev *et al.*, 1989) is an example of such synergism. In mice diassociated with *C. perfringens* and only one of the inhibitory strains, *C. perfringens* counts were only slightly reduced compared with those obtained in mice monoassociated with *C. perfringens*. *E. coli* often exerts a synergistic effect in bacterial antagonism. As shown by Ducluzeau *et al.* (1977) and Hudault *et al.* (1982), *E. coli* was acting with clostridial strains to exert an antagonistic effect against *Sh. flexneri* and *C. perfringens*. The mechanism of such effects has not been elucidated.

Bacterial synergism can also lead to an increase in pathological disorders. Dabard *et al.* (1979) have shown that when a strain of *C. perfringens* or of *C. tertium* or both was associated with *C. difficile*, diarrhoea and death of young gnotobiotic hares occurred more rapidly than in hares monoassociated with *C. difficile*. In these experiments, neither *C. perfringens* nor *C. tertium* shared with *C. difficile* in the pathology observed in gnotobiotic hares, but each of the former strains allowed the latter to reach a population level up to 10^8 g^{-1} of faeces and to kill animals within 2 to 11 days post-inoculation instead of 12 to 17 days, when *C. difficile* was inoculated alone. Conversely, *C. difficile* allowed *C. perfringens* to become established in gnotobiotic hares. The mechanism of another bacterial synergistic effect has been elucidated by Dubos *et al.* (1985). A strain of *C. perenne* was unable to become established in germ-free mice fed a semi-synthetic diet containing casein unless a strain of *C. difficile* became established. The effect of the latter strain was to hydrolyse β-aspartic ϵ-lysin, a dipeptide probably produced by heating casein. It accumulated within the gut of germ-free mice because it was not hydrolysed by the host peptidases. The dipeptide was not inhibitory by itself against *C. perenne* but became inhibitory when chelated with copper in the gut. It is likely that similar bacterial synergisms can be responsible for the sequential establishment of bacteria in the gut of conventional animals. 'Pioneer' strains first implanted could clear the gut of inhibitory compounds, thus allowing other bacterial strains to become established. 'Pioneer' strains can also lower the redox potential of the gut to a level compatible with the development of strictly anaerobic oxygen-susceptible bacteria. Growth promoters produced by 'pioneer' strains can also be responsible for the establishment of subsequent strains. This was the case in the experimental model described by Ducluzeau *et al.* (1986) where some strains of *Clostridium* were able to release diaminopimelic acid (DAP) in the gut of gnotobiotic mice, thus permitting the development of a DAP-dependant mutant of *E. coli*. Fonty *et al.* (1983) have also shown that the development of cellulolytic strains, like *Bact. succinogenes*, *Ruminococcus flavefaciens* or *R. albus*, was dependent upon the prior establishment of other unknown bacterial strains in gnotobiotic lambs.

2.3.6 Bacterial interactions affecting metabolic activities

Bacterial interactions not only affect bacterial population levels in the gut, they also affect the bacterial metabolism resulting in increased or decreased activity. ω-Muricholic acid is a bacterial metabolite derived by epimerization from β-muricholic acid, a bile acid produced by the rat liver, mainly in the form of a taurine conjugate. Sacquet *et al.* (1979) have shown that a *Clostridium* sp. was unable to produce ω-muricholic acid unless a strain of *Bact. vulgatus* was already established. In that case, the synergistic mechanism was the hydrolysis of taurine conjugate, rendering it accessible to the enzyme system of the *Clostridium*. Volatile fatty acid (VFA) production can also depend on bacterial interactions. Using gnotobiotic chickens fed a 4% lactose diet, Szylit *et al.* (1988) have shown that the caecal concentration of total VFAs (mainly propionate) was increased seven times in chickens diassociated with a strain of *L. acidophilus* and a strain of *Veillonella alcalescens*, when compared with chickens monoassociated with each strain. This increase was due to the hydrolysis of lactose into lactic acid by the *Lactobacillus* and a concomittant production of propionic acid from lactic acid by the *Veillonella* in the chicken caeca. It is also known that methanogenic bacteria fail to produce methane in the rumen unless other bacteria provide hydrogen and carbon precursors.

Another type of bacterial interaction affecting metabolic activity is illustrated by the experiments of Corthier *et al.* (1985). Using gnotobiotic mice these authors have demonstrated that the production of cytotoxin in the gut by *C. difficile* was deeply affected by the concomittant presence of other bacteria such as some *Bifidobacterium* strains, although *C. difficile* counts were not significantly decreased. As a consequence of such an interaction, all gnotobiotic mice harbouring *Bifidobacterium* and *C. difficile* survived whereas all those dosed with *C. difficile* alone died. By quantifying cytotoxin, these authors showed that its titre in the gut was 1000 times lower in mice diassociated with *Bifidobacterium* and *C. difficile* than in mice monoassociated with *C. difficile* alone. Duval-Iflah *et al.* (1983) have observed similar results in gnotobiotic piglets dosed with a non-toxigenic strain of *E. coli*, which protected the piglets against a toxigenic *E. coli* strain, without bringing about a major diminution in the population level of the toxigenic strain. An interesting result was also obtained by Castex *et al.* (1990), who showed that *Saccharomyces boulardii* protected gnotobiotic mice against *C. difficile*, provided that the yeast was continuously given at a high dose to the mice. In that case, no enterotoxin could be detected in the gut, and the cytotoxin titre was lowered by 100- to 1000-fold, whereas both toxins were present in a high amount in dying mice challenged with *C. difficile* alone. Here also, *C. difficile* counts were not significantly lowered in the caecum of protected gnotobiotic mice (Table 2.4).

It is likely that synergism between various bacterial strains is also required for fully stimulating the host intestinal immune system. As shown by Moreau *et al.* (1982), plasma cells synthetizing IgA were ten times less abundant in adult germ-free mice than in conventional ones. Adult germ-free mice, dosed with the whole gut flora of suckling 7–23-day-old conventional mice, were partially stimulated, whereas those dosed with the whole gut flora of a 25-day-old conventional mouse were fully stimulated. This suggests that a synergistic bacterial effect is involved in this very important function of the gut flora.

Table 2.4 Modulation of *Clostridium difficile* toxin production by *Saccharomyces boulardii* continuously administered *per os* to gnotobiotic mice. (After Castex *et al.*, 1990.)

Sac. boulardii in drinking water	*C. difficile* mean counts in caecum	Toxins in caecum (mean lg g^{-1})	
		Cytotoxin	Enterotoxin
+	2×10^8	0.08	< 0.1
–	4×10^8	2.0	1.0

All mice without *Sac. boulardii* were dying; all those with *Sac boulardii* were healthy. *Sac. boulardii* counts were around 10^9 g^{-1} of caecum. Counts of *C. perfringens* were not significantly different. The amounts of toxins were significantly different.

2.3.7 Bacterial interactions resulting in transfer of bacterial genes

The expression of bacterial genes in the gut can be modified by environmental factors. But transfers of genes can also occur between bacterial species within the gut. Duval-Iflah *et al.* (1981) have demonstrated such transfers by conjugation between strains belonging to the same species. It has also been demonstrated that transfers can occur by conjugation between bacteria belonging to different species. Indeed, plasmid transfer from *L. reuteri* to *Ent. faecalis* has been demonstrated in the gut of gnotobiotic mice (Morelli *et al.*, 1988). Duval-Iflah *et al.* (1980) have shown that a *Serratia liquefaciens* strain harbouring a conjugative R plasmid responsible for resistance to 14 antibiotics can transfer this plasmid to an *E. coli* strain in gnotobiotic mice dosed with a whole human flora, even though the *Serratia* strain was only transient in the gut, and without any antibiotic selection pressure. The result of such a mating was that the *E. coli* strain gained resistance to 14 antibiotics instead of 4. This points out the danger of ingesting bacteria harbouring conjugative R plasmids.

2.4 CONCLUSIONS

Bacterial interactions involve multiple mechanisms that are poorly understood. Such mechanisms are involved either in the size of subdominant bacterial populations or in the metabolic activities of predominant populations. They can differ according to the host species, to the individual host, and to the location of bacterial populations in the gut. Diet and perhaps other environmental factors, such as stress, can modify their expression. It is very unlikely that mechanisms demonstrated under *in vitro* conditions have the same activity in the gut. Some bacterial interactions involve synergism between more than one predominant indigenous microrganism. Consequently, great efforts are required to simplify complex floras playing a role in a given interaction, because the number of combinations to be examined dramatically increases when the number of involved strains increases. For example, if 2 of 50 strains are involved in a synergic interaction, the number of combinations possible is as much as 1225; it is 19 600 if 3 of 50 strains are involved. Great effort and expense are also required to examine such interactions because gnotobiotic animals seem to be the best experimental model for assessing both the interactions and the effects of environmental factors on them. Thus, considerable work remains to be done if we are to fully understand and further control bacterial interactions in the gut.

REFERENCES

Andrieux, C., Gadelle, D., Leprince, C. and Sacquet, E. (1989) Effects of some poorly digestible carbohydrates on bile acid bacterial transformation in the rat. *Brit. J. Nutr.*, **62**, 103–19.

Castex, F., Corthier, G., Jouvert, S. *et al.* (1990) Prevention of *Clostridium difficile* induced experimental pseudomembranous colitis by *Saccharomyces boulardii*: a scanning electron microscopic and microbiological study. *J. Gen. Microbiol.*, **136**, 1085–9.

Cole, C. B. and Fuller, R. (1984) A note on the effect of host specific fermented milk on the coliform population of the neonatal rat gut. *J. Appl. Bacteriol.*, **56**, 495–8.

Contrepois, M. and Gouet, P. (1969) Utilisation d'une technique microbiologique pour la mesure de la vitesse de transit des microparticules dans le tractus digestif des ruminants. *C.R. Acad. Sci. (Paris)*, **268**, 1757–9.

Corthier, G., Dubos, F. and Raibaud, P. (1985) Modulation of cytotoxin production by *Clostridium difficile* in the intestinal tracts of gnotobiotic mice inoculated with various human intestinal bacteria. *Appl. Environ. Microbiol.*, **49**, 250–2.

Corthier, G. and Muller, M. C. (1988) Emergence in gnotobiotic mice of nontoxinogenic clones of *Clostridium difficile* from a toxinogenic one. *Infect. Immun.*, **56**, 1500–4.

Dabard, J., Dubos, F., Martinet, L. and Ducluzeau, R. (1979) Experimental reproduction of neonatal diarrhea in young gnotobiotic hares simultaneously associated with *Clostridium difficile* and other *Clostridium* strains. *Infect. Immun.*, **24**, 7–11.

Dubos, R.J. (1963) Staphylococci and infection immunity. *Amer. J. Dis. Child.*, **105**, 643–5.

Dubos, F., Pelissier, J.P., Andrieux, C. *et al.* (1985) Inhibitory effect of a copper-dipeptide complex on the establishment of a *Clostridium perenne* strain in the intestinal tract of gnotobiotic mice. *Appl. Environ. Microbiol.*, **50**, 1258–61.

Ducluzeau, R. and Galinha, A. (1967) Recombinaison *in vivo* entre une souche Hfr et une souche F-d'*Escherichia coli* K12, ensemencées dans le tube digestif de souris axéniques. *C.R. Acad. Sci. (Paris)*, **264**, 177–80.

Ducluzeau, R. (1967) Equilibre entre deux souches bactériennes, *Escherichia coli* et *Staphylococcus pyogenes* selon les conditions de leur ensemencement dans le tube digestif de souris axéniques. *C.R. Acad. Sci. (Paris)*, **265**, 1657–60.

Ducluzeau, R. and Raibaud, P. (1968) Lyse '*in vitro*' de certains types de *Lactobacillus* par une souche de *Streptococcus liquefaciens*: Incidence du phénomène sur l'équilibre entre ces bactéries dans le tube digestif de souris 'gnotoxéniques'. *C.R. Acad. Sci. (Paris)*, **266**, 1332–4.

Ducluzeau, R., Bellier, M. and Raibaud, P. (1970) Transit digestif de divers inoculums bactériens introduits 'per os' chez des souris axéniques et 'holoxéniques' (conventionnelles): effet antagoniste de la microflore du tractus gastro-intestinal. *Zbl. Bakt. I. Orig.*, **213**, 533–48.

Ducluzeau, R., Dubos, F. and Raibaud, P. (1971) Effet antagoniste d'une souche de *Lactobacillus* sur une souche de *Ristella* sp. dans le tube digestif de souris 'gnotoxéniques' absorbant du lactose. *Ann. Inst. Pasteur*, **121**, 777–94.

Ducluzeau, R. and Raibaud, P. (1973) Effet d'une souche de *Lactobacillus* sur la cinétique d'établissement de *Shigella flexneri* et d'*Escherichia coli* dans le tube digestif de souriceaux 'gnotoxéniques'. Rôle de l'immunisation des mères. *Can. J. Microbiol.*, **19**, 1021–30.

Ducluzeau, R., Raibaud, P. and Ladiré, M. (1974) Cinétique de l'établissement d'une microflore anaérobie stricte dans le tube digestif de souriceaux nés de mères gnotoxéniques entre la naissance et le sevrage. Effet du régime alimentaire des mères. *Ann. Microbiol. (Inst. Pasteur)*, **125A**, 57–68.

Ducluzeau, R. and Raibaud, P. (1974) Interaction between *Escherichia coli* and *Shigella flexneri* in the digestive tract of 'gnotobiotic' mice. *Infect. Immun.*, **9**, 730–3.

Ducluzeau, R., Dubos, F., Raibaud, P. and Abrams, G.D. (1976) Inhibition of *Clostridium perfringens* by an antibiotic substance produced by *Bacillus licheniformis* in the digestive tract of gnotobiotic mice: effect of other bacteria from the digestive tract. *Antimicrob. Agents Chemother.*, **9**, 20–5.

Ducluzeau, R., Ladiré, M., Callut, C. *et al.* (1977) Antagonistic effect of extremely oxygen-sensitive clostridia from the microflora of conventional mice and of *Escherichia coli* against *Shigella flexneri* in the digestive tract of gnotobiotic mice. *Infect. Immun.*, **17**, 415–24.

Ducluzeau, R., Rapine, P., Courvalin, C. and Raibaud, P. (1978a) Transfert de la flore microbienne fécale de porcelets et de porcs adultes holoxéniques à des souris adultes et des porcelets axéniques: effet de l'animal hôte et du régime alimentaire sur le faciès microbien du tube digestif des divers animaux. *Ann. Microbiol. (Inst. Pasteur)*, **129B**, 597–612.

Ducluzeau, R., Dubos, F., Raibaud, P. and Abrams, G.D. (1978b), Production of an antibiotic substance by *Bacillus licheniformis* within the digestive tract of gnotobiotic mice. *Antimicrob. Agents Chemother.*, **13**, 97–109.

Ducluzeau, R., Ladiré, M. and Raibaud, P. (1986), Implantation d'un mutant de *Escherichia coli* exigeant en acide diaminopimélique dans le tube digestif de souris gnotoxéniques. *Ann. Microbiol. (Inst. Pasteur)*, **137A**, 79–87.

Duval-Iflah, Y., Raibaud, P., Tancrède, C. and Rousseau, M. (1980) R-plasmid transfer from *Serratia liquefaciens* to *Escherichia coli* in vitro and in vivo in the digestive tract of gnotobiotic mice associated with human fecal flora. *Infect. Immun.*, **28**, 981–90.

Duval-Iflah, Y., Raibaud, P. and Rousseau, M. (1981), Antagonisms among isogenic strains of *Escherichia coli* in the digestive tracts of gnotobiotic mice. *Infect. Immun.*, **34**, 957–69.

Duval-Iflah, Y., Ouriet, M.F., Moreau, C. *et al.* (1982) Implantation précoce d'une souche de *Escherichia coli* dans l'intestin de nouveau-nés humains: effet de barrière vis-à-vis de souches de *E. coli* antibiorésistantes. *Ann. Microbiol. (Inst. Pasteur)*, **133A**, 393–408.

Duval-Iflah, Y., Chappuis, J.P., Ducluzeau, R. and Raibaud, P. (1983) Intraspecific interactions between *Escherichia coli* strains in human newborns and in gnotobiotic mice and piglets. *Prog. Food Nutr. Sci.*, **7**, 107–16.

Freter, R. (1956) Experimental enteric shigella and vibrio infection in mice and guinea pigs. *J. Exp. Med.*, **104**, 411–18.

Freter, R., Brickner, H., Botney, M. *et al.* (1983) Mechanisms that control bacterial populations in continuous-flow culture models of mouse large intestinal flora. *Infect. Immun.*, **39**, 676–85.

Fonty, G., Gouet, P., Jouany, J.P. and Senaud, J. (1983) Ecological factors determining establishment of cellulolytic bacteria and protozoa in the rumen of meroxenic lambs. *J. Gen. Microbiol.*, **129**, 213–23.

Hazenberg, M.P., Bakker, M. and Verschoor-Burggraaf, A. (1981) Effects of the human intestinal flora on germ-free mice. *J. Appl. Bacteriol.*, **50**, 95–106.

Hudault, S., Ducluzeau, R., Dubos, F., *et al.* (1976) Elimination du tube digestif d'un enfant 'gnotoxénique' d'une souche de *Lactobacillus casei* issue d'une préparation commerciale: démonstration chez des souris 'gnotoxéniques' du rôle antagoniste d'une souche de *Escherichia coli* d'origine humaine. *Ann. Microbiol. (Inst. Pasteur)*, **127B**, 75–82.

Hudault, S., Raibaud, P., Ducluzeau, R. and Bridonneau, C. (1982) Effet antagoniste à l'égard de *Clostridium perfringens* exercé par des souches de *Clostridium* isolées de la microflore de souris holoxéniques dans le tube digestif de souris gnotoxéniques. *Ann. Microbiol. (Inst. Pasteur)*, **133A**, 443–59.

Hudault, S., Bewa, H., Bridonneau, C. and Raibaud, P. (1985) Efficiency of various bacterial suspensions derived from cecal floras of conventional chickens in reducing the population level of *Salmonella typhimurium* in gnotobiotic mice and chicken intestines. *Can. J. Microbiol.*, **31**, 832–8.

Ikari, N.S., Kenton, D.M. and Young, V.M. (1969) Interaction in the germfree mouse intestine of colicinogenic and colicin sensitive microorganisms. *Proc. Soc. Exp. Biol. Med.*, **130**, 1280–4.

Impey, C.S., Mead, G.C. and George, S. (1982) Competitive exclusion of salmonellas from the chick caecum using a defined mixture of bacterial isolates from the caecal microflora of an adult bird. *J. Hyg. (Camb.)*, **89**, 479–90.

Itoh, K. and Freter, R. (1989) Control of *Escherichia coli* populations by a combination of indigenous clostridia and lactobacilli in gnotobiotic mice and continuous-flow cultures. *Infect. Immun.*, **57**, 559–65.

Lloyd, A.B., Cumming, R.B. and Kent, R.D. (1977) Prevention of *Salmonella typhimurium* infection in poultry by pretreatment of chickens and poults with intestinal extracts. *Austral. Vet. J.*, **53**, 82–7.

Mallett, A.K., Bearne, C.A., Rowland, I.R. *et al.* (1987) The use of rats associated with a human faecal flora as a model for studying the effects of diet on the human gut microflora. *J. Appl. Bacteriol.*, **63**, 39–45.

Moreau, M.C., Raibaud, P. and Muller, M.C. (1982) Relation entre le développement du système immunitaire intestinal à IgA et l'établissement de la flore microbienne dans le tube digestif du souriceau holoxénique. *Ann. Immunol. (Inst. Pasteur)*, **133D**, 29–39.

Morelli, L., Sarra, P.G. and Bottazzi, V. (1988) *In vivo* transfer of p AM ß1 from *Lactobacillus reuteri* to *Enterococcus faecalis*. *J. Appl. Bacteriol.*, **65**, 371–5.

Nugon-Baudon, L., Szylit, O. and Raibaud, P. (1988) Production of toxic glucosinolate derivatives from rapeseed meal by intestinal microflora of rat and chicken. *J. Sci. Food Agric.*, **43**, 299–308.

Ratcliffe, B., Cole, C.B., Fuller, R. and Newport, M.J. (1986) The effect of yoghurt and milk fermented with a porcine intestinal strain of *Lactobacillus reuteri* on the performance and gastrointestinal flora of pigs weaned at two days of age. *Food Microbiol.*, **3**, 203–11.

Raibaud, P., Ducluzeau, R., Dubos, F. **et al.** (1980) Implantation of bacteria from the digestive tract of man and various animals into gnotobiotic mice. *Amer. J. Clin. Nutr.*, **33**, 2440–7.

Sacquet, E., Raibaud, P., Mejean, C. *et al.*, (1979) Bacterial formation of ω-muricholic acid in rats. *Appl. Environ. Microbiol.*, **37**, 1127–31.

Sarra, P.G., Vescovo, M. and Bottazzi, V. (1989) Antagonism and adhesion among isogenic strains of *Lactobacillus reuteri* in the caecum of gnotobiotic mice. *Microbiologica*, **12**, 69–74.

Schneitz, C., Hakkinen, M., Nuotio, L., *et al.* (1990) Droplet application for protecting chicks against salmonella colonisation by competitive exclusion. *Vet. Rec.*, 126–510.

Su, W.J., Waechter, M.J., Bourlioux, P. *et al.*, (1987) Role of volatile fatty acids in colonization resistance to *Clostridium difficile* in gnotobiotic mice. *Infect. Immun.*, **55**, 1686–91.

Szylit, O., Dabard, J., Durand, M. *et al.*, (1988) Production of volatile fatty acids as a result of bacterial interactions in the cecum of gnotobiotic rats and chickens fed a lactose-containing diet. *Reprod. Nutr. Develop.*, **28**, 1455–64.

van der Waaij, D., Berghuis de Vries, J.M. and Lekkerkerk van der Wees, J.E.C. (1971) Colonization resistance of the digestive tract in conventional and antibiotic-treated mice. *J. Hyg. (Camb.)*, **69**, 405–11

Veilleux, B.G. and Rowland, I.R. (1981) Simulation of the rat intestinal ecosystem using a two-stage continuous culture system. *J. Gen. Microbiol.*, **123**, 103–15.

Yurdusev, N., Nicolas, J.L., Ladiré, M. *et al.* (1986) Antagonistic effect exerted by three strictly anaerobic strains against various strains of *Clostridium perfringens* in gnotobiotic rodent intestines. *Can. J. Microbiol.*, **33**, 226–31.

Yurdusev, N., Ladiré, M., Ducluzeau, R. and Raibaud, P. (1989) Antagonism exerted by an association of a *Bacteroides thetaiotaomicron* strain and a *Fusobacterium necrogenes* strain against *Clostridium perfringens* in gnotobiotic mice and in fecal suspensions incubated in vitro. *Infect. Immun.*, **57**, 724–31.

Chapter Three

Metabolic interactions in the gut

IAN R. ROWLAND

3.1 INTRODUCTION

One of the most important ways in which a probiotic organism may exert a beneficial effect on its host is to modify metabolic processes, particularly those occurring in the gut. Such a beneficial effect could be achieved in theory by a variety of mechanisms:

1. By suppressing reactions which result in the generation of toxic or carcinogenic metabolites.
2. By stimulating enzymic reactions involved in detoxification of potentially toxic substances, either ingested or formed endogenously.
3. By stimulating mammalian enzymes involved in the digestion of complex nutrients, or where such enzymes are absent (due to genetics or disease) providing a bacterial source of these enzymes.
4. By synthesizing vitamins and other essential nutrients not provided in sufficient quantities in the diet.

This review discusses enzymic reactions in the gut, both mammalian and bacterial in origin, the role of those enzymes in nutrition and toxic events in man and animals and the potential consequences of modification of enzyme activity by probiotic organisms. Before embarking on this discussion a number of general points need to be made. It is important to consider the gut microflora, not as a separate entity segregated from the host within the gut walls, but as an ecosystem which has a complex and intimate interaction with its host and in particular with that host's own metabolic processes. Thus metabolic reactions occurring in the gut can have consequences both locally – for example, on the gut mucosa – or systemically. An example of a remote consequence of a metabolite produced in the gut is the generation by gut bacteria of amines and phenols from amino acids, which can have effects on the central nervous system, the vascular system and potentially on tumorigenesis in various organs of the body (Bakke, 1969; Boutwell and Bosch, 1959; Drasar and Hill, 1974). There are examples, discussed below (Mizutani and Mitsuoka, 1979), of changes in gut microflora influencing cancer in various tissues. The mechanism involved is unknown but is quite likely to involve metabolism. Finally, it should be pointed out that in this review I shall include studies not just on probiotic organisms themselves, but also on food supplements, particularly indigestible sugars, designed to promote the growth of beneficial organisms in the colon and thus have a similar effect to the administration of 'true' probiotics.

3.2 MAMMALIAN INTESTINAL METABOLISM

3.2.1 Digestive enzymes

In humans, deficiencies can occur in most of the enzymes involved in the digestion of the major macromolecular components of the diet. The causes of enzyme deficiencies can be congenital or due to a variety of clinical conditions causing decreased synthesis or output. Although deficiencies in metabolism of proteins and lipids can occur, the most well-known clinical manifestation of an enzyme deficiency is lactose intolerance. The latter is due to low jejunal lactase concentration which is insufficient to hydrolyse a physiological lactose load. The disaccharide is therefore unable to be absorbed and is fermented by colonic bacteria to short-chain fatty acids, hydrogen and carbon dioxide causing diarrhoea and abdominal pain (Cook, 1973; Dahlqvist, 1984; Gray, 1984). Lactose intolerance can be caused, rarely, by congenital lactase deficiency (Dahlqvist and Asp, 1975) and can occur in patients with diseases or conditions in which the small intestinal mucosa is damaged, for example, Crohn's disease, ileostomy and bacterial overgrowth in the jejunum/duodenum (Jones *et al.*, 1977). Hypolactasia is, however, the normal state for the majority of the world's population, particularly non-caucasians. In children, lactase activity in the jejunum is highest in the neonate and then in most populations declines during the first few years of life to about 5–10% of its initial level (Cook, 1968). It seems that the high lactase concentrations that persist in adults in Europe and North America are the result of a mutation in a regulatory gene for the enzyme.

(a) Influence of probiotics on lactose intolerance

The gastrointestinal symptoms evident after the consumption of lactose by lactase-deficient subjects do not appear to occur after ingestion of yoghurt and other fermented milks (Savaiano *et al.*, 1984). The mechanisms involved are not entirely clear. In some cases, it may be attributable to a lower lactose concentration in the milk product as a consequence of its degradation during the fermentation process (Kolars *et al.*, 1984; Hitchins and McDonough, 1989). However, many yoghurts have comparable lactose concentrations to unfermented milks yet still do not induce deleterious effects in lactase-deficient individuals. There is some evidence from studies in humans and rats to suggest that the ingestion of lactic acid bacteria possessing β-galactosidase activity that are present in fermented dairy products raises intestinal lactase activity thus assisting lactose digestion *in vivo* (Kolars *et al.*, 1984; Goodenough and Kleyn, 1976). However, mixtures of heat-treated yoghurt and lactose were well tolerated by lactase-deficient subjects despite the virtual absence of lactase activity in the yoghurt (Savaiano *et al.*, 1984).

3.2.2 Mucosal oxidative enzymes

In addition to the enzymes involved in digestion of macromolecular food components, the mammalian gut also possesses a variety of enzymes, located in the mucosa, capable of oxidizing endogenous and dietary compounds including drugs and other foreign substances. The enzymes involved include cytochromes P450-mediated mixed function oxidases (e.g. arylhydrocarbon hydroxylase, dimethylnitrosamine *N*-demethylase) and phase II conjugating enzymes such as glutathione-*S*-transferase and UDP-glucuronyltransferase, analogous to those found in the liver (Combes, 1989). The oxidation reactions can lead to the conversion, to active genotoxins and carcinogens, of a variety of foreign compounds including benzo(a)pyrene, dimethylhydrazine, aflatoxin B1 and nitrosamines (Combes, 1989; Thies and Siegers, 1989).

The phase I biotransformation enzymes (mixed function oxidases) are found throughout the small and large intestines of both rat and man. The enzyme activity, which comprises about 5–20% of the specific monooxygenase activity in the liver, declines from the proximal to distal ends of the small and large intestines (Thies and Siegers, 1989). Mixed function oxidase activities in the intestine are susceptible to modulation by inducers and inhibitors, such as phenobarbitone and 3-methylcholanthrene, known to influence the corresponding enzymes in the liver. In addition, dietary components such as fat, protein, vitamins, flavones, indoles and various phenolics, can substantially modify intestinal enzyme activities (Combes, 1989).

(a) Influence of probiotics on gut mucosal enzymes

It seems likely that gut bacterial activities can result in metabolites capable of modulating the activity of mucosal oxidative enzymes in the light of similar effects observed on hepatic drug metabolizing enzymes.

For example, the plant glycoside rutin is hydrolysed in the gut by bacterial β-glycoside enzymes releasing the aglycone quercetin (MacDonald *et al.*, 1984; Lasker *et al.*, 1984; see 3.3.1(a) below). Quercetin has been shown to act as an inducer of carcinogen activating enzymes in the liver (Alldrick *et al.*, 1989) and would presumably have similar effects on gut mucosal enzymes. It seems reasonable to surmise that modification of the gut microflora and its activities by ingestion of a probiotic could, in turn, modify mucosal activity, although this has not yet been studied.

3.3 GUT BACTERIAL METABOLISM

When studying metabolism by the gut microflora, information on the metabolic reactions performed by individual species which comprise the flora is often of little use. The colonic flora of man and other mammals is a highly complex ecosystem comprising over 400 different species. The abilities of these species to metabolize nutrients and foreign compounds are not known in any detail and even if they were, it would be difficult to predict whether a reaction that occurs *in vitro* with a pure culture would proceed when the organism was surrounded by, and interacting with, the other members of its ecosystem *in vivo*. Some limited information on metabolic reactions *in vitro* and *in vivo* has, however, been obtained using gnotobiotic animals monoassociated with specific gut microorganisms (Cole *et al.*, 1985). A better approach to understanding the role of the gut microflora in nutrition and toxic events in man is to treat the flora as an entity, or indeed as another 'organ' of the body, ignoring its multi-organism composition. This approach has been used with some success by a number of researchers using a variety of functional assays on faeces or gut contents (Goldin and Gorbach, 1976; Midtvedt, 1989; Rowland, 1989).

The metabolic activities of the gut microflora can have wide-ranging implications for the health of the host, resulting in both detrimental and beneficial effects (Rowland *et al.*, 1985; Rowland and Walker, 1983) summarized in Table 3.1. By modifying these reactions, a probiotic may influence the generation of potentially toxic compounds in the gut. Some examples of microbial metabolism leading to changes in toxicity of ingested or endogenous substances are described below, together with evidence for the effect of probiotics on the reactions.

The region of the gut which harbours the greatest number of bacteria is the colon; indeed, other areas of the human gastrointestinal tract are very sparsely populated (Drasar, 1988). This does not mean, however, that only poorly absorbed ingested chemicals encounter the colonic flora. Substances, and their metabolites, may partition across the intestinal wall from the blood or may reach the colon after excretion in the bile (see section 3.3.1(b)). Thus there is ample opportunity for a wide variety of materials in the diet to encounter and be metabolized by the colonic microflora.

Table 3.1 Health implications of gut flora metabolism.

1. Production of toxic, carcinogenic or mutagenic metabolites from substances derived from diet or produced endogenously.
2. Detoxification of dietary toxicants.
3. Enterohepatic circulation of drugs, food additives and steroids.
4. Alteration in susceptibility of host to tumour induction.

3.3.1 Bacterial xenobiotic-metabolizing enzymes

(a) β-Glycosidases

A wide variety of glycosidic compounds are produced by plants (Brown, 1988). Their presence, often in large quantities, in edible fruits and vegetables and in beverages, such as tea and wine, derived from plant extracts results in significant human intake of around 1 g per day. Most of these ingested glycosides are harmless, *per se* but are poorly absorbed and pass into the colon. There they are subjected to the action of β-glycosidases associated with the resident microbial flora, which cleaves the sugar moiety releasing aglycones, which may be toxic, carcinogenic or mutagenic (Table 3.2). For example, amygdalin, a glycoside found in several drupes and pomes including apricot, apple, plum and almond, is hydrolysed in the gut to mandelonitrile which subsequently decomposes to cyanide. Germ-free rats, unlike their conventional flora counterparts, do not exhibit cyanide toxicity after amydalin exposure, indicating the critical role played by the gut microflora in the metabolic reaction (reviewed by Rowland and Walker, 1983). Similarly, cycasin, when fed to conventional flora rats, is hydrolysed by bacterial β-glucosidase in the colon releasing the aglycone methylazoxymethanol which induces colon tumours. No such tumours are found in germ-free rats fed the glycoside (Laqueur and Spatz, 1968).

In some cases microbial metabolism of glycosides can have more subtle toxicological consequences. Rutin, a glycoside found in many vegetables and in tea and wine, is hydrolysed to the flavonol quercetin, which is mutagenic (Elliger *et al.*, 1984; MacDonald *et al.*, 1984). Quercetin has also been shown, however, to affect the activity of some drug-metabolizing enzymes in the liver (Lasker *et al.*, 1984) and, furthermore, to increase the capacity of liver preparations to convert to active mutagenic species, a number of substances formed during the heating of meat and fish, so-called pyrolysate products, or cooked food mutagens (Alldrick *et al.*, 1989). Clearly, the generation of free quercetin in the gut by bacterial hydrolysis of ingested rutin could have similar consequences on hepatic metabolism. Mallett *et al.*, (1989a) confirmed that this could occur by demonstrating that hepatic preparations derived from rats fed rutin (1% w/w) in the diet, metabolized pyrolysate products to mutagens more effectively than preparations from rats fed a rutin-free diet.

Table 3.2 Some plant glycosides hydrolysed to toxic derivatives by the intestinal microflora.

Glycoside	Aglycone	Toxicity
Cycasin	Methylazoxymethanol	Carcinogen
Amygdalin	Mandelonitrile	Cyanide toxicity
Franguloside	Emodin	Mutagen
Rutin	Quercetin	Mutagen
Stevioside	Steviol	Mutagen

Based on data from Brown (1988); Laqueur and Spatz (1968).

(b) β-Glucuronidase

Many xenobiotics and endogenously produced compounds such as steroids are metabolized in the liver and conjugated to glucuronic acid before being excreted *via* the bile into the small intestine (Smith, 1966). In the colon, bacterial β-glucuronidase can hydrolyse the glucuronide linkage releasing the parent compound, or its hepatic metabolite. This may result in an enterohepatic circulation as the compound is reabsorbed, returning to the liver where it can be subjected to further metabolism and conjugation and then once more secreted in the bile. Enterohepatic circulation results in increased retention of xenobiotics and steroids in the body with a concomitant potentiation of their pharmacological and physiological effects. In some cases, ingested carcinogens may be conjugated and secreted into the intestine. β-Glucuronidase action on such conjugates can release the parent carcinogen in the colon. For example, 1,2-dimethylhydrazine (DMH) is converted to a carcinogen methylazoxymethanol (MAM) in the liver, then conjugated to glucuronic acid and excreted in the bile. In the colon, MAM is released after hydrolysis of the conjugate. As might be expected, germ-free rats have fewer colon tumours after DMH treatment than do conventional animals (Reddy *et al.*, 1974). The carcinogen benzo(a)pyrene, a contaminant of the human diet, undergoes a similar sequence of reactions to that of DMH. Gut bacteria have been shown to release reactive metabolites which covalently bind to DNA and are genotoxic (Renwick and Drasar, 1976; Chipman *et al.*, 1983).

(c) Nitrate reductase

Nitrate is a widely distributed environmental contaminant and is present in both the diet and drinking water of man. Although of very low toxicity, it is readily converted by the bacterial population of the mammalian gastrointestinal tract to its more reactive and toxic reduction product, nitrite. The latter reacts with oxyhaemoglobin in the blood yielding methaemoglobin. The reaction is of clinical importance since the methaemoglobin is unable to transport oxygen to the tissues (de Bruin, 1976). Infants are at risk, particularly in areas where drinking water nitrate is very high, since they drink more than adults on a bodyweight basis and have a less efficient methaemoglobin reductase system, which restores the function of the respiratory pigment (Green and Tannenbaum, 1982). The activity of bacterial nitrate reductase activity has been shown to be important in determining the degree of methaemoglobinaemia after exposure to nitrate (Wise *et al.*, 1982).

A more important consequence of nitrate reduction is the reaction of nitrite with nitrogenous compounds in the body to produce *N*-nitroso compounds, many of which are highly carcinogenic (Rowland, 1988). Although acidic conditions are necessary for the reaction to occur chemically, it can be catalysed at neutral pH by bacteria (Klubes *et al.*, 1971; Leach *et al.*, 1985). Using germ-free rats, we have demonstrated the importance of the gut microflora for the nitrosation reaction (Massey *et al.*, 1988) and recently have confirmed that the reaction can occur in man by measuring *N*-nitroso compounds in faeces (Rowland *et al.* 1991). Ingestion of nitrate was found to be important for the formation of *N*-nitroso compounds in the gut.

(d) Azoreductase

A number of dyes used in food, cosmetics and for textiles and leather are based on azo compounds. These compounds are reduced to varying degrees in the gut by the intestinal flora to produce, ultimately, amines. The reduction products are often toxic – for example, the food dye Brown FK, which, after bacterial reduction, causes vacuolar myopathy in cardiac and skeletal muscle in rats given high doses (Grasso and Goldberg, 1968). Workers exposed to Direct Black 38, a dye used in the leather and textile industry, have an elevated risk of bladder cancer, which has been attributed to the reduction of the dye by the gut microflora to benzidine, a known human bladder carcinogen (Powell *et al.*, 1979; Cerniglia *et al.*, 1982). Some of the azo dyes used in the food industry possess mutagenic activity, which is detectable only after metabolism by gut microflora preparations (Venitt and Bushell, 1976; Combes and Haveland-Smith, 1982).

(e) Nitroreductase

Heterocyclic and aromatic nitro compounds are extensively used in industry and medicine. They are important intermediates in the manufacture of thousands of consumer products (Hartter, 1984) and clinically are used as antibiotic, antiparasitic and radiosensitizing drugs. Nitroaromatics are also ubiquitous environmental pollutants resulting from combustion of fossil fuels and have been found in diesel exhaust, cigarette smoke and airborne particulates. Many of these compounds have been shown to possess toxic, mutagenic and carcinogenic activity and so may contribute to the environmental cancer risk in man (Rosenkranz and Mermelstein, 1985). Reduction of the nitro group, which occurs in a series of steps involving formation of nitroso and hydroxylamino intermediates, leading ultimately to an amine, is usually required for the pharmaceutical and toxicological activity of these compounds to be expressed, for example, for the antitrichomonad activity and mutagenicity of nitroimidazoles (Lindmark and Muller, 1976) and for the induction of methaemoglobinaemia by nitrobenzenes (Reddy *et al.*, 1976). Although reduction of the nitro group can be effected by both mammalian and bacterial reductases, in most cases nitroreduction by the gut microflora appears to play a more important role than hepatic enzymes. Conclusive evidence for the importance of gut flora reductases in the toxicity of nitrobenzene has been obtained by exposing conventional flora, antibiotic-treated and germ-free rats to the compound. Methaemoglobin levels of 30–40% were induced in the conventional rats within 2 hours of exposure to nitrobenzene, whereas no increase was detected in the animals without a gut flora (Reddy *et al.*, 1976). Similarly, the toxicity of nitrotoluenes, important intermediates in the manufacture of plastics and dyes, is largely dependent on the reductive activity of the intestinal microflora. In this case, genotoxicity and the ability to bind covalently to macromolecules (thought to be an early step in tumour induction) was decreased in germ-free rats (Doolittle *et al.*, 1983; Mirsalis *et al.*, 1982). Gut bacterial nitroreduction is also believed to play a crucial role in the activation of nitrated polycyclic hydrocarbons, such as 6-nitrochrysene, to carcinogenic derivatives (Rickert, 1988).

(f) Decarboxylation of amino acids

Decarboxylation of several amino acids by bacteria in the gut can produce pharmacologically active monoamines such as tyramine, from tryrosine, and tryptamine, from tryptophan (Melnykowycz and

Johansson, 1955). Diamines can also be produced by decarboxylation of lysine and histidine. Such amines are vasoactive and some have been implicated in migraine attacks (Anon., 1968). In healthy individuals, these metabolites are usually detoxified by mammalian amine oxidases, but in the presence of amine oxidase inhibitor drugs, or in patients with impaired hepatic function, increased levels of the toxic amines may be present in body tissues and fluids.

(g) Deamination of amino acids

Deamination of amino acids derived from dietary or endogenously produced protein is the major source of ammonia in the colon (Wrong *et al.*, 1982). Ammonia generated in the colon readily passes across the gut wall, thereby gaining access to other tissues of the body. High concentrations of ammonia have been associated with a number of toxic events *in vivo*, including damage to circulating erythrocytes (Dang and Visek, 1968), increased susceptibility to viral infection (Anderson *et al.*, 1964) and induced neuropsychiatric disturbances in man (Dawson, 1978). In addition, ammonia has been postulated to stimulate the development of neoplastic change in normal tissues, especially the large intestinal mucosa (reviewed by Clinton, 1991). In the cases of some aromatic amino acids, bacterial deamination can lead to the production of phenols and cresols, which have been found to possess tumour-promoting and convulsant activity (Bakke, 1969; Angel and Rogers, 1968). A recent paper has provided some preliminary evidence that elevated levels of faecal *p*-cresol may be associated with hyperactive activity in children (Adams *et al.*, 1985)

It should be noted that amino acid metabolism by gut microflora has been demonstrated both *in vitro* and *in vivo*; indeed, bacterial decarboxylation is responsible for the majority of amines in human urine (Drasar and Hill, 1974).

(h) Sulphamatase

The sodium and calcium salts of cyclohexylsulphamic acid (cyclamates) were widely used as artificial sweeteners until feeding studies of sweetener mixtures at high doses in rats suggested that they may induce bladder tumours (Oser *et al.*, 1975). Cyclamate was also shown to be converted *in vivo* to cyclohexylamine, a compound reported to induce testicular damage in rats following chronic treatment (Renwick, 1988). The formation of cyclohexylamine *in vivo* is inhibited by coadministration of antibiotics in laboratory animals (Bickel *et al.*, 1974) indicating the importance of gut bacteria. Furthermore, the conversion was shown to be inducible: prior exposure to cyclamate increases the rate of transformation.

(i) Bile acid metabolism

The two major bile acids secreted by the liver – cholic acid and chenodeoxycholic acid – are subject to a variety of biotransformations in *in vitro* incubations with the faecal microflora in the form of mixed and pure cultures. These reactions have been reviewed by MacDonald *et al.* (1983) and Hill (1986). The most important of these reactions *in vivo* is 7-α-dehydroxylation, which results in the conversion of cholic acid and chenodeoxycholic acid to deoxycholic acid and lithocholic acid, respectively. In man, over 80% of the bile acids in faeces are dehydroxylated.

Bacterial metabolism of bile acids has been postulated to play an important role in the aetiology of colon cancer (Hill, 1986; Reddy *et al.*, 1975). It is suggested that the secondary bile acids (i.e. those resulting from bacterial metabolism) can act as promotors of the tumorigenic process in the colon. In addition, Hill and his co-workers have proposed that dehydrogenation of the steroid nucleus to produce delta 1 and delta 4 bonds in association with 3-keto groups is particularly important in relation to colon cancer. Certain strains of clostridia can perform this reaction *in vitro* although the significance of the reaction *in vivo* is questionable (Lombardi *et al.*, 1978). Early studies, which revealed an association of the 'nuclear dehydrogenating clostridia' in faeces with colon cancer risk (Goddard *et al.*, 1975), were not confirmed in subsequent studies (IARC, 1977; Crowther *et al.*, 1976).

Toxicity of secondary bile acids

In general it would appear that the secondary bile acids, e.g. deoxycholic and lithocholic acids, are more toxic than the primary bile acids (Kelsey, 1983). A wide variety of *in vitro* and *in vivo* studies provide evidence that the bacterially degraded bile acids possess toxic, comutagenic and cocarcinogenic activity. For example, dehydroxylated acid steroids, particularly lithocholic acid, induce proliferation of liver and bile duct epithelia, cause cell necrosis, inflammation and interfere with mitochondrial function (Kelsey, 1983). It is important to note that bile acids are not carcinogenic *per se*, nor are they mutagenic in the Ames bacterial mutagenicity assay (Silverman and Andrews, 1977; Kawalek and Andrews, 1977). However, there is considerable evidence that they have comutagenic and cocarcinogenic activity (summarized by Kelsey, 1983). For example, secondary bile acids have been shown to stimulate binding of the carcinogen benzo(a)pyrene to DNA in cultured human colon cells (Autrup *et al.*, 1978) and to enhance the mutagenicity of benzo(a)pyrene and 2-aminoanthracene (Silverman and Andrews, 1977). Narisawa *et al.* (1974) showed that intrarectal administration of secondary bile acids increased the incidence of colon adenomas induced by N-methyl-N-nitro-N-nitrosoguanidine (MNNG) in rats. Subsequently, Reddy *et al.* (1977) provided evidence for the importance of bacterial metabolism of bile acids for this tumour-promoting activity by demonstrating that the primary bile acids act as promotors of MNNG-induced colon tumours only in conventional flora rats, whereas the secondary bile acids exert tumour-promoting activity in both germ-free and conventional flora animals.

Thus, these and other studies (summarized by Rowland *et al.*, 1985) lead to the conclusion that in humans, bacterial metabolism of bile acids may lead to the formation in the colon of agents capable of stimulating the development or growth of tumours in the mucosa.

(j) Short-chain fatty acids

The short-chain fatty acids (SCFAs), acetic, propionic and butyric acids, are the major anionic products of bacterial intermediary metabolism and are largely the end-products of fermentation of carbohydrates (dietary complex polysaccharides, poorly digestible sugars and endogenous carbohydrate such as mucins) (Cummings and Branch, 1986). Fermentation of protein and lipid reaching the large intestine also contributes to SCFAs in the colon, particularly the branched-chain SCFAs. The mucosa of the colon readily absorbs SCFA from the lumen, particularly acetic and propionic acids, which may then contribute to the available energy pool of the host (Rombeau *et al.*, 1990). In addition, some SCFAs may protect against pathological changes in the colonic mucosa since butyric acid has been shown to inhibit the expression of certain neoplastic characteristics in mammalian cell cultures (Leavitt *et al.*, 1978). The SCFA concentration is an important factor determining the pH of the colonic lumen. Since the expression of many bacterial enzymes is influenced by the pH of the medium, changes in amounts of SCFA produced could have important implications for foreign compound and carcinogen metabolism in the gut (Mallett *et al.*, 1989b).

3.3.2 Effect of probiotics on foreign compound metabolism by gut microflora

The effect of administration of probiotic organisms on bacterial enzymes of toxicological importance has been addressed in a number of papers. There is evidence from one of the early papers by Goldin and Gorbach (1977) that the modulating effect of probiotic organisms was dependent on the type of diet fed. In rats, no significant effects on enzyme activities were seen when a grain-based diet was supplemented with *Lactobacillus acidophilus* (to provide 10^{10} organisms per rat per day). When the animals were transferred to a meat-based diet (72%, w/w beef), a marked (>2-fold) increase in faecal azoreductase and nitroreductase activities occurred. This increase was prevented in those animals given the *Lactobacillus* supplement with the meat diet. A similar rise in β-glucuronidase activity associated with meat feeding was also inhibited, but to smaller extent, by administration of *L. acidophilus* (Goldin and Gorbach, 1977). In a subsequent study, using the same meat diet and the same *Lactobacillus* strain (10^9–10^{10} day^{-1}) a decrease of 43% in faecal β-glucuronidase was observed together with a more marked decrease in nitroreductase activity (Goldin and Gorbach, 1984a).

These studies have been extended to humans who were given milk supplemented with 10^9 viable lactobacilli per day for one month (Goldin and Gorbach, 1984a). The *Lactobacillus* strains were of human intestinal origin and not from fermented milks or used in commercial yoghurt production. Two strains of lactobacilli were used and were said to give identical results, so data were combined. Prior to lactobacillus feeding, faecal levels of β-glucuronidase ranged between 1.74 and 2.14 units mg^{-1} faecal protein. After ingestion of the lactobacilli for 30 days, the activity of β-glucuronidase declined in all 21 subjects, reaching a mean value of 1.12 units mg^{-1} protein. On stopping the feeding of the lactobacilli, values returned to the baseline value after 10 days. More marked reductions were seen in nitro and azo reductase activities. In the case of nitroreductase, the enzyme activity in faeces decreased from approximately 4.5 units mg^{-1} protein during the control period to 1.2 units mg^{-1} protein after 30 days exposure to the lactobacilli. Control values were not restored until 30 days after lactobacillus supplementation ceased. Faecal azoreductase activity was similarly decreased from control range of 3.9–5.1 units mg^{-1} protein to a mean level of 1.2 units mg^{-1} protein after 30 days of lactobacillus ingestion. Again 30 days were required to regain pre-dosing levels after lactobacillus ingestion was stopped.

Thus the kinetics of the changes in the activities of the three enzymes were similar: namely, a progressive decline in enzymic activity after lactobacillus administration, which levelled out after 3–4 weeks. The changes in activity were not sustained when the ingestion of lactobacilli ceased, suggesting that the intestinal tract was not permanently colonized.

Analogous results to those of Goldin and Gorbach have been obtained recently by Cole *et al.* (1989) who used one of the *L. acidophilus* strains of Goldin and Gorbach and investigated the effect of its administration on bacterial metabolic activities in germ-free rats colonized with a human faecal microflora. Although the lactobacilli were administered for only 3 days (albeit at a higher dose per kilogram of body weight than in the human study of Goldin and Gorbach), a significant reduction in β-glucosidase and β-glucuronidase activities was observed with the effects persisting for 7 days after dosing ceased.

Table 3.3 Possible mechanisms of interaction of probiotics with bacterial metabolism in the gut.

1. Probiotics displace or dilute normal gut flora organisms which activate ingested substances to toxic or carcinogenic derivatives.
2. Probiotics provide enzymes which detoxify ingested substances or their active metabolites.
3. Probiotics generate conditions in the gut which alter the rate of bacterial activation of ingested chemicals, e.g. lowering of pH affects ammonia production, bile acid metabolism.

The mechanisms involved in the reduction in enzyme activity by the ingestion of lactobacilli have not been investigated although several can be advanced (Table 3.3). Of these, the most likely would appear to be displacement, by lactobacilli, of some members of the normal colonic microflora. *L. salivarius* ingestion is known to cause qualitative changes in gut microflora (Cole and Fuller, 1984) suggesting that it may decrease the population of bacteria with high β-glucuronidase activity in the gut. This is supported by studies of the enzymic activity of pure cultures of colonic bacteria *in vitro* (Table 3.4) and *in vivo* (germ-free rats monocontaminated with the organisms (Cole *et al.*, 1985). These investigations suggest that lactobacilli and bifidobacteria have lower enzyme activities than coliforms, clostridia and bacteroides. Thus, a decrease in the latter groups and an increase in lactobacilli or bifidobacteria after ingestion would result in the observed changes in enzyme activities.

In the various studies discussed above, the effects of probiotic organisms have been investigated in healthy individuals. In contrast, Tamura *et al.* (1983) administered a bifidobacterial preparation comprising *Bifidobacterium breve* and *Bif. bifidum* ($>10^9$ organisms per d for 2–3 weeks) to patients with gastrointestinal disorders and monitored pH, ammonia concentration and urease activity in faeces. Statistically significant decreases reported although the changes were small and of questionable biological significance.

The decrease in activity of intestinal azo- and nitroreductases and β-glucaronidase, seen after feeding lactobacilli to animals, results in corresponding changes in the metabolism of potentially toxic substrates of those enzymes (Goldin and Gorbach, 1984b). For example, the activity of nitroreductase and the production of the bladder carcinogen, 2-napthylamine, were both decreased by lactobacillus feeding.

The effect of ingestion by human subjects of *L. casei* on urinary excretion of potentially toxic amino acid metabolites has been studied by Tohyama *et al.* (1981). The urinary concentration of indican (from tryptophan) and *p*-cresol (from tyrosine) was significantly decreased by feeding 10^{10} organisms per day for 5 weeks with mean reductions of 29 and 43% respectively in the seven subjects.

Table 3.4 Enzyme activities of intestinal organisms *in vitro*.

Species	Enzyme activities (lmol h^{-1} per 10^{10} cells)			
	Azoreductase	Glucuronidase	Glucosidase	Nitroreductase
Bifidobacterium				
longum	ND	0.001	0.203	NE
adolescentis	ND	ND	0.648	NE
breve	ND	ND	3.67	NE
infantis	ND	ND	2.73	NE
Bacteroides				
fragilis	0.011	0.007	0.514	NE
vulgatus	0.032	0.012	0.064	NE
thetaiotaomicron	0.022	ND	0.815	NE
Clostridium				
perfringens	0.343	0.018	0.841	NE
paraputrificum	7.53	0.026	8.20	NE
Eubacterium				
aerofaciens	0.348	0.012	0.166	NE
Streptococcus sp.	NE	0.04	0.03	0.0004
Lactobacillus				
salivarius	NE	ND	0.10	ND

NE = Not estimated.
ND = Not detected.
After Cole *et al.* (1985) and Saito and Rowland (unpublished observations, 1991).

Final confirmation that these changes in metabolism can result in significant alterations in toxicity of ingested chemicals has come from a study of carcinogenesis induced by 1,2-dimethylhydrazine in rats.

Dimethylhydrazine (DMH) is thought to induce colon tumours after metabolism involving oxidation in the liver by mixed-function oxidase enzymes, conjugation to glucuronic acid and excretion in bile, then deconjugation in the colon by bacterial β-glucuronidase releasing the active carcinogen (see section 3.1(b)). DMH-treated rats given *L. acidophilus* supplements were found to have significantly fewer colon tumours at 20 weeks than control animals given DMH alone (Goldin and Gorbach, 1980).

3.3.3 Effect of milk (fermented and non-fermented) and yoghurt on metabolism by gut microflora

Ayebo *et al.* (1980) have assessed the effect of ingestion of non-fermented milk containing *L. acidophilus* (about 2×10^6 organisms ml^{-1} on activity of β-glucuronidase and β-glucosidase in faeces in a cross-over study in elderly (>65 yr) humans. The control group were given low fat milk (3 cups per day) for 4 weeks and the treated group the same volume of milk containing lactobacilli. The volunteers were allowed to consume their own diets.

During the period of consumption of the lactobacillus milk, the viable count of lactobacilli in faeces rose by one order of magnitude to about 3×10^5 g^{-1}. β-Glucuronidase activity decreased slightly after 4 weeks of lactobacillus feeding. The effects on β-glucosidase were inconsistent – during one period of lactobacillus exposure, the activity decreased from 0.9 units to about 0.45 units, whereas during another period on the same diet there was no change in activity.

In an analogous study to that of Ayebo *et al.* (1980), Marteau *et al.* (1990) studied the effect of ingestion of a fermented milk product on a number of reductive and hydrolytic faecal enzyme activities in human volunteers. The subjects were asked to consume 300g day^{-1} of a fermented milk product containing *L. acidophilus* (10^7 g^{-1}), *Bif. bifidum* (10^8 g^{-1}) and *Streptococcus lactis* (*Lactococcus lactis* subsp. *lactis*) and *S. cremoris* (*Lact. lactis* subsp. *cremoris*) (both at $10^8 g.^{-1}$) Azoreductase and β-glucuronidase activities did not change in response to consumption of the fermented milk product, but nitroreductase activity was decreased by 38% and remained depressed for at least 3 weeks. In contrast, β-glucosidase increased after ingestion of the fermented milk – a change attributed to the high activity of this enzyme in *Bif. bifidum*.

In the various studies described above, performed using either lactobacilli or milk products containing lactobacilli, the most consistent finding is a decrease in nitroreductase activity in faeces. The inconsistency of the other changes in metabolism may be due to differences in the types of probiotic fed or their concentrations in the product.

(a) In vivo *studies*

Takano *et al.* (1985) studied the influence of sour milk (prepared by fermenting skimmed milk with *L. helveticus* subsp. *jurguti* and *Candida utilis*) on colon tumours induced by DMH. Only small numbers of animals were used in each group and although the number of rats with colon tumours was decreased in those fed sour milk, the difference was not significant (Table 3.5). However, the total number of tumours per rat was significantly lower in the sour-milk fed group. Interestingly, animals fed milk acidified with lactic acid, or fed cells of *L. helveticus* and *Can. utilis* did not exhibit lower tumour incidence than the control animals. The mechanisms involved were not investigated, although numbers of bifidobacteria were higher, by about one order of magnitude in the sour-milk fed rats by comparison with rats fed whole milk or control diet.

Table 3.5 Incidence of DMH-induced colon tumours in rats fed a diet containing sour milk, acidified milk or 'starter' bacteria.

Group	Animals with colon tumours	Colon tumours per rat‡		
		Total	Adenoma	Adenocarcinoma
Control*	9/9	2.6	1.3	1.3
Sour milk	6/9	1.0†	0.4	0.6
Acidified milk	9/9	3.4	2.0	1.4
Starter bacteria	7/9	2.3	1.8	0.5

* Control commercial diet.
† Significantly different from control ($p < 0.05$).
‡ Rats (6 wk) fed milks (28% in diet). DMH given i.p. 20 mg kg^{-1} wk^{-1} for 16 weeks. Tumours assessed 10 weeks later.
After Takano *et al.* (1985).

3.3.4 Indigestible sugars

A number of companies (particularly in Japan) have isolated various oligosaccharides, which are not substrates, for mammalian hydrolytic enzymes and so pass undegraded into the colon where they may be fermented by colonic bacteria. Unlike other indigestible sugars such as lactulose, which is hydrolysed by a wide variety of gut bacteria, these oligosaccharides are usually only fermented by a limited range of microorganisms and so in theory can selectively stimulate the growth of chosen organisms. They are often co-administered with a specific organism (usually a *Bifidobacterium* species) in order to encourage its multiplication in the gut and hence potentiate the beneficial 'probiotic' effects of the organism. Thus, although not strictly probiotics, these indigestible sugars have been included in this review since they may exhibit properties analogous to those of true probiotics.

The main oligosaccharides that have been studied for beneficial effects on the consumer are:

1. Transgalactosylated oligosaccharides (TOS; Tanaka *et al.* 1983). TOS is a mixture of tri-, tetra-, penta- and hexasaccharides (of galactose and glucose) obtained when lactose is enzymatically hydrolysed by *Aspergillus oryzae*. This sugar can be utilized by all *Bifidobacterium* species tested and by some lactobacilli, bacteroides, streptococci and enterobacteria (Tanaka *et al.*, 1983).
2. Soybean oligosaccharides extract (SOE; Hayakawa *et al.*, 1990). SOE is prepared from defatted soybean whey and comprises a mixture of sucrose (44%), stachyose (23%), raffinose (7%) and monosaccharides. The extract has also been used in a refined form (SOR) in which the stachyose and raffinose contents are inceased to 71 and 20% respectively. The efficiency of utilization of SOR by intestinal bacteria *in vitro* was greatest with *Bifidobacterium* species although other genera, including *Lactobacillus* and *Bacteroides*, were capable of fermenting the sugar.
3. Fructooligosaccharides (Hidaka *et al.*, 1986). The commercial preparation is a mixture of tri-, tetra- and pentasaccharides (glucose and fructose) obtained by the action of the enzyme β-fructofuranoside, derived from *A. niger*, on sucrose. It is utilized by most bifidobacteria, bacteroides and some streptococci, lactobacilli and enterobacteria, but not *Escherichia coli* (Hidaka *et al.*, 1986).

(a) Effect of indigestible sugars on bacterial numbers and metabolism

The effect of ingestion of TOS (3 or 10 g day^{-1}), *Bif. breve*, or both, on faecal bacterial counts and faecal ammonia has been investigated in human volunteers (Tanaka *et al.*, 1983). In general, significant effects were seen only during periods when both TOS and *Bif. breve* were ingested simultaneously. For example, during these periods, the viable counts of bacteroides and enterobacteria in faeces declined and faecal ammonia concentration decreased markedly in 4 out of the 5 volunteers. Ingestion of TOS alone, even at 10 g day^{-1}, gave inconsistent effects on these parameters.

Ingestion of a commercial preparation (8 g day^{-1} for 14 days) by senile patients (50–90 years) led to an slight increase in total viable count in faeces and an average increase of about 10-fold in bifidobacteria (Hidaka *et al.*, 1986). However, the increase in bifidobacteria was seen only in those individuals who originally had low numbers ($<10^8 g^{-1}$) of the organisms in the faeces. Similar individual differences in response were seen in gut bacterial metabolism when the commercial preparation was administered for two months at a dose rate of 8 g day^{-1}. An increase in bifidobacteria was accompanied by a rise in SCFA concentration in faeces and a decrease in *p*-cresol and indole. In the same person, however, faecal ammonia was increased nearly two-fold. Changes induced by the commercial sugar in these metabolic profiles were not seen in a subject whose bifidobacteria count was initially high (Hidaka *et al.*, 1986). These results need to be interpreted with caution since only two individuals were studied.

Further experiments were performed in rats fed a purified diet containing tyrosine and tryptophan. Incorporation of a sugar preparation into the diet at 0.4–10% appeared to reduce the *p*-cresol concentration in faeces, especially at the highest dose of the oligosaccharide (Hidaka *et al.*, 1986). Concentrations of 10 and 20% sugar preparation in the diet of rats also markedly increased the SCFA concentration in faeces (Tokunaga *et al.*, 1986). The total daily excretion of neutral steroids was also raised, with a smaller increase in bile acid excretion, seen only at the higher (20%) dose, which would not be achieved in human exposures. The changes in faecal neutral steroid excretion in rats were not reflected in serum cholesterol concentrations suggesting that the commercial sugar stimulates cholesterol synthesis rather than increases faecal excretion (Tokunaga *et al.*, 1986).

The influence on the gut microflora of ingestion of oligosaccharides derived from soybeans (SOE) has been studied in human volunteers. The subjects consumed their normal diet with the addition of 10 g SOE per day with or without 6×10^9 c.f.u. *Bif. longum* for 3 weeks (Hayakawa *et al.*, 1990). During the periods of ingestion of SOE the viable count of bifidobacteria in faeces was slightly but significantly higher (10^{10} g^{-1}) than during control periods ($10^{9.4}$–$10^{9.7}$ g^{-1}). Intake of *Bif. breve* with SOE appeared to have little additional effect on bifidobacteria numbers in faeces. The biological significance of such changes has yet to be determined. In the same study, faecal pH and amino acid degradation products in faeces (indole, *p*-cresol and phenol) were also determined, but no consistently significant differences were observed between the various dietary periods.

The refined soybean oligosaccharide SOR has been studied in an *in vitro* continuous flow culture system which models the human colonic microflora (Bearne *et al.*, 1990). Numbers of bifidobacteria were determined and various bacterial enzyme activities assayed before and after SOR was incorporated into the growth medium at a concentration of 0.1% (w/v). Following addition of SOR, there was an increase in concentration of bifidobacteria in the culture from 8×10^7 to $4 \times 10^8 g^{-1}$ (Saito, Y. and Rowland, I.R., unpublished observation, 1991). Thus the response of the colonic flora to soybean oligosaccharides in this *in vitro* study was analogous to that seen in human volunteers (Hayakawa *et al.*, 1990). Changes were also seen in the activity of some bacterial enzymes. Azoreductase activity decreased significantly ($p<0.05$) by 40% and β-glucuronidase and β-glucosidase activities also decreased (by 38 and 32%, respectively), although the results were not significant. Clearly, further *in vivo* metabolic studies in animals and humans are necessary before conclusions can be drawn.

3.4 CONCLUSIONS

The evidence that at least certain strains of *Lactobacillus* can modify intestinal bacterial metabolism is strong and the biological and toxicological significance of the changes seen has been established and indicates that ingestion of such probiotic organisms may have beneficial effects in humans. It is also clear however, that diet may modulate the effects of probiotics on the gut microflora. Other probiotic organisms such as bifidobacteria have been studied less extensively with regard to their influence on the intestinal bacterial metabolism. Their effects on the proportions of the various bacterial types in the gut have been investigated in great detail, but such studies, while of value in establishing that a treatment can modifiy the flora, give little information on the biological consequences of such changes for the host animal. Therefore, further metabolic and toxicological studies are urgently needed in this area. The use of indigestible oligosaccharides as specific modifiers of the intestinal flora holds great promise and presumably these will continue to be refined in terms of the specificity of their effects. Again, metabolic and toxicological studies will be necessary to assess their beneficial consequences for the consumer.

REFERENCES

Adams, R. F., Murray, K. E. and Earl, J. W. (1985) High levels of faecal *p*-cresol in a group of hyperactive children. *Lancet*, **ii**, 1313.

Alldrick, A. J., Lake, B. G. and Rowland, I. R. (1989) Modification of *in vivo* heterocyclic amine genotoxicity by dietary flavonoids. *Mutagenesis*, **4**, 365–70.

Anderson, D. P., Beard, C. W. and Hansen, R. P. (1964) The adverse effect of ammonia on chickens including resistance to infection with Newcastle disease virus. *Avian Dis.*, **8**, 369.

Angel, A. and Rogers, K. J. (1968) Convulsant action of polyphenols. *Nature*, **217**, 84–5.

Anon. (1968) Headache, tyramine serotonin and migraine. *Nutr. Rev.*, **26**, 40–4.

Autrup, H., Harris, C. C., Trump, B. I. and Jeffrey, A. M. (1978) Metabolism of benzo(a)pyrene and identification of the major benzo(a)pyrene-DNA adducts in cultured human colon. *Cancer Res.*, **38**, 3689.

Ayebo, A. D., Angelo, I. A. and Shahani, K. M. (1980) Effect of ingesting *Lactobacillus acidophilus* milk upon fecal flora and enzyme activity in humans. *Milchwiss.*, **35**, 730–3.

Bakke, O.M. (1969) Studies on the degradation of tyrosine by rat caecal contents. *Scand. J. Gastroenterol.*, **4**, 603–8.

Bearne, C.A., Mallett, A.K., Rowland, I.R. and Brennan-Craddock, W.E. (1990) Continuous culture of human faecal bacteria as an *in vitro* model for the colonic microflora. *Toxicol. in Vitro*, **4**, 522–5.

Bickel, M.H., Burkard, B., Meier-Strasser, E. and van den Broek-Boot, M. (1974) Enterobacterial formation of cyclohexylamine in rats ingesting cyclamate. *Xenobiotica*, **4**, 425–39.

Boutwell, R.K. and Bosch D.K. (1959) The tumour-promoting action of phenol and related compounds for mouse skin. *Cancer Res.*, **19**, 413–24.

Brown, J.P. (1988) Hydrolysis of glycosides and esters, in *Role of the Gut Flora in Toxicity and Cancer* (ed. I. R. Rowland), Academic Press, London, pp. 109–44.

Cerniglia, C.E., Freeman, J.P., Franklin, W. and Pack, L.D. (1982) Metabolism of benzidine and benzidine-congener based dyes by human, monkey and rat intestinal bacteria. *Biochem. Biophys. Res. Commun.*, **107**, 1224–9.

Chipman, J.K., Millburn, P. and Brooks, T.M.(1983) Mutagenicity and *in vivo* disposition of biliary metabolites of benzo(a)pyrene. *Toxicol, Lett.*, **17**, 233–40.

Clinton, S.K. (1991) Dietary protein and carcinogenesis, in *Nutrition, Toxicity, and Cancer* (ed. I.R. Rowland), CRC Press, Boca Raton. pp 455–75.

Cole, C.B. and Fuller, R. (1984) A note on the effect of host specific fermented milk on the coliform population of the neonatal rat gut. *J. Appl. Bacteriol.*, **56**, 495–8.

Cole, C.B., Fuller, R. and Carter S.M. (1989) Effect of probiotic supplements of *Lactobacillus acidophilus* and *Bifidobacterium adolescentis* 2204 on β-glucosidase and β-glucuronidase activity in the lower gut of rats associated with a human faecal flora. *Microb. Ecol. Health Dis.*, **2**, 223–5.

Cole, C.B., Fuller, R., Mallett, A.K. and Rowland, I.R. (1985) The influence of the host on expression of intestinal microbial enzyme activities involved in metabolism of foreign compounds. *J. Appl. Bacteriol.*, **58**, 549–53.

Combes, R.D. (1989) Cytochrome P-450, mixed function oxidases and formation of genotoxic metabolites by the intestinal tract. *Prog. Pharm. Clin. Pharm.*, **7**, 119–45.

Combes, R.D. and Haveland-Smith, R.B. (1982) A review of the genotoxicity of food, drug and cosmetic colours and other azo, triphenylmethane and xanthene dyes. *Mutation Res.*, **98**, 101.

Cook, G.C. (1968) Defects of sugar absorption. Some observations on racial lactase deficiency. *Proc. R. Soc. Med.*, **61**, 1102–4.

Cook, G.C. (1973) Incidence and clinical features of specific hypolactasia in adult man, in *Intestinal Enzyme Deficiencies and their Nutritional Implications* (eds A. Dahlqvist and L. Hambracus), Swedish Nutrition Foundation, pp. 52–73.

Crowther, J.S., Drasar, B.S., Hill, M.J. *et al.* (1976) Faecal steroids and bacteria and large bowel cancer in Hong Kong by socio-economic groups. *Brit. J. Cancer*, **34**, 191–7.

Cummings, J.H. and Branch, W.J. (1986) Fermentation and the production of short chain fatty acids in the human large intestine, in *Dietary Fiber : Basic and Clinical Aspects* (eds G. Vahouny and D. Kritchevsky), Plenum Press, New York and London, pp. 131–49.

Dahlqvist, A. (1984) Lactose intolerance. *Nutr. Abstr. Rev. (Series A)*, **54**, 649–58.

Dahlqvist, A. and Asp, N.–G. (1975) Specific disaccharidase deficiency in adults. *Trans. Biochem. Soc.*, **3**, 227–32.

Dang, H.C. and Visek, W.J. (1968), Some characteristics of blood in normal and immune rabbits after urease injection. *Amer. J. Physiol.*, **215**, 502.

Dawson, A.M. (1978) The regulation of blood ammonia. *Gut*, **19**, 504.

De Bruin, A. (1976) *Biochemical Toxicology of Environmental Agents*, Elsevier, Amsterdam.

Doolittle, D.J. Sherrill, J.M. and Butterworth, B.E. (1983) Influence of intestinal bacteria, sex of animal, and position of the nitro group on the hepatic genotoxicity of nitrotoluene isomers *in vivo*. *Cancer Res.*, **43**, 2846.

Drasar, B.S. (1988) The bacterial flora of the intestine, in *The Role of the Gut Flora in Toxicity and Cancer* (ed. I.R. Rowland.), Academic Press, London, pp. 23–38.

Drasar, B.S. and Hill, M.J. (1974) *Human Intestinal Flora*, Academic Press, London.

Elliger, C., Henika, P. R. and MacGregor, J.T. (1984) Mutagenicity of flavones, chromones and acetophenones in *Salmonella typhimurium*. New structure – activity relationships. *Mutation Res.*, **135**, 77–86.

Goddard, P., Fernandez, F., West, B. *et al.* (1975) The nuclear dehydrogenation of steroids by intestinal bacteria. J. Med. Microbiol. **8**, 429–35.

Goldin, B.R. and Gorbach, S.L. (1976) The relationship between diet and rat fecal enzymes implicated in colon cancer. *J. Natl. Cancer Inst.*, **57**, 371–5.

Goldin, B.R. and Gorbach, S.L. (1977) Alterations in fecal microflora enzymes related to diet, age, *Lactobacillus* supplements, and dimethylhydrazine. *Cancer*, **40**, 2421–6.

Goldin, B.R. and Gorbach, S.L. (1980) Effect of *Lactobacillus acidophilus* supplements on 1,2-dimethylhydrazine dihydrochloride-induced intestinal cancer in rats. *J. Natl Cancer Inst.*, **64**, 263–5.

Goldin, B.R. and Gorbach, S.L. (1984a) The effect of milk and lactobacillus feeding on human intestinal bacterial enzyme activity. *Amer. J. Clin. Nutr.*, **39**, 756–61.

Goldin, B.R. and Gorbach, S.L. (1984b) Alterations of the intestinal microflora by diet, oral antibiotics, and *Lactobacillus*: decreased production of free amines from aromatic nitro compounds, azo dyes, and glucuronides. *J. Natl Cancer Inst.*, **73**, 689–95.

Goodenough, E.R. and Kleyn, D.H. (1976) Influence of viable yogurt microflora on digestion of lactose by the rat. *J. Dairy Sci.*, **59**, 601–6.

Grasso, P. and Goldberg, L. (1968) Problems confronted and lessons learnt in the safety evaluation of Brown FK. *Food Cosmet. Toxicol.*, **8**, 539.

Gray, G.M. (1984) Carbohydrate absorption and malabsorption, in *The Role of the Gastrointestinal Tract in Nutrient Delivery* (eds M Green and H.L. Greene) Academic Press, London, pp. 133–44.

Green, L.C. and Tannenbaum, S.R. (1982) Nitrate, nitrite and N-nitroso-compounds: Biochemistry metabolism toxicity and carcinogenicity, in *Human Nutrition* (eds A. Neuberger and T.H. Jukes), MTP Press, Lancaster, pp. 87–140.

Hartter, D.R. (1984) Use and importance of nitroaromatic chemicals in the chemical industry, in *Toxicity of Nitroaromatic Compounds* (ed. E.D. Rickert), Hemisphere Publishing Co., Washington, DC, pp. 1–13.

Hayakawa, K., Mizutani, J., Wada, K. *et al.* (1990) Effects of soybean oligosaccharides on human faecal microflora. *Microb. Ecol. Health Dis.*, **3**, 293–303.

Hidaka, H., Eida, T., Takizawa, T. *et al.* (1986) Effects of fructooligosaccharides on intestinal flora and human health. *Bifidobacteria Microflora*, **5**, 37–50.

Hill M.J. (1986) *Microbes and Human Carcinogenesis*, Arnold, London.

Hitchins, A.D. and McDonough, F.E. (1989) Prophylactic and therapeutic aspects of fermented milk. *Amer. J. Clin. Nutr.*, **49**, 675–84.

IARC Intestinal Microecology Group (1977) Dietary fibre, transit time, fecal bacteria, steroids and colon cancer in two Scandinavian populations. *Lancet*, **i**, 207–11.

Jones, A., Flanagan, P.R. and Forstner, G.C. (1977) Pathogenesis of mucosal injury in the blind loop syndrome. Brush border enzyme activity and glycoprotein degradation. *J. Clin. Invest.*, **60**, 1321–30.

Kawalek, J.C. and Andrews, A.W. (1977) The effect of bile acids on the metabolism of benzo(a)pyrene and 2-aminoanthracene to mutagenic products. *Fed. Proc.*, **36**, 844.

Kelsey, M.I. (1983) *In vitro* effect of bile acids, in *Experimental Colon Carcinogenesis* (eds H. Autrup and G.M. Williams), CRC Press, Boca Raton, p. 241.

Klubes, P, Cerna, I, Rabinowitz, A.D., and Jondorf, W.R. (1971) Factors affecting dimethylnitrosamine formation from simple precursors by rat intestinal bacteria. *Food Cosmet, Toxicol.*, **10**, 757–67.

Kolars, J.C., Levitt, M.D., Aouji, M. and Savaiano, D.A. (1984) Yogurt, an autodigesting source of lactose. *New Engl. J. Med.*, **310**, 1–3.

Laqueur, G.L. and Spatz, M. (1968) The toxicology of cycasin. *Cancer Res.*, **28**, 2262.

Lasker, J. M., Huang M.T. and Conney, A. H. (1984) *In vitro* and *in vivo* activation of oxidative drug metabolism by flavonoids. *J. Pharmacol. Exp. Ther.*, **229**, 162–70.

Leach, S.A., Challis, B., Cook, A.R. *et al.* (1985) Bacterial catalysis of the N-nitrosation of secondary amines. *Biochem. Soc. Trans.*, **13**, 380.

Leavitt, J., Barrett, J.C., Crawford, B.D. and Ts'o, P.O.P. (1978) Butyric acid supression on the *in vitro* neoplastic state of Syrian hamster cells. *Nature*, **271**, 262.

Lindmark, D.G. and Muller, M. (1976) Antitrichomonad action, mutagenicity, and reduction of metronidazole and other nitroimidazoles. *Antimicrob. Agents Chemother.*, **10**, 476.

Lombardi, P., Goldin, B, Boutin, E. and Gorbach, S.L. (1978) Metabolism of androgens and estrogens by human fecal microorganisms. *J. Steroid Biochem.*, **9**, 795–801.

MacDonald, I.A., Bokkenheuser, V.D., Winter, J. *et al.* (1983) Degradation of steroids in the human gut. *J. Lipid Res.*, **24**, 675–700.

MacDonald I. A., Bussard, R. G., Hutchinson D. M. and Holdeman, L. V. (1984) Rutin-induced β-glucosidase activity in *Streptococcus faecium* VGH-1 and *Streptococcus* sp. strain FRP-17 isolated from human feces: formation of the mutagen quercetin from rutin. *Appl. Environ. Microbiol.*, **47**, 350–5.

Mallett, A.K., Bearne, C.A., Lake, B.G. and Rowland, I.R. (1989a) Modified mutagen activation in hepatic fractions from rats fed dietary rutin – interaction between gut microflora and host metabolism. *Food Chem. Toxicol.*, **27**, 607–11.

Mallett, A.K., Bearne, C.A. and Rowland, I.R. (1989b) The influence of incubation pH on the activity of rat and human gut flora enzymes. *J. Appl. Bacteriol.*, **66**, 433–7.

Marteau, P., Pochart, P., Flourie, B. *et al.* (1990) Effect of chronic ingestion of a fermented dairy product containing *Lactobacillus acidophilus* and *Bifidobacterium bifidum* on metabolic activities of the colonic flora in humans. *Amer. J. Clin. Nutr.*, **52**, 685–8.

Massey, R.C., Key, P.E., Mallett, A.K., Rowland, I.R. (1988) An investigation of the endogenous formation of apparent total *N*-nitroso compounds in conventional microflora and germ-free rats. *Food Chem. Toxicol.*, **26**, 595–600.

Melnykowycz, J. and Johansson, K.R. (1955) Formation of amines by intestinal microorganisms and influence of chlortetracycline. *J. Exp. Med.*, **101**, 507–17.

Midtvedt, T. (1989) In *Recent Advances in Microbial Ecology* (eds T. Hattori *et al.*), Japan Scientific Societies Press, Tokyo, pp. 515–9

Mirsalis, J.C., Hamm, T.E., Sherrill, J.M. and Butterworth, B.E. (1982) Role of gut flora in genotoxicity of dinitrotoluene. *Nature*, **295**, 322–3.

Mizutani, T. and Mitsuoka, T. (1979) Effect of intestinal bacteria on incidence of liver tumours in gnotobiotic C3H/He male mice. *J. Natl Cancer Inst.*, **63**, 1365–70.

Narisawa, T., Magadia, N.E., Weisburger, J.H. and Wynder, E.L. (1974) Promoting effect of bile acids on colon carcinogenesis after intrarectal instillation of *N*-methyl-*N*-nitro-*N*-nitrosoguanidine in rats. *J. Natl Cancer Inst.*, **53**, 1093–7.

Oser, B.L., Carson, S., Cox, G.E. *et al.* (1975) Chronic toxicity of cyclamate:saccharin (10:1) in rats. *Toxicology*, **4**, 315.

Powell, R., Murray, M., Chen, C. and Lee, A. (1979) Survey of the manufacture, import and uses for benzidine, related substances and related dyes and pigments. *E.P.A. Report 560/13–79–005*, Environmental Protection Agency, Washington, DC.

Renwick, A.G. (1988) Intense sweeteners and the gut microflora, In *The Role of the Gut Flora in Toxicity and Cancer* (ed. I.R. Rowland), Academic Press, London, pp. 175–206.

Reddy, B.G., Pohl, L.R. and Krishna, G. (1976) The requirement of the gut flora in nitrobenzene-induced methemoglobinemia in rats. *Biochem. Pharmacol.*, **25**, 1119

Reddy, B.S., Watanabe, K., Weisburger, J.H. and Wynder, E. (1977) Promoting effect of bile acids in colon carcinogenesis in germ free and conventional F344 rats. *Cancer Res.*, **37**, 3238.

Reddy, B.S., Weisburger, J.H., Narisawa, T. and Wynder, E.L. (1974) Colon carcinogenesis in germfree rats with 1, 2-dimethylhydrazine and *N*-methyl-*N*-nitro-*N*-nitrosoguanidine. *Cancer Res.*, **74**, 2368–72.

Reddy, B.S., Weisburger, J.H. and Wynder, E.L. (1975) Effects of high risk and low risk diets for colon carcinogenesis on faecal microflora and steroids in man. *J. Nutr.*, **105**, 878–84.

Renwick, A.G. and Drasar, B.S. (1976) Environmental carcinogens and large bowel cancer. *Nature*, **263**, 234–5.

Rickert, D.E. (1988) Metabolism of nitro compounds, in *Role of the Gut Flora in Toxicity and Cancer* (ed. I.R. Rowland), Academic Press, London, pp. 145–52.

Rombeau, J.L., Kripke, S.A. and Settle, R.G. (1990) Short chain fatty acids. Production, absorption, metabolism and intestinal effects, in *Dietary Fiber Chemistry, Physiology, and Health Effects* (eds D. Kritchevsky *et al.*), Plenum Press, New York and London, pp. 317–37.

Rosenkranz, H.S. and Mermelstein, R. (1985) The genotoxicity, metabolism and carcinogenicity of nitrated polycyclic aromatic hydrocarbons. *J. Environ. Sci. Health*, **3**, 221–72.

Rowland, I.R. (1988) The toxicology of *N*-nitrosocompounds, in *Nitrosamines Toxicolgy and Microbiology* (ed. M.J Hill), Ellis Horwood, Chichester, pp. 117–41.

Rowland, I.R. (1989) Metabolic profiles of intestinal floras, in *Recent Advances in Microbial Ecology* (eds. T. Hattori *et al.*), Japan Scientific Societies Press, Tokyo. pp 510–4

Rowland IR Granli T Bøckman OC *et al.* (1991) Endogneous – nitrosation in man assessed by measurement of apparent total – mtroso compounds in faeces. *Carinogenesis*, **12**, 1395–1401

Rowland, I.R., Mallett, A.K. and Wise, A. (1985) The effect of diet on the mammalian gut flora and its metabolic activities. *CRC Crit. Rev. Toxicol.*, **16**, 31–103.

Rowland I.R. and Walker, R. (1983) The gastrointestinal tract in food toxicology, A.B.G. in *Toxic Hazards in Food* (eds D.M Conning, D.M. and A.B.G. Lansdown), Croom Helm, London, pp. 183–274.

Savaiano, D.A., Abou El Anouar, A., Smith, D.E. and Levitt, M.D. (1984) Lactose malabsorption from yogurt, pasteurized yogurt, sweet acidophilus milk, and cultured milk in lactase-deficient individuals. *Amer. J. Clin. Nutr.*, **40**, 1219–23.

Silverman, S.J. and Andrews, A.W. (1977) Bile acids: comutagenic activity in the *Salmonella*-mammalian microsome mutagenicity test: brief communication. *J. Natl Cancer Inst.*, **59**, 1557–9.

Smith, R.L. (1966) The biliary excretion and enterohepatic circulation of drugs and other organic compounds, in *Progressive Drug Research* (ed. J. Tucker) Birkhauser, Basel.

Takano, T., Arai, K., Murota, I. *et al.* (1985) Effects of feeding sour milk on longevity and tumorigenesis in mice and rats. *Bifidobacteria Microflora*, **4**, 31–7.

Tamura, N., Norimoto, M., Yoshida, K. *et al.* (1983) Alteration of fecal bacterial flora following oral administration of bifidobacterial preparation. *Gastroenterologia Japonica*, **18**, 47–55.

Tanaka, R., Takayama, H., Morotomi, M. *et al.* (1983) Effects of administration of TOS and *Bifidobacterium breve* 4006 on the human fecal flora. *Bifidobacteria Microflora*, **2**, 17–24.

Thies, E. and Siegers, C.–P. (1989) Metabolic activation and tumourigenesis. *Prog. Pharm. Clin. Pharm.*, **7**, 199–214.

Tohyama, K., Kobayashi, Y., Kan, T. *et al.* (1981) Effect of lactobacilli on urinary indican excretion in gnotobiotic rats and man. *Microbiol. Immunol.*, **25**, 101–12.

Tokunaga, T, Oku, T. and Hoysoya, N. (1986) Influence of chronic intake of new sweetener fructooligosaccharide (Neosugar) on growth and gastrointestinal function of the rat. *J. Nutr. Sci. Vitaminol.*, **32**, 111–21.

Venitt, S. and Bushell, C.T. (1976) Mutagenicity of the food colour Brown FK and its constituents in *Salmonella typhimurium*. *Mutation Res.*, **40**, 309.

Wise, A., Mallett, A.K. and Rowland, I.R. (1982) Dietary fibre, bacterial metabolism and toxicity of nitrate in the rat. *Xenobiotica*, **12**, 111–18.

Wrong, O.M., Vince, A.J. and Waterlow, J.C. (1982) The origins and bacterial metabolism of faecal ammonia, in *Falk Symposium 32 – Colon and Nutrition*, (eds H. Kaspar and H. Goebell), MTP Press Ltd, Lancaster, Chap. 14.

Chapter Four

Translocation and the indigenous gut flora

RODNEY D. BERG

4.1 INTRODUCTION

In 1885, investigators examined the internal organs of apparently healthy animals for viable bacteria (reviewed in Ford, 1900). Although the early results are inconclusive because of the variation in culturing methods, Ford (1901) reported that approximately 70% of the internal organs of domestic cats, dogs, rabbits and guinea-pigs contained viable bacteria. He even found viable bacteria in the livers, spleens and kidneys of human patients. There was not much interest in this phenomenon, however, until Ellis and Dragstedt (1930) confirmed the report of Wolbach and Saiki (1909) that the livers of presumably healthy dogs contain viable anaerobes. A series of investigations then ensued utilizing various dog models in attempts to determine the source of these anaerobic bacteria (Trusler and Reeves, 1934; Chau *et al.*, 1951; Schatten, 1954). These studies suggested that certain bacteria can pass from the gastrointestinal tract to the portal blood and liver when the permeability of the intestinal mucosa is increased.

Table 4.1 Terms and references.

Term	Reference
Absorption	Wilson and Walzer (1935)
Persorption	Volkheimer (1974)
Transmural migration	Frank *et al.* (1948)
Phage translocation	Keller and Engley (1958); Hildebrand and Wolochow (1962).
Bacterial translocation	Wolochow, *et al.* (1966); Fuller and Jayne-Williams (1970); Berg and Garlington (1979)

Various terms have been used to describe the passage of substances, including viruses and bacteria, from the gastrointestinal tract across the mucosal barrier to extraintestinal sites and the bloodstream (Table 4.1). *Absorption* is the term commonly used to describe the passage of intact molecules across the intestinal epithelium (Wilson and Walzer, 1935; Cornell *et al.*, 1972). Substances are absorbed by passive diffusion, enzymatic active transport or pinocytosis through enterocytes. Volkheimer coined the term *persorption* to describe the transport of large solid particles in the micrometre range across the mucosal barrier (reviewed in Volkheimer, 1974). A variety of substances, such as pollen, spores, cellulose particles and starch granules, were given orally and then detected in the blood and urine of mice, rats, guinea-pigs, chickens and dogs. Volkheimer also ingested these substances himself and demonstrated their persorption to his blood and urine. It was suggested that persorption occurred between epithelial cells (paracellular route) since many of these particles have a greater diameter than the lumenal border of an individual enterocyte. Apparently, only 1 out of 50 000 ingested particles was actually persorbed. The mechanisms suggested for persorption were loosening of the epithelial cell tight junctions in the immediate vicinity of goblet cells, spontaneous loosening of tight junctions as a result of epithelial cell desquamation, and mechanical loosening of the tight junctions of epithelial cells located at the base of villi caused by the entry of these large particles into the enterocyte. The route of passage of the particles after entering the mucosal epithelial cells appeared to be *via* the chyle or portal circulation.

Frank *et al.* (1948) used the term *transmural migration* to describe the passage of *Escherichia coli* and other enteric bacteria from the gastrointestinal tract through the bowel wall to the peritoneal cavity of acute renal failure patients receiving peritoneal irrigation. Schweinburg *et al.* (1949) were able to repeat these observations in uremic dogs and then to demonstrate the transmural migration of radioactive-labelled *E. coli* from the gastrointestinal tract to the peritoneal cavity of dogs exhibiting chemically induced peritonitis (Schweinburg *et al.*, 1950). Keller and Engley (1958), however, coined the term *translocation* to describe the passage of orally inoculated *Bacillus megaterium* bacteriophage from the intestinal lumen to the bloodstream of mice. They speculated that the route of passage may have been *via* the blood capillaries but no experiments were performed to distinguish between the blood and lymphatic routes. Hildebrand and Wolochow (1962) used the term *translocation* for the passage of orally administered *E. coli* bacteriophage to the lymph. Wolochow *et al.* (1966) appear to be the first to have used the term *translocation* to describe the passage of bacteria from the gastrointestinal tract. They found *Serratia marcescens* in the lymph from the cisterna chyli but not in the blood of rats after intraduodenal inoculation of *Ser. marcescens*. Fuller and Jayne-Williams (1970) detected indigenous staphylococci, streptococci, lactobacilli and other unidentified Gram-positive and Gram-negative anaerobic rod-shaped bacteria in the livers of newly hatched, conventional chicks. These indigenous bacteria most likely originated from the gastrointestinal tract, since in other experiments *Enterococcus faecalis* translocated to the livers of ex-germ-free chicks monoassociated with *Ent. faecalis*. Fuller and Jayne-Williams also used the earlier term *translocation* to describe this passage of indigenous bacteria from the gastrointestinal tract to the liver.

Unaware of the above studies, I found viable indigenous *E. coli* in the mesenteric lymph node complex (MLN) of gnotobiotic mice monoassociated for 1 or 2 weeks with *E. coli* (Berg, 1975). Further investigation revealed that indigenous bacteria, such as *E. coli*. are not normally cultured from the MLN, spleen or liver of so-called specific pathogen-free (SPF) mice (Berg and Garlington, 1979). Indigenous bacteria, however, appeared in the MLN of gnotobiotic mice intragastrically inoculated with a whole caecal flora obtained from a donor SPF mouse. Indigenous bacteria did not appear in the MLN of SPF mice similarly inoculated with this whole caecal flora preparation. *E. coli* and *Lactobacillus acidophilus* also were cultured from the MLN of gnotobiotic mice monoassociated with either of these bacteria. I also adopted the term *bacterial translocation* to describe this phenomenon since this term simply describes the relocation of bacteria from one site to another without implying anything concerning the actual mechanisms, which undoubtedly would prove to be very complex (Berg and Garlington, 1979). A search of the literature then revealed that earlier investigators had observed this phenomenon of the passage of viable bacteria from the gastrointestinal tract to extraintestinal sites but did not attract attention by attaching a term, such as translocation, to the process (Gordon *et al.*, 1955; Canada and Strong, 1965; van der Waaij *et al.*, 1972; Maejima and Tajima, 1973).

Bacteria of the indigenous microflora are not normally found in extraintestinal sites, such as the MLN, spleen, liver or blood of SPF mice housed under barrier conditions (Berg and Garlington, 1979). However, indigenous bacteria are most likely continuously translocating across the mucosal epithelium in these mice at a very low rate but are killed by the host immune defences either *in transit* through the *lamina propria* or *in situ* in reticuloendothelial organs, such as the MLN. On the rare occasions that MLN of healthy SPF mice contain translocating bacteria, they are usually identified as lactobacilli or enterococci. The Enterobacteriaceae, on the other hand, are the bacterial types most often found translocating from the gastrointestinal tract in experimental models designed to promote bacterial translocation (Berg and Garlington,

1979; Berg, 1980b, 1981b, 1983b, 1985). Most likely lactobacilli and enterococci, rather than Enterobacteriaceae are the bacterial types 'spontaneously' translocating in healthy SPF mice because lactobacilli and enterococci normally colonize the gastrointestinal tract at higher population levels than do the Enterobacteriaceae. On the other hand, obligate anaerobes populate the gastrointestinal tract at the highest levels but are rarely found to translocate spontaneously in SPF mice or in animal models designed to promote bacterial translocation. Sensitivity to oxygen may be one mechanism inhibiting the translocation of these obligate anaerobes to the MLN and other sites (Itoh and Berg, unpublished observations). However, there is very little information concerning the mechanisms whereby bacteria translocate across the presumably healthy intestinal mucosa.

The translocation efficiency of *E. coli* C25 to the MLN was compared in ten inbred mouse strains (Maejima *et al.*, 1984c). The mice were antibiotic-decontaminated and then monoassociated with *E. coli* C25. The caecal populations of *E. coli* C25 were similar in all the mouse strains. There were no differences in the incidence or magnitude of *E. coli* C25 translocation to the MLN of the various mouse strains. Thus, genetic differences among the mouse strains did not influence *E. coli* C25 translocation.

Salmonella typhimurium and enteropathogenic *E. coli* translocate primarily through the mucosal epithelial cells (an intracellular or transcellular route) rather than between the epithelial cells (an intercellular or paracellular route) (Takeuchi, 1967; Staley *et al.*, 1968, 1969). Consequently, I have suggested that the relatively nonpathogenic, indigenous bacteria also translocate through enterocytes rather than interrupting tight junctions to pass between enterocytes (Owens and Berg, 1982; Berg, 1985). Recent studies have demonstrated that *Candida albicans* and *E. coli* translocate intracellularly from Thiry-Vella intestinal loops of thermally injured guinea-pigs and rats (Alexander *et al.*, 1990). Once into the *lamina propria* both *Can. albicans* and *E. coli* were found either engulfed by macrophages or free in the lymphatics and blood vessels. Wells *et al.* (1990) demonstrated that *Ent. faecalis* also translocates intracellularly through enterocytes. If the intact mucosal barrier is disrupted by inflammation or injury, then indigenous bacteria can easily pass through the denuded or ulcerated areas of the mucosa, and perhaps even through loosened tight junctions. It is not known whether indigenous bacteria translocate with equal efficiencies from all sites in the gastrointestinal tract or whether they translocate more efficiently from certain regions, such as the ileum or caecum.

The primary defence mechanisms preventing indigenous bacteria from translocating from the gastrointestinal tract are: (a) a stable gastrointestinal microflora preventing bacterial overgrowth by certain indigenous bacteria or colonization by more pathogenic exogenous bacteria (Berg and Owens, 1979; Berg, 1980a, b, 1981c; Steffen and Berg, 1983); (b) the host immune defences (Berg, 1983a; Owens and Berg, 1980, 1982); and (c) an intact mucosal barrier (Morehouse *et al.*, 1986; Deitch *et al.*, 1989b). More than one of these defence mechanisms can be involved, depending upon the animal model or clinical situation. These defence mechanisms are examined in more detail in the following sections.

4.2 DEFENCE AGAINST BACTERIAL TRANSLOCATION

4.2.1 A stable gastrointestinal microflora

Gnotobiotic rodents have been the animal models most often used to demonstrate that intestinal bacterial overgrowth promotes bacterial translocation. Hale and Hill (1973) and Maejima and Tajima (1973) found viable bacteria in the MLN, spleens, livers and kidneys of gnotobiotic rodents. Berg and Garlington (1979) intragastrically inoculated germ-free or SPF mice with the whole caecal microflora obtained from an SPF mouse. In order of frequency, *E. coli*, *L. acidophilus*, *Ent. faecalis*, *Klebsiella pneumoniae* and *Proteus mirabilis* were cultured 1 week later from the MLN of the ex-germ-free mice, but no bacteria were cultured from the MLN of the SPF mice. The SPF mice were intragastrically inoculated with the same caecal microflora preparation and housed under the same conditions in germ-free isolators as the ex-germ-free mice. Indigenous *E. coli* and *L. acidophilus* also translocated to the MLN of gnotobiotic or antibiotic-decontaminated SPF mice intragastrically inoculated with either of these bacteria. *E. coli* translocated at a greater incidence than did *L. acidophilus*. In another experiment, *E. coli* was found translocating to the MLN 112 days after monoassociation of germ-free mice, the longest period tested. In gnotobiotic mice diassociated with indigenous *E. coli* and *Bacteroides fragilis*, *E. coli* translocated to 100% of the MLN whereas *Bact. fragilis* was not found in any MLN cultures.

The results of these early experiments suggestes several hypotheses. Hypothesis 1: Intestinal bacterial overgrowth promotes bacterial translocation. In the gnotobiotic mice tested 1 week after intragatrically inoculation with the whole caoecal microflora, bacterial translocation occurred because ecologic equilibrium had not been established as yet and various indegenous bacteria attained abnormally high population levels. Bacterial translocation did not occur in SPF mice intragastrically innoculated with this same ceacal microflora because the indignous bacteria did not attain abnormally high population levels; the various ecologic niches were already occupied by indigenous bacteria. The high rate of translocation observed in gnotobiotic mice monassociated with *E. coli* or *L. acidophilus* also supports this hypothesis. Hypothesis 2: The indigenous translocating bacteria are continuously seeded from the gastrointestinal tract and are not simply multiplying in the MLN. This hypothesis is supported by the observation that *E. coli*. The rate of *E. coli*translocation to the MLN does not increase overtime but maintains a dynamic steady state. Furthermore, when the intestinal population levels of *E. coli* are decreased to less than $10^8 g^{-1}$, *E. coli* ceases to translocate with equal efficiencies from the gastrointestinal tract. For example, indienous *Bact. fragilis* did not translocate to the MLN in gnotobiotic mice inoculated with the entire caecal microflora or in gnotobiotic mice diassociated with *E. coli* plus *Bact. fragilis*. These hypotheses are further tested in the following studies.

To further strengthen the hypothesis that increased intestinal population levels promote bacterial translocation, *E. coli* population levels and *E. coli* translocation were compared between gnotobiotic mice monoassociated with *E. coli* and gnotobiotic mice colonized with *E. coli* plus an indigenous microflora (Berg and Owens, 1979). *E. coli* C25 (obtained from R. Freter, Univ. of Mich., Ann Arbor, MI) was used since it can be easily isolated due to its resistance to streptomycin, whereas the indigenous *E. coli* strains normally present in SPF mice are streptomycin-sensitive. Two groups of germ-free mice were monoassociated with *E. coli* C25. After 1 week, one group was intragastrically inoculated with the whole caecal contents obtained from three SPF mice. At various intervals after inoculation with the caecal contents, mice from both the experimental and control groups were sacrificed and the bacterial population levels and bacterial translocation assessed. In the control group colonized (monoassociated) with only *E. coli* C25, caecal populations of *E. coli* C25 remained greater than 10^{11} g^{-1} during the entire test period and *E. coli* C25 translocated to 100% of the MLN. In the gnotobiotes colonized with *E. coli* C25 plus a whole indigenous microflora, there was a 1000-fold reduction in *E. coli* C25 caecal populations by 48 h Concomitant with this decrease in *E. coli* C25 population levels were decreases in both the incidence and magnitude of *E. coli* C25 translocation. Thus, the incidence and magnitude of *E. coli* C25 translocation to the MLN were related to the caecal population levels of *E. coli* C25. Bacterial antagonism by the obligate anaerobes appears to be the defence mechanism inhibiting the translocation of Enterobacteriaceae, such as *E. coli*. Factors other than an increase in *E. coli* caecal population levels possibly could have promoted *E. coli* translocation from the gastrointestinal tract to the MLN in the above experiments. For example, biochemical factors in the caecal flora inoculum, other than the antagonistic bacteria, might inhibit the caecal populations of *E. coli* and thereby reduce *E. coli* translocation to the MLN. The thicker *lamina propria* of the conventional mouse as compared with the germ-free mouse might provide a more efficient anatomical barrier against bacterial translocation. The *lamina propria* of the germ-free mouse contains only one-tenth the number of secretory IgA-producing lymphocytes as that of the conventional mouse. Therefore, the *lamina propria* of the conventional mouse might be a more efficient immunologic barrier to bacterial translocation than that of germ-free mice. Bammann *et al.* (1978) reported that streptomycin-resistant bacterial mutants do not colonize the gastrointestinal tract as well as do parental strains. Consequently, it also is possible that *E. coli* C25, a streptomycin-resistant mutant, does not colonize germ-free mice to the same population levels as would streptomycin-sensitive parental strains.

Further studies were conducted to provide information concerning these questions (Berg, 1980a). Germ-free mice were monoassociated with an indigenous strain of *E. coli* susceptible to streptomycin rather than streptomycin-resistant *E. coli* C25. After 1 week, one group of mice again was inoculated intragastrically with a whole caecal microflora while the other group served as the *E. coli* monoassociated control. The caecal population levels of the indigenous *E. coli* strain remained at high levels (10^{10-11} g^{-1}) in the monoassociated gnotobiotes and *E. coli* translocated to 100% of the MLN. Similarly to the previous experiment, the indigenous *E. coli* levels in the caeca of the gnotobiotes colonized with the indigenous *E. coli* strain plus the whole caecal flora decreased over time with a concomitant reduction in the translocation of indigenous *E. coli* to the MLN. Thus, bacterial antagonism of either a streptomycin-resistant mutant of *E. coli* or of a streptomycin-sensitive, indigenous strain of *E. coli* resulted in decreased translocation to the MLN. Neither the *E. coli* caecal populations nor the *E. coli* translocation were decreased when heat-treated, formalin-treated, or filter-sterilized caecal contents were given intragastrically to gnotobiotes monoassociated with *E. coli*. Thus, viable bacteria in the caecal inoculum, most likely antagonistic obligate anaerobes, are required to reduce *E. coli* populations and inhibit *E. coli* translocation from the gastrointestinal tract.

Only two weeks of colonization of gnotobiotic mice by the indigenous microflora might not be sufficient time to stimulate an effective immunologic or anatomic barrier to *E. coli* translocation. Consequently, the above experiments utilizing germ-free mice were repeated in SPF mice, which already possess a thick *lamina propria* compared with germ-free mice (Berg, 1980c). Adult SPF mice were given bacitracin plus streptomycin in their drinking water for 7 days to remove the antagonistic indigenous microflora. The SPF mice exhibited approximately 10^{10}/g caecal populations of obligately bacteria before antibiotic treatment and no anaerobic bacteria after 4 or 7 days of ntibiotic treatment (Figure 4.1). The mice were inoculated intragastrically with streptomycin-resistant *E. coli* C25 on day 4; the *E. coli* indigenous to mice are streptomycin sensitive. By day, the caecal populations of *E. coli* C25 had increased to approximately 10^{11}/g since there were no obligate anaerobes present to antagonize the *E. coli* C25 levels. As expected, the increase in *E. coli* C25 caecal populations promoted the translocation of *E. coli* C25 to the MLN (100% positive incidence; 10/10). The SPF mice then were reinoculated intragastrically on day 7 with a whole caecal microflora obtained from SPF mice. The caecal levels of the obligate anaerobes returned to approximately 10^{10}/g by day 14 and the levels of *E. coli* C25 fell from 10^{11} to 10^{6-7}/g. Accompanying the decrease in *E. coli* C25 caecal populations was a decrease in *E. coli* C25 translocation to the MLN (40% positive incidence). These results suggest a 'cause and effect' relationship between *E. coli* C25 caecal population levels and *E. coli* C25 translocation to the MLN.

A direct relationship between caecal population levels and bacterial translocation was demonstrated by comparing the caecal populations and translocation to the MLN of three strains of Enterobacteriaceae in gnotobiotic mice (Steffen and Berg, 1983). Germ-free mice were triassociated with indigenous *E. coli*, *K. pneumoniae*, and *Pr. mirabilis*. At 1, 3, 5, 8 and 12 weeks, the caecal populations and translocation rates of each of the three species were determined. Minimal translocation occurred to the spleen, liver, kidney or peritoneal cavity. However, the three species translocated readily to the MLN. The total caecal population (consisting of all three species) was determined and the proportion contributed by each of the three bacterial species was calculated. The proportion of each species contributing to the total bacteria translocating to the MLN was also determined. The proportion of each species in the total caecal populations was then compared to the proportion of that species translocating to the MLN. In nearly all cases (18 of 21 comparisons), the proportion of a particular species (e.g. *E. coli*) in the total caecal population was the same as the proportion of that species (e.g. *E. coli*) translocating to the MLN. Thus, a direct relationship was demonstrated between the numbers of a particular bacterial species populating the caeca of the triassociated gnotobiotic mice and the numbers of that same species translocating to the MLN.

These observations in animal models are relevant to the clinical situation. Decontamination of SPF mice with oral antibiotics to allow bacterial overgrowth by certain antibiotic-resistant bacteria is similar to the ecologic disruption that occurs when patients receive oral antibiotic therapy and certain indigenous bacteria then overgrow their intestines. Because of the reduction in 'colonization resistance', these patients also are often colonized by antibiotic-resistant, exogenous bacteria. To mimic this clinical condition more closely, bacterial translocation and caecal population levels were determined in mice that were orally treated with only one type of antibiotic for only a short period of time (Berg, 1981c). SPF mice were given penicillin-G

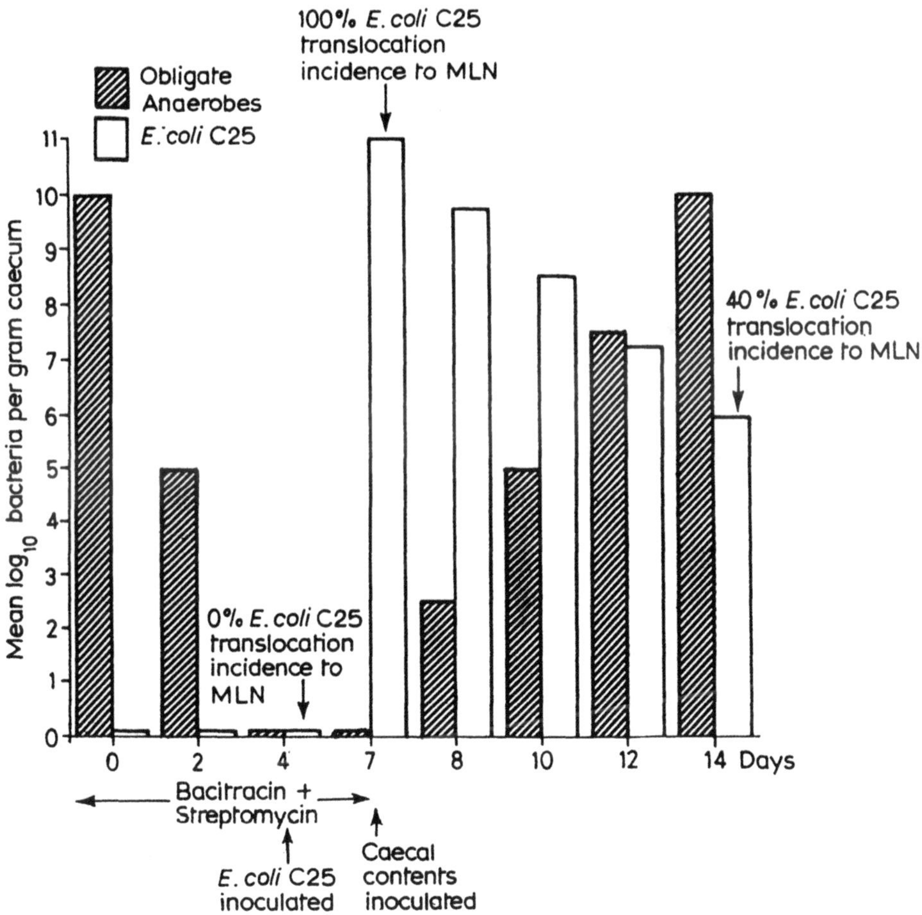

Figure 4.1 Promotion of *Escherichia coli* C25 translocation to the MLN by increased *Escherichia coli* C25 caecal populations.

(500 U ml^{-1}) in their drinking water for 4 days and the penicillin-G was then discontinued. The mice were sacrificed at various intervals and the caeca and MLN cultured for aerobic, facultatively anaerobic and obligately anaerobic bacteria. The numbers of obligate anaerobes decreased 1000-fold whereas the numbers of Enterobacteriaceae increased 10 000-fold to $> 10^{10}$ g^{-1} caecum (Figure 4.2). The penicillin treatment removed the portion of the indigenous microflora that was antagonistic to the Enterobacteriaceae. Concomitant with the increase in Enterobacteriaceae population levels was an increase in Enterobacteriaceae translocation to the MLN; 100% of the MLN cultures were positive by day 4. The translocating bacteria were identified as *Enterobacter cloacae*, *Eb. aerogenes*, *K. pneumoniae*, *E. coli*, *Pr. morganii*, and *Pr. mirabilis*. When penicillin-G was discontinued on day 4, the obligate anaerobes began to increase back to their normal population levels and the Enterobacteriaceae populations began to decrease. However, it a took considerable time for ecologic equilibrium to re-establish in the gastrointestinal tract. Not until day 35 (31 days after pencillin-G was discontinued), did the MLN cultures become negative for translocating bacteria. Thus, oral penicillin treatment for only 4 days is enough to disrupt the gastrointestinal ecology to such a degree that bacterial translocation continues long after antibiotic treatment is stopped. Similar results were obtained using oral clindamycin or oral metronidazole (Berg, 1981c). These results suggest that patients receiving oral antibiotic therapy, especially those that are immunocompromised, are at increased risk to bacterial translocation from the gastrointestinal tract.

The adverse effects of oral antibiotic therapy, such as antibiotic-associated diarrhoea and increased risk of septicaemia due to intestinal overgrowth by antibiotic-resistant bacteria, have stimulated efforts to devise alternative therapies. For example, in a double-blind clinical trial, patients taking lyophilized cultures of viable *Saccharomyces boulardii* (1 g day^{-1} orally) exhibited decreased incidence of antibiotic-associated diarrhoea (Surawicz *et al.*, 1989). Investigations are currently underway to determine whether *Sac. boulardii* exerts this protection by microbial antagonism, by inactivation of bacterial toxins, or by enhancement of non-specific immunological defences.

4.2.2 Host immune system

(a) Immunoglobulins

The various facets of the host immune system, such as serum immunity, secretory (mucosal) immunity, and cell-mediated immunity, are most likely all important to various degrees in protecting the host against bacterial translocation. Secretory IgA may inhibit the close association of bacteria with the gut mucosa that must occur prior to bacterial translocation across the mucosal barrier. However, the role of secretory IgA in the defence against bacterial translocation has not been tested to date. Serum immunoglobulins also may be important in clearing translocating bacteria once they have entered the *lamina propria*, lymph, blood or reticuloendothelial organs, such as the MLN. Again, very little information is available in this area.

Mice injected once intraperitoneally with immunosuppressive agents, such as cyclophosphamide, prednisone, methotrexate, 5-fluorouracil or cytosine arabinoside, exhibit increased translocation of *L. acidophilus*,

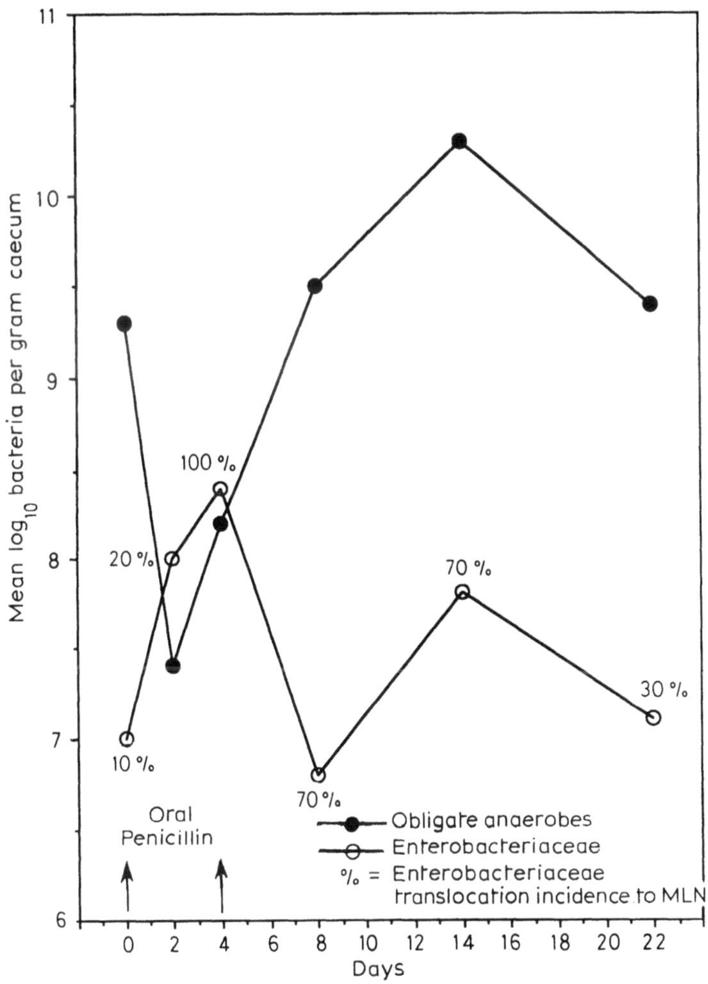

Figure 4.2 Promotion of Enterobacteriaceae translocation to the MLN by Enterobacteriaceae caecal overgrowth in SPF mice.

E. coli, *K. pneumoniae*, *Pr. mirabilis*, *Ent. faecalis* and *Staphylococcus aureus* to the MLN, spleen or liver (Berg, 1983a). Cyclophosphamide, an alkylating agent, and prednisone, a steroid, were more effective in promoting bacterial translocation than were the various antimetabolites. These mice also exhibited decreased serum immune responses following *E. coli* vaccination. These results provide indirect evidence that serum immunity may be effective in clearing translocating bacteria once they have penetrated the mucosa. Serum immunity appears to be especially effective in preventing the systemic spread of translocating bacteria from the MLN to other organs, such as the spleen and liver. Again, however, there is very little information available concerning the contribution of serum immunity in the defence against bacterial translocation from the gastrointestinal tract.

(b) T-cell-mediated immunity

Most of the studies attempting to delineate host immune defences against bacterial translocation have concentrated on cell-mediated immunity, especially the roles of macrophages and T-cells. The availability of congenitally athymic (nu/nu) mice provided a unique animal model for determining whether T-cell-dependent immunity is involved in the immune defence against bacterial translocation. Athymic (nu/nu) and heterozygous euthymic (nu/+) BALB/c mice, 6 months old, were tested for 'spontaneous' translocation of aerobic, facultative or obligately anaerobic bacteria to the MLN, spleen, liver or kidney (Table 4.2) (Owens and Berg, 1980). One-half of these organs (50/100) from athymic nu/nu mice contained translocated bacteria, predominantly *E. coli* and *L. acidophilus*, compared to only 5.2%

Table 4.2 Effect of lack of T-cell-mediated immunity on the incidence of bacterial translocation.

Organs	nu/nu (%)	nu/+ (%)	Thymus-grafted nu/nu (%)	Thymectomized +/+ (%)	Sham-thymectomized +/+ (%)
MLN	44	20	18	46	5
Spleen	56	0	6	29	0
Liver	40	0	0	24	5
Kidney	60	0	6	12	0
All organs	50	5	8	28	2.5

positive organs (5/96) from euthymic nu/+ mice. Bacterial translocation to the spleen, liver and kidney was greater in nu/nu mice than in nu/+ mice; however, translocation to the MLN was similar in both groups. Thus, T-cell-mediated immunity appears to be particularly important in preventing the spread of translocating bacteria from the MLN to other sites, such as the spleen, liver and kidney.

Deitch *et al.* (1986) confirmed the presence of 'spontaneous' translocation in athymic mice compared to euthymic mice. Penn *et al.* (unpublished observations) also found increased *E. coli* C25 translocation to the spleen and liver of BALB/c athymic mice monoassociated with *E. coli* C25 compared to *E. coli* C25 monoassociated heterozygotes. Maddus *et al.* (1988), however, in a more recent study did not detect increased bacterial translocation in athymic nu/nu mice compared with heterozygotes. However, their translocation assay required at least ten translocating bacteria to produce a positive culture result, whereas our translocation assay theoretically requires only one viable bacterium. Also, they used a different mouse strain which most likely harbours a different microflora than the mouse strain we used.

We also grafted nu/nu mice with thymuses from nu/+ mice to provide additional support for the role of T-cell-dependent immunity in the defence against bacterial translocation (Owens and Berg, 1980). Nu/nu recipient mice, 21 days old, were grafted with neonatal thymuses from 1–2-day-old donor nu/+ mice. The grafted mice were tested for their immune responsiveness to T-cell-dependent antigens (sheep erythrocyte vaccination) 4 weeks after receiving the thymic grafts. Both the thymus-grafted (nu/nu) mice and the heterozygous (nu/+) mice exhibited increased serum haemagglutinins compared with nu/nu mice. Histologic examination of recovered thymus grafts at sacrifice also demonstrated normal thymus architecture with well-defined cortex and medulla regions. At 8 weeks of age, the total incidence of bacterial translocation to the MLN, spleen, liver and kidney had decreased from 50% in the nu/nu mice to 8% (5/64 organs) in the thymus-grafted nu/nu mice; similar to the 5% translocation incidence (5/96 organs) noted in nu/+ mice (Table 4.2).

Congenitally athymic (nu/nu) mice might possess unrecognized genetic abnormalities in addition to their athymic condition that could influence bacterial translocation. Consequently, SPF mice, 12–24 h old, were thymectomized and then tested for bacterial translocation at 9 weeks of age. The thymectomized mice demonstrated a reduced haemagglutinin response to sheep erythrocytes as compared with sham-thymectomized mice. Thymectomized mice also exhibited 46% positive MLN cultures (19/41) versus 5% positive MLN cultures (1/20) from sham-thymectomized mice (Table 4.2). Of the total MLN, spleen, liver and kidney cultures from the thymectomized mice, 2%

were positive compared with only 2.5% positive organs from the sham-thymectomized mice. The predominant translocating species was *E. coli*, although *Staph. aureus* and *Ent. faecalis* were also present. Cantrell and Jutila (1970) in an earlier study also found bacteria, that presumably translocated from the GI tract, in the liver, spleen and blood of thymectomized BALB/c mice receiving injections of rabbit antimouse thymocyte sera.

These results suggest that T-cell-mediated immunity contributes to the host defence against bacterial translocation and, in particular, inhibits the spread of translocating bacteria from the MLN to other sites, such as the spleen and liver. It is well known that athymic mice are more susceptible than euthymic mice to a variety of bacterial, mycotic and viral infections. Consequently, it is not surprising that athymic mice also are more susceptible to the systemic spread of translocating bacteria. Since secretory immunity is T-cell dependent, it is possible that the lack of secretory IgA on the intestinal mucosal surfaces of athymic mice allows increased bacterial translocation from the gastrointestinal tract. Future research will focus on the role of secretory IgA and the identification of specific T-cell subpopulations important in the host defence against bacterial translocation.

(c) Macrophages

In most animal models demonstrating bacterial translocation, the translocating bacteria first appear in the MLN prior to their appearance in other organs, such as the liver or spleen (Berg, 1981a, 1983b). Thus, resident macrophages in the MLN are ideally situated to clear bacteria translocating from the gastrointestinal tract. Consequently, we determined whether bacterial translocation would be reduced in mice given immunomodulators known to activate macrophages. Since we cannot predict which of the many species of bacteria in the gastrointestinal tract will translocate in any particular clinical condition, it would be useful to develop an immunological regimen that would inhibit the translocation of a broad range of bacterial species. Therefore, we examined three immunomodulators known to non-specifically activate macrophages – namely, glucan, muramyl dipeptide and *Propionibacterium acnes* – for their abilities to inhibit bacterial translocation to the MLN.

Glucan, a polyglycan derived from the cell wall of *Sac. cerevisiae*, increases the activation and proliferation of macrophages (Kimura *et al.*, 1983) and stimulates both cell-mediated and humoral immunity (Hassid *et al.*, 1941; Wooles and DiLuzio, 1962). Muramyl dipeptide (MDP), a small molecular weight glycopeptide responsible for the adjuvant action of *Mycobacterium*, activates macrophages directly (Pabst and Johnson, 1980). *P. acnes* (formerly classified as *Corynebacterium parvum*) exerts many effects on the immune system, the most important being the non-specific activation of macrophages (Hebert *et al.*, 1983).

SPF mice were antibiotic-decontaminated with oral streptomycin and penicillin-G in their drinking water. One group of mice was injected intraperitoneally once with 1.4 mg of formalin-killed *P. acnes* (Burroughs-Welcome, Research Triangle, NC), a second group injected twice with 0.4 mg of particulate glucan-P (Accurate Chemical and Scientific Corp., Westbury, NY), and a third group injected twice with muramyl dipeptide. The mice were challenged by giving viable *E. coli* C25 ($1 \times 10^9 ml^{-1}$) in their drinking water. The mice were sacrificed on day 14 and the MLN cultured for translocating *E. coli* C25. Vaccination with *P. acnes*, but not glucan or muramyl dipeptide, inhibited *E. coli* C25 translocation to the MLN. The vaccination stimulated a lymphoreticular response in all three groups of mice, as demonstrated by splenomegaly, a commonly used indicator of macrophage activation (Burgaleta and Golde, 1977; Patchen and McVittie, 1983). Interestingly, even though glucan vaccination did not decrease *E. coli* C25 translocation to the MLN, glucan vaccination did reduce mortality following intraperitoneal challenge with 10^{10} viable *E. coli* C25. It is surprising that non-specific immunostimulation by particulate glucan reduces lethal *E. coli* C25 peritonitis caused by large numbers of *E. coli* C25 but does not inhibit the very low numbers of *E. coli* C25 translocating from the gastrointestinal tract to the MLN. Perhaps glucan activates neutrophils more effectively than macrophages.

Studies were continued with *P. acnes* but not with MDP or glucan since neither of these immunomodulators inhibited *E. coli* C25 translocation in these initial experiments. *P. acnes* vaccination of antibiotic-decontaminated mice monoassociated with *E. coli* C25 reduces both the incidence and numbers of *E. coli* C25 translocating to the MLN (Table 4.3) (Fuller and Berg, 1985). Interestingly, *P. acnes* vaccination did not inhibit either the incidence or magnitude of *E. coli* C25 translocation to the MLN of gnotobiotic mice monoassociated with *E. coli* C25. The gnotobiotic mice were immunologically 'stimulated' by the *P. acnes* vaccination since they exhibited a 7.1 splenic index that was even greater than the 5.7 splenic index of the SPF mice. Also, differences in *E. coli* C25 population levels were not a factor since the caecal levels of *E. coli* C25 were the same in both the control, non-vaccinated gnotobiotes and the vaccinated gnotobiotes. Thus, it appears that the mouse immune system must be 'primed' by antigens of the indigenous microflora before a 'second stimulation' by *P. acnes* vaccination can activate macrophages sufficiently to inhibit bacterial translocation.

Table 4.3 Inhibition of *Escherichia coli* C25 translocation by *Propionibacterium acnes* vaccination in antibiotic-decontaminated SPF or gnotobiotic mice monoassociated with *Escherichia coli* C25.

Mice	Vaccine	Spleen weight (g)	Splenic index	CFU g^{-1} caecum	Translocation incidence to MLN(%)	CFU g^{-1} MLN
Antibiotic-treated SPF	Control	0.1	5.7	1.7×10^9	75	1860
	P. acnes	0.6		2.2×10^9	41*	304*
Gnotobiotes	Control	0.1	7.1	4.6×10^8	100	2650
	P. acnes	0.7		5.4×10^8	100	1730

*$p < 0.01$.

To test this hypothesis, adult germ-free mice were colonized with either *E. coli* C25 or the whole caecal microflora for 8 weeks prior to *P. acnes* vaccination. *P. acnes* vaccination did not reduce *E. coli* C25 translocation to the MLN in these adult gnotobiotes monoassociated with *E. coli* C25 or colonized with the whole indigenous microflora (Berg and Itoh, 1986). However, if the germ-free mice were colonized with the whole indigenous microflora within 1 week after birth, then a subsequent *P. acnes* vaccination given at 8 weeks of age reduced *E. coli* C25 translocation (Berg and Itoh, 1986). Other investigators also have reported that prior exposure to the indigenous microflora increases the effectiveness of *P. acnes* vaccination (Scott, 1972; Bash, 1978; Wells and Balish, 1979). It is not surprising that the indigenous microflora affects the development of the host immune response since the indigenous microflora so profoundly influences the anatomic and physiological development of the host. However, the immunostimulating effect of the indigenous microflora usually is demonstrated with T-cell-dependent sheep erythrocytes rather than with T-cell-independent bacterial antigens.

P. acnes vaccination decreases bacterial translocation when given at the time of bacterial challenge or even when given after the bacteria are already translocating to the MLN. *P. acnes* vaccination also non-specifically protects against translocation to the MLN of either indigenous *E. coli*, *Pr. mirabilis* or *Eb. cloacae*. However, it has not been determined as yet whether *P. acnes* vaccination also will inhibit the translocation of Gram-positive bacterial species.

These studies demonstrate that non-specifically activated macrophages inhibit bacterial translocation from the gastrointestinal tract to the MLN. Other investigators, however, have suggested that macrophages actually may help transport bacteria and particles from the gastrointestinal tract. Harmsen *et al.* (1985) demonstrated that fluorescent microspheres could be transported by macrophages from the lungs to the tracheobronchial lymph nodes. The microspheres did not travel to the lymph nodes to be phagocytized subsequently by lymph node macrophages, but instead the microspheres were first engulfed by lung macrophages and then the macrophages travelled to the nodes. In similar experiments, Wells *et al.* (1988) found that fluorescent microspheres injected into ligated intestinal loops were taken up by macrophages. However, since ligated intestinal loops are subjected to internal pressure, it is not known if macrophage transport of engulfed bacteria also occurs in the normal situation. Wells *et al.* (1987) also have reported that polymorphonuclear leukocytes can transport bacteria from the gastrointestinal tract to experimental abdominal abscesses.

Our results suggest that macrophages may be important effector cells in the host defence against bacterial translocation since bacterial translocation is inhibited by vaccination with killed *P. acnes*, a non-specific macrophage activator. This hypothesis is further strengthened by our observations that the adoptive transfer of MLN or spleen cells obtained from *P. acnes*-vaccinated donor mice inhibits *E. coli* C25 translocation to the MLN in non-vaccinated recipient mice (Gautreaux and Berg, 1990). However, there are many questions yet to be answered. For example: what are the relative contributions of secretory immunity (e.g. secretory IgA at mucosal surfaces), systemic immunity (serum IgG and IgM), and cell-mediated immunity (macrophages and T-cells) in the host immune defence against bacterial translocation? Are T-cells required for the activation of effector macrophages? Which subpopulations of T-cells are most important? Elucidation of the mechanisms whereby the host immune system inhibits bacterial translocation from the gastrointestinal tract is required in order to devise preventative immunological strategies.

4.2.3 Intact mucosal barrier

The intact intestinal mucosa provides a physical or mechanical barrier to prevent bacteria colonizing the gastrointestinal tract from translocating to invade extraintestinal organs and tissues. Thus, the integrity of the gut mucosa is an important host defence mechanism and indigenous bacteria are not normally cultured from the MLN or other extraintestinal sites of healthy adult rodents (Garlington and Berg, 1979). Ricinoleic acid (12-hydroxy-9-octadecenoic acid), the pharmacologically active constituent of castor oil, given orally to mice severely damages the mucosal barrier and allows bacteria to translocate from the gastrointestinal tract (Morehouse *et al.*, 1986). Mice were given a single intragastric inoculation of ricinoleic acid at the same physiologic dose by weight usually given therapeutically to humans. Massive exfoliation of columnar and goblet cells in the proximal small intestine was observed by 2 hours after the ricinoleic acid administration. Because of this loss of the mucosal barrier, both facultative and obligate anaerobic bacteria of the indigenous microflora translocated to the MLN, spleen and liver. The incidence of bacterial translocation was greatest at 4 days following the administration of ricinoleic acid but ceased completely by 7 days. Bacterial translocation would most likely have had more serious consequences if the mice had been colonized by more pathogenic bacteria, such as *Pseudomonas aeruginosa*, or if the mice had been immunocompromised. Perhaps castor oil, a commonly used purgative, should not be given to immunocompromised patients who are already at risk to bacterial septicaemia.

Large amounts of endotoxin, the lipopolysaccharide component of Gram-negative bacteria, are naturally present in the intestines since the gastrointestinal tract normally harbours tremendous populations of Gram-negative indigenous bacteria. Endotoxin is released during bacterial cell growth and particularly upon bacterial cell death and lysis. However, only very small amounts of endotoxin are absorbed from the healthy intestines. The small amounts of endotoxin that are absorbed are detoxified by Kuppfer cells in the liver (Nolan, 1981). However, in severely ill or injured patients, bacteria and endotoxin can cross the gut mucosal barrier and gain access to tissues and the systemic circulation.

A relationship among endotoxaemia, intestinal ischaemia and shock was first proposed by Ravin and Fine (1962). More recently, it is hypothesized that gut barrier failure, in conjunction with hepatic dysfunction, promotes or potentiates the newly recognized clinical syndrome called multiple organ failure (Carrico *et al.*, 1986). Multiple organ failure is a common final pathway leading to death in a variety of patients. Conditions such as shock, infection or immunosuppression increase gut mucosal permeability resulting in increased translocation of bacteria and bacterial products (e.g. endotoxin) from the gastrointestinal tract to the portal and systemic circulations. Translocated endotoxin then activates various plasma protein cascades, resident macrophages and circulating neutrophils to release monokines and proteins that, in turn, further increase gut mucosal permeability.

Since endotoxaemia is common in a variety of patients, studies were performed in rodent models to determine whether endotoxin also could promote the translocation of indigenous bacteria from the gastrointestinal tract. SPF mice were injected once intraperitoneally with *E. coli* 026 : B6 endotoxin and the peritoneum, MLN, spleen, liver and blood tested 24 h later for translocating indigenous bacteria (Deitch *et al.*, 1987a). Endotoxin promoted bacterial translocation to the MLN in a dose-dependent fashion; the spleen and liver cultures were negative. Thus, one intraperitoneal injection of endotoxin promotes bacterial translocation to the MLN, but the translocating bacteria remain confined to the MLN and do not spread systemically to other organs, such as the spleen or liver.

Endotoxin is composed of a core polysaccharide, a lipid-A component containing long-chain fatty acids linked to a glucosamine backbone, and polysaccharide side chains (the O-antigens). To determine which components of the endotoxin structure are important in promoting bacterial translocation, mice were injected with endotoxin from six R-mutant strains of *Salmonella* (Ra, Rb, Rc, Rd, Re or lipid A) differing in their endotoxin compositions. Intact *Salmonella* endotoxin (wild type) and the Ra and Rb endotoxin fragments promoted bacterial translocation to the MLN whereas the *Salmonella* endotoxin fragments lacking the terminal-3 sugars of the core polysaccharide (Rc, Rd, Re or lipid A) did not promote bacterial translocation (Deitch *et al.*, 1989c). Thus, the terminal portion of the core polysaccharide of *Salmonella* endotoxin appears to be required to promote bacterial translocation. Only the endotoxin fragments that promoted bacterial translocation were also associated with increased caecal population levels of Enterobacteriaceae. Also, injection of the translocation-promoting Ra fragment increased ileal xanthine oxidase and xanthine dehydrogenase activities, suggesting mucosal injury, whereas the non-promoting Rc and Re fragments did not increase these enzymatic activities. Histological examination of gastrointestinal tissue from endotoxin-challenged mice revealed physical disruption of the mucosal barrier. The ileal and caecal *lamina propria* were oedematous with separation of the epithelium from the *lamina propria* in certain areas, such as at villus tips. The duodenal, jejunal and colonic mucosa appeared normal. Thus, endotoxin-induced bacterial translocation appears to be due primarily to damage to the gut mucosal barrier. It is suggested that both lipid A and endotoxin produce their toxic manifestations by stimulating host cells, especially macrophages, to release mediator substances that then act as second messengers to disrupt various homeostatic systems.

Parks *et al.* (1982) have implicated xanthine oxidase-generated, oxygen-free radicals as mediators of intestinal injury. Consequently, it was determined whether inhibition of xanthine oxidase activity by allopurinol or inactivation of xanthine oxidase activity by a sodium tungstate diet would prevent mucosal damage and the subsequent bacterial translocation that occurs after endotoxin challenge (Deitch *et al.*, 1989b, c). SPF mice were antibiotic-decontaminated, monoassociated with *E. coli* C25 and injected once intraperitoneally with *E. coli* 026 :

B6 endotoxin. One group of mice was given allopurinol by gastric lavage ($50\ mg^{-1}kg^{-1}$) 48 and 24 h prior to endotoxin challenge. Another group of mice was placed on a diet supplemented with sodium tungstate ($0.7g^{-1}kg^{-1}$ diet), but restricted in molybdenum (ICN Biochemical, Cleveland, OH). This diet contains normal levels of protein, vitamins, minerals and other trace elements. To deplete intestinal levels of xanthine oxidase, the mice were maintained on the sodium tungstate diet a minimum of 14 days prior to endotoxin challenge. Ileal xanthine oxidase activities were reduced in the mice fed the tungstate diet or treated with allopurinol prior to endotoxin challenge compared to control mice. There also was no significant intestinal mucosal damage in the endotoxin-challenged mice pre-treated with allopurinol or pre-fed the tungstate diet. Furthermore the incidence and magnitude of bacterial translocation to the MLN promoted by endotoxin challenge were reduced in the mice pre-treated with allopurinol or fed the tungstate diet.

To further strengthen these findings, the effect of allopurinol pre-treatment on bacterial translocation was tested in a bacterial overgrowth model in which translocation is promoted by an increase in caecal population levels rather than by damage to the gut mucosal barrier. SPF mice were antibiotic-decontaminated, monoassociated with *E. coli* C25, treated with allopurinol and tested for *E. coli* C25 translocation to the MLN. As expected, there was no difference in *E. coli* C25 translocation between the allopurinol-treated mice and the non-treated control group. Allopurinol pre-treatment should not have decreased bacterial translocation in this overgrowth model since bacterial translocation was promoted by bacterial overgrowth and not mucosal injury. Thus, endotoxin-induced bacterial translocation in the earlier experiments is due, in part at least, to mucosal damage caused by xanthine oxidase activity (Deitch *et al.*, 1989c, d).

Interestingly, the caecal populations of Enterobacteriaceae, such as *E. coli*, increased 100-fold 24 h following endotoxin injection (Deitch *et al.*, 1987a). By 48 h following endotoxin injection, the caecal levels of these bacteria returned to normal and concomitantly both the incidence and magnitude of bacterial translocation to the MLN also decreased. The mechanisms whereby endotoxin injection influences caecal population levels of indigenous bacteria are unknown, although endotoxin is known to reduce gastrointestinal motility and to cause transient intestinal ischaemia. Zymosan, a yeast cell wall product unrelated to endotoxin but inflammatory due to its complement activating activity, injected intraperitoneally also increased the caecal population levels of indigenous Enterobacteriaceae and promoted their translocation to the MLN (Deitch *et al.* 1990b, Mainous *et al.*, 1991).

Haemorrhagic shock also produces intestinal mucosal damage and increased bacterial translocation (Baker *et al.*, 1987, 1988). Rats submitted to haemorrhagic shock for 90 min exhibit necrosis of the ileal mucosa and increased bacterial translocation to the MLN. Mucosal damage in the haemorrhagic shock model also is mediated by oxidants generated by the xanthine oxidase system. Oxygen-free radicals generated during the period of intestinal reperfusion are particularly important. Allopurinol administered orally prior to haemorrhagic shock reduced the mucosal damage and decreased bacterial translocation (Deitch *et al.*, 1988, 1989a). Rats fed the tungsten-supplemented, molybdenum-free diet to inactivate xanthine oxidase prior to haemorrhagic shock also exhibit reduced mucosal damage and decreased bacterial translocation compared with control rats fed a regular diet (Deitch *et al.*, 1988, 1989a).

These experimental models demonstrate that the intact intestinal mucosa is important in the host defence against bacterial translocation. It is likely that patients suffering from severe trauma with associated haemorrhagic shock, thermal injury or endotoxaemia also exhibit intestinal mucosal damage and are at risk to systemic infection by bacteria translocating from their gastrointestinal tracts.

4.3 BACTERIAL TRANSLOCATION IN ANIMAL MODELS WITH MULTIPLE DEFICIENCIES IN HOST DEFENCES

The disruption of the ecological balance in the gastrointestinal tract by oral antibiotics allows both bacterial overgrowth and increased translocation. Immunosuppressive agents also promote the translocation of indigenous bacteria from the gastrointestinal tract. Since immunocompromised patients often receive antibiotic therapy, it was of interest to determine whether the combination treatment of an oral antibiotic plus an immunosuppressive agent could synergistically promote bacterial translocation.

SPF mice were given either penicillin-G or clindamycin in their drinking water *ad libitum*. Groups of these mice also were injected four times with either cyclophosphamide or prednisone. Other groups were given combinations of these drugs. Oral penicillin or oral clindamycin alone promoted the translocation of enteric bacilli to 100% and 85% of the MLN respectively (Table 4.4) (Berg *et al.*, 1988). The translocating bacteria did not spread systemically since none of the peritoneal cultures became positive. Prednisone or cyclophosphamide injection also promoted bacterial translocation to the MLN, although to a lesser degree than the oral antibiotics. Again, the peritoneal cultures were negative. However, the combination treatment of an antibiotic plus an immunosuppressive agent not only promoted the translocation of Enterobacteriaceae to the MLN, but 39 – 75% of the peritoneal cultures also were positive. By 14 days after treatment with clindamycin plus prednisone, all the mice had died of septicaemia caused by their own indigenous microflora. Thus, the combination treatment of an oral antibiotic plus an immunosuppressive agent synergistically promotes the systemic spread of translocating bacteria to the peritoneal cavity.

Table 4.4 Caecal overgrowth plus immunosuppression synergistically promotes translocation of Enterobacteriaceae.

Oral antibiotic	Immunosuppressive agent	Translocation incidence (%) to MLN	Peritoneal cultures (% positive)
Penicillin	None	35	0
Clindamycin	None	100	0
None	Cyclophosphamide	17	1.7
None	Prednisone	46	0
Penicillin	Cyclophosphamide	80	39
Penicillin	Prednisone	95	75
Clindamycin	Cyclophosphamide	89	50
Clindamycin	Prednisone	94	55

The synergistic effect of multiple deficiencies in the defence against bacterial translocation has been demonstrated in a variety of other animal models. For example, protein malnutrition alone produces histologic atrophy of the mucosa of the small bowel and caecum, but the mucosal barrier to translocation remains intact (Deitch and Berg, 1987). However, the combination of protein malnutrition plus one endotoxin injection produces mucosal ulceration with a concomitant increase in bacterial translocation (Ma *et al.*, 1989). Similarly, endotoxin injection acts synergistically to promote mucosal damage and bacterial translocation after a 30% total body surface area burn (Deitch and Berg, 1987). Studies are in progress to delineate the relationships among endotoxic shock, increased mucosal permeability, mucosal injury, bacterial translocation and lethal sepsis.

4.4 CONCLUSION

The defence mechanisms preventing bacterial translocation from the gastrointestinal tract include: (a) an ecologically balanced, gastrointestinal microflora exhibiting bacterial antagonism to prevent bacterial overgrowth, (b) the host immune system, and (c) an intact intestinal mucosa providing a physical barrier. Bacterial translocation has been studied in a variety of animal models, including rodents subjected to oral antibiotic treatment (Berg, 1981c; Deitch *et al.*, 1985), streptozotocin-induced diabetes (Berg, 1985), thermal injury (Maejima *et al.*, 1984a, b), solid tumours (Penn *et al.*, 1985), leukaemia (Penn *et al.*, 1986), endotoxaemia (Deitch and Berg, 1987; Deitch *et al.*, 1987a, 1989c), haemorrhagic shock (Baker *et al.*, 1987; Deitch *et al.*, 1990d), protein malnutrition (Deitch *et al.*, 1987b, 1990e), intestinal obstruction (Deitch *et al.*, 1990a), parenteral nutrition (Spaeth *et al.*, 1990) and bile duct ligation (Deitch *et al.*, 1990e). The mechanisms responsible for promoting bacterial translocation in these animal models are presented in Table 4.5 (Berg, 1980b, 1981a, b, 1983b, 1985). In some of these models, the host exhibits multiple deficiencies in defence mechanisms resulting in bacteraemia and lethal sepsis by translocating indigenous bacteria.

Table 4.5 Defence mechanisms against bacterial translocation in animal models.

Animal model	Stable gastrointestinal microflora	Host immune system	Intact intestinal mucosa
Oral antibiotics	+	−	−
Immunosuppressive agents	−	+	±
Athymic (nu/nu) mice	−	+	−
Thymectomized mice	−	+	−
Endotoxaemia	±	+	+
Haemorrhagic shock	−	?	+
Thermal injury	−	+	±
Intestinal obstruction	+	?	+
Protein malnutrition	±	±	±
Leukaemia	−	+	−
Diabetes	±	±	±

The results obtained from the various animal models suggest that the pathogenesis of bacterial translocation occurs in several discrete stages. In the healthy animal, 'spontaneous' bacterial translocation is likely, occurring continuously at a very low rate, but these low numbers of translocating bacteria are killed by the host immune defence. In the first stage of translocation, as occurs during oral antibiotic treatment, the bacteria translocate to the MLN but usually do not spread to other organs. The host does not appear ill and the immune system apparently can clear this low level of translocating bacteria. Of course, if the host is colonized by more virulent exogenous bacteria, then the consequences can be more serious. The second stage of translocation occurs when the translocating bacteria spread from the MLN to other organs, such as the liver, spleen or kidney. This second stage readily occurs when the host immune system is compromised. Again, depending upon the virulence properties of the translocating bacteria, the host may be able to confine the infection. The third and final stage of translocation occurs when the translocating bacteria spread systemically to cause septicaemia. The host can either recover from this systemic infection or succumb to septic shock, depending on the degree of immunosuppression, the degree of mucosal damage and the pathogenic properties of the translocating bacteria.

A few human studies have been conducted suggesting that the results from these animal models may be applicable to the clinical situation. Surveillance cultures of faecal samples from humans with leukaemia or other immunosuppressive disorders have demonstrated an association between the bacterial biotype or serotype present at the highest level in the patient's faeces and the bacterial biotype or serotype causing septicaemia (Tancrede and Andremont, 1985). These studies provide indirect evidence that bacteria translocating from the patient's own gastrointestinal tract can cause septicaemia. Interestingly, in our animal models, the bacterial species most likely to translocate from the gastrointestinal tract, such as *E. coli*, *K. pneumoniae*, *Pr. mirabilis*, *Ps. aeruginosa* and *Ent. faecalis*, are the same species of bacteria that cause a large proportion of septicaemias in debilitated patients (Steffen *et al.*, 1988).

Direct evidence for the translocation of microorganisms from the gastrointestinal tract of humans is provided by the translocation of *Can. albicans* to the blood and urine of a volunteer who ingested large quantities of viable *Can. albicans* (Krause *et al.*, 1969). Recently, translocating bacteria also have been cultured from the MLN of patients with bowel obstruction (Deitch, 1989), colorectal cancer (Vincent *et al.*, 1988) or Crohn's disease (Ambrose *et al.*, 1984). Thus, evidence is accumulating that bacterial translocation from the gastrointestinal tract is an early step in the pathogenesis of opportunistic infections in debilitated patients.

More knowledge is required concerning the mechanisms whereby bacterial translocation is prevented in the healthy host and how these defence mechanisms are compromised in the debilitated patient. The results described here demonstrating that vaccination with killed *P. acnes* inhibits the translocation of various Enterobacteriaceae from the gastrointestinal tract to the MLN suggest that macrophages are important immune effector cells in host defence. The fact that macrophages can be activated non-specifically to inhibit the translocation of a variety of bacteria is of practical significance, since we cannot be predict with certainty which of the many indigenous bacterial species will translocate from the gastrointestinal tract in any given clinical situation. The studies utilizing genetically athymic mice and thymectomized euthymic mice suggest that T-cell-mediated immunity also may play a role in the defence against translocation. Studies are in progress to determine the relative roles in preventing bacterial translocation of the various compartments of the host immune system, namely secretory immunity on mucosal surfaces, cell-mediated immunity (macrophages, neutrophils and T-cell subpopulations) and systemic immunity (serum IgG and IgM).

Oral and even systemic antibiotics must be employed with caution, since bacterial overgrowth in the intestines is such an efficient mechanism of promoting bacterial translocation from the gastrointestinal tract. The results of bacterial translocation studies support the usefulness of the concept of selective antibiotic decontamination to remove only certain bacteria, such as the Enterobacteriaceae, but leave intact the obligate anaerobes to exert colonization resistance. The Gram-negative enteric bacilli (Enterobacteriaceae) translocate at a much greater rate than do the strictly anaerobic bacteria believed to be primarily responsible for colonization resistance. Also, non-indigenous bacteria generally translocate at greater efficiencies from the gastrointestinal tract than do indigenous bacteria. Selective antibiotic decontamination of debilitated patients may eventually prove beneficial in preventing bacterial translocation. The results of bacterial translocation studies also support the probiotic concept of the oral introduction of certain bacteria to correct particular gastrointestinal problems. For example, *L. acidophilus* does not translocate at a high rate from the gastrointestinal tract even when it is present at high population levels. *L. acidophilus* does not cause severe disease even when it does translocate from the gastrointestinal tract under special conditions. It is even possible that a low level of translocation by certain Gram-positive bacteria is helpful in stimulating the host immune response. For example, *P. acnes*, a Gram-positive bacterium, non-specifically stimulates host macrophages to inhibit translocation by Gram-negative bacteria, such as the Enterobacteriaceae.

Bacterial translocation involves complex interactions between the host defence mechanisms and the abilities of bacteria to translocate mucosal barriers and to survive in hostile environments. Delineation of the defence mechanisms inhibiting bacterial translocation in animal models will provide information for devising rational approaches for the control of opportunistic infections originating from the gastrointestinal tract in debilitated patients, such as those with endotoxaemia, thermal injury, trauma, and immunosuppressive disorders such as AIDS.

REFERENCES

Alexander, J. W., Boyce, S. T., Babcock, G. F., *et al.*, (1990). The process of microbial translocation. *Arch. Surg.*, **212**, 496–512.

Ambrose, M.S., Johnson, M., Burdon, D.W. and Keighley, M.R.B. (1984) Incidence of pathogenic bacteria from mesentric lymph nodes and ileal serosa during Crohn's disease surgery. *Brit. J. Surg.*, **71**, 623–5.

Baker, J.W., Deitch, E.A., Berg, R.D. and Ma, L. (1987) Hemorrhagic shock impairs the mucosal barrier resulting in bacterial translocation from the gut and sepsis. *Surg. Forum*, **37**, 73–4.

Baker, J.W., Deitch, E.A., Ma, L. and Berg, R. (1988) Hemorrhagic shock promotes the systemic translocation of bacteria from the gut. *J. Trauma*, **28**, 896–906.

Bammann, L.L., Clark, W.B. and Gibbons, R.G. (1978) Impaired colonization of gnotobiotic and conventional rats by streptomycin-resistant strains of *Streptococcus mutans*. *Infect. Immun.*, **22**, 721–6.

Bash, J.A. (1978) Suppression of rat T-cell proliferation by *Corynebacterium parvum*: T-cell requirement for induction. *J. Reticuloendothelial Soc.*, **23**, 63–73.

Berg, R.D. (1975) Immunological responsiveness of mice to indigenous and nonindigenous microbes. PhD Dissertation, Univ. of Texas, Austin, Texas.

Berg, R.D. (1980a) Inhibition of *Escherichia coli* translocation from the gastrointestinal tract by normal cecal flora in gnotobiotic or antibiotic-decontaminated mice. *Infect. Immun.*, **29**, 1073–81.

Berg, R.D. (1980b) Mechanisms confining indigenous bacteria to the gastrointestinal tract. *Amer. J. Clin. Nutr.*, **33**, 2472–84.

Berg, R.D. (1981a) The translocation of normal flora bacteria from the gastrointestinal tract to the mesenteric lymph nodes and other organs. *Microecol. Ther.*, **11**, 27–34.

Berg, R.D. (1981b) Factors in influencing the translocation of bacteria from the gastrointestinal tract, in *Recent Advances in Germfree Research* (eds S. Sasaki, *et al*), Tokai University Press, Tokyo, pp. 411–18.

Berg, R.D. (1981c) Promotion of the translocation of enteric bacteria from the gastrointestinal tracts of mice by oral treatment with penicillin, clindamycin, or metronidazole. *Infect. Immun.*, **33**, 854–61.

Berg, R.D. (1983a) Bacterial translocation from the gastrointestinal tracts of mice receiving immunosuppressive chemotherapeutic agents. *Curr. Microbiol.*, **8**, 285–92.

Berg, R.D. (1983b) Translocation of the indigenous bacteria from the intestinal tract$_2$ in *The Intestinal Microflora in Health and Disease* (ed. D.), Hentges Academic Press, New York, pp. 333–52.

Berg, R.D. (1985) Bacterial translocation from the intestines. *Exptl Animals*, **34**, 1–16.

Berg, R.D. and Garlington, A.W. (1979) Translocation of certain indigenous bacteria from the gastrointestinal tract to the mesenteric lymph nodes and other organs in a gnotobiotic mouse model. *Infect. Immun.*, **23**, 403–11.

Berg, R.D. and Itoh, K. (1986) Bacterial translocation from the gastrointestinal tract. Immunologic aspects. *Microecol. Ther.*, **16**, 131–45.

Berg, R.D. and Owens, W.E. (1979) Inhibition of translocation of viable *Escherichia coli* from the gastrointestinal tract to the mesenteric lymph nodes and other organs in a gnotobiotic mouse model. *Infect. Immun.*, **25**, 820–7.

Berg, R.D., Wommack, E. and Deitch, E.A. (1988) Immunosuppression and intestinal bacterial overgrowth synergistically promote bacterial translocation. *Arch. Surg.*, **123**, 1359–64.

Burgaleta, C. and Golde, D.L. (1977) Effect of glucan on granulopoiesis and macrophage genesis in mice. *Cancer Res.*, **37**, 1739–42.

Canada, J.C. and Strong, D.H. (1965) Incidence of *Clostridium perfringens* in the livers of conventional and gnotobiotic mice. *J. Bacteriol.*, **89**, 1623–4.

Cantrell, J.C. and Jutila, J.W. (1970), Bacteriologic studies on wasting disease induced by neonatal thymectomy. *J. Immunol.*, **104**, 79–85.

Carrico, C.J., Meakins, J.L., Marshall, J.C. *et al.* (1986) Multiple organ failure syndrome. *Arch. Surg.*, **121**, 196–208.

Chau, A.Y.S., Goldbloom, V.C. and Gurd, F.N. (1951) Clostridia infection as a cause of death after ligation of the hepatic artery. *Arch. Surg.*, **63**, 390–402.

Cornell, R., Walker, W.A. and Isselbacher, K.J. (1972) Small intestinal absorption of horseradish peroxidase: A cytochemical study. *Lab. Invest.*, **25**, 42–8.

Deitch, E.A. (1989) Simple intestinal obstruction causes bacterial translocation in man. *Arch. Surg.*, **123**, 699–701.

Deitch, E.A. and Berg, R.D. (1987) Endotoxin but not malnutrition promotes bacterial translocation of the gut flora in burned mice. *J. Trauma*, **27**, 161–6.

Deitch, E.A., Maejima, K. and Berg, R.D. (1985) Effect of oral antibiotics and bacterial overgrowth on the translocation of the GI tract microflora in burned rats. *J. Trauma*, **25**, 385–92.

Deitch, E.A., Winterton, J. and Berg, R.D. (1986) Thermal injury promotes bacterial translocation from the gastrointestinal tract in mice with impaired T-cell-mediated immunity. *Arch. Surg.*, **121**, 97–101.

Deitch, E.A., Berg, R.D. and Specian, R. (1987a) Endotoxin promotes the translocation of bacteria from the gut. *Arch. Surg.*, **122**, 185–90.

Deitch, E.A., Winterton, J., Ma, L. and Berg, R.D. (1987b) The gut as a portal of entry for bacteremia: The role of protein malnutrition. *Ann. Surg.*, **20**, 8–16.

Deitch, E.A., Bridges, W.R., Baker, J. *et al.*, (1988) Hemorrhagic shock-induced bacterial translocation is reduced by xanthine oxidase inhibition or inactivation. Surgery, **104**, 191–8.

Deitch, E.A., Bridges, W.R., Ma, L. *et al.* (1989a) Role of xanthine oxidase and neutrophil generated oxy radicals in shock-induced bacterial translocation. *J. Trauma*, **29**, 1027.

Deitch, E.A., Ma, J.W., Ma, L. *et al.* (1989b) Endotoxin-induced bacterial translocation: A study of mechanism. Surgery, **106**, 292–300.

Deitch, E.A., Ma, L., Ma, J.W. *et al.* (1989c) Inhibition of endotoxin-induced bacterial translocation in mice. *J. Clin. Invest.*, **84**, 36–42.

Deitch, E.A., Taylor, M., Grisham, M. *et al.* (1989d) Endotoxin induces bacterial translocation and increases xanthine oxidase activity. *J. Trauma*, **29**, 1679–83.

Deitch, E.A., Bridges, W.R., Ma, J.W. *et al.* (1990a) The obstructed intestine as a reservoir for systemic infection. *Amer. J. Surg.*, **159**, 394–401.

Deitch, E.A., Kemper, A.C., Specian, R.D. and Berg, R.D. (1990b) A study of the relationships between bacterial translocation, gut-origin sepsis and survival in a model of systemic inflammation. *J. Trauma*, (in press).

Deitch, E.A., Ma, J.W., Ma, L. *et al.* (1990c) Protein malnutrition predisposes to inflammatory-induced gut origin septic states. *Ann. Surg.*, **211**, 560–8.

Deitch, E.A., Morrison, J.D., Berg, R. and Specian, R. (1990d) The effect of hemorrhagic shock on bacterial translocation, intestinal morphology, and intestinal permeability in conventional and antibiotic-decontaminated rats. *Crit. Care Med.*, **18**, 529–36.

Deitch, E.A., Sittig, K., Ma, L. *et al.* (1990e) Obstructive jaundice promotes bacterial translocation from the gut. *Amer. J. Surg.*, **159**, 79–84.

Ellis, J.C. and Dragstedt, L.R. (1930) Liver autolysis *in vivo*. *Arch. Surg.*, **20**, 8–16.

Ford, W.W. (1900) The bacteriology of healthy organs. *Trans. Assoc. Amer. Physicians*, **15**, 389–415.

Ford, W.W. (1901) On the bacteriology of normal organs. *J. Hyg.* (Camb.), **1**, 277–84.

Frank, H.A., Seligman, A.M. and Fine, J. (1948) Further experiences with peritoneal irrigation for acute renal failure. *Ann. Surg.*, **128**, 561–608.

Fuller, K.G. and Berg, R.D. (1985) Inhibition of bacterial translocation from the gastrointestinal tract by nonspecific immunostimulation, in Germfree Research: Microflora Control and its Application to the Biomedical Sciences (eds B.S. Wostmann, *et al.*), Alan R. Liss, New York, pp. 195–8.

Fuller, R. and Jayne-Williams, D.J. (1970) Resistance of the fowl (*Gallus domesticus*) to invasion by its intestinal flora. II. Clearance of translocated intestinal bacteria. *Res. Vet. Sci.*, **11**, 368–74.

Gautreaux, M.D., Deitch, E.A and Berg, R.D. (1990) Immunologic mechanisms preventing bacterial translocation from the gastrointestinal tract. in *Proceedings of the 10th Int. Symp. Gnotobiology* (ed. P. Heidt). Leiden, The Netherlands. (In press).

Gordon, L.E., Ruml, D., Hahne, H.J. and Miller, C.P. (1955) Studies on susceptibility to infection following ionizing radiation. IV. The pathogenesis of the endogenous bacteremias of mice. *J. Exp. Med.*, **102**, 413–24.

Hale, P. and Hill, A. (1973) The recovery of *Lactobacillus* sp. from the livers of healthy mice. *Lab. Animals*, **7**, 119–24.

Harmsen, A.G., Muggenburg, B.A., Snipes, M.B. and Bice, D.E. (1985) The role of macrophages in particle translocation from lungs to lymph nodes. *Science*, **230**, 1277–80.

Hassid, W.Z., Joslyn, M.A. and McCready, R.M. (1941) The molecular constitution of an insoluble polysaccharide from yeast, *Saccharomyces cerevisiae*. *J. Amer. Chem. Soc.*, **63**, 295–8.

Hebert, J.C., Gamelli, R.L., Foster, R.S. Jr *et al.* (1983) Improved survival after pneumococcus in splenectomized and nonsplenectomized mice with *Corynebacterium parvum*. *Arch. Surg.*, **118**, 328–32.

Hildebrand, G.J. and Wolochow, H. (1962) Translocation of bacteriophage across the intestinal wall of the rat. *Proc. Soc. Exp. Biol. Med.*, **109**, 183–5.

Keller, R. and Engley, F.B. Jr (1958) Fate of bacteriophage particles introduced into mice by various routes. *Proc. Soc. Exp. Biol. Med.*, **98**, 577–80.

Kimura, A., Sherwood, R.L. and Goldstein, E. (1983) Glucan alteration of pulmonary antibacterial disease. *J. Reticuloendothelial Soc.*, **34**, 1–11.

Krause, W., Matheis, H. and Wulf, K. (1969) Fungaemia and funguria after oral administration of *Candida albicans*. *Lancet*, **i**, 598–9.

Ma, L., Specian, R.D., Berg, R.D. and Deitch, E.A. (1989) Effects of protein malnutrition and endotoxin on the intestinal musocal barrier to the translocation of indigenous flora in mice. *J. Parenteral Enteral Nutr.*, **13**, 572–8.

Maddus, M.A., Wells, C.L., Platt, J.L. *et al.* (1988) Effect of T-cell modulation on the translocation of bacteria from the gut and mesenteric lymph node. *Ann. Surg.*, **207**, 387–98.

Maejima, K., Deitch, E.A. and Berg, R.D. (1984a) Promotion by burn stress of the translocation of bacteria from the gastrointestinal tracts of mice. *Arch. Surg.*, **119**, 166–72.

Maejima, K., Deitch, E.A. and Berg, R.D. (1984b) Bacterial translocation from the gastrointestinal tract of rats receiving thermal injury. *Infect. Immun.*, **43**, 6–10.

Maejima, K., Shimoda, K. and Berg, R.D. (1984c) Assessment of mouse strain on bacterial translocation from the gastrointestinal tract. *Exp. Animals*, **33**, 345–9.

Maejima, K. and Tajima, Y. (1973) Association of gnotobiotic mice with various organisms isolated from conventional mice. *Jap. J. Exp. Med.*, **43**, 289–96.

Mainous, M., Tso, P., Berg, R.D. Deitch, E.A. (1991) Studies of the route, magnitude, and time course of bacterial translocation in model of systemic inflammation. Arch. of Surg., **126**, 33–7.

Morehouse, J., Specian, R., Stewart, J. and Berg, R.D. (1986) Promotion of the translocation of indigenous bacteria of mice from the GI tract by oral ricinoleic acid. *Gastroenterology*, **91**, 673–82.

Nolan, J.P. (1981) Endotoxin, reticuloendothelial function, and liver injury. *Hepatology*, **1**, 458–65.

Owens, W.E. and Berg, R.D. (1980) Bacterial translocation from the gastrointestinal tract of athymic (nu/nu) mice. *Infect. Immun.*, **24**, 308–12.

Owens, W.E. and Berg, R.D. (1982) Bacterial translocation from the gastrointestinal tracts of thymectomized mice. *Curr. Microbiol.*, **7**, 169-74.

Pabst, M.J. and Johnson, R.B. (1980) Increased production of superoxide anion by macrophages exposed *in vitro* to muramyl dipeptide or lipopolysaccharide. *J. Exp. Med.*, **151**, 101–14.

Parks, D.A., Bulkley, G.B., Granger, D.N. *et al.* (1982) Ischemic injury in the cat small intestine: role of superoxide radicals. *Gastroenterology*, **82**, 9–15.

Patchen, M.L. and MacVittie, T.J. (1983) Dose-dependent responses of murine pluripotent stem cells and myeloid and erythroid progenitor cells following administration of the immunomodulating agent glucan. *Immunopharmacology*, **5**, 303–13.

Penn, R.L., Maca, R.D. and Berg, R.D. (1985) Increased translocation of bacteria from the gastrointestinal tract of tumor-bearing mice. *Infect. Immun.*, **47**, 793–8.

Penn, R.L., Maca, R.D. and Berg, R.D. (1986) Leukemia promotes the translocation of indigenous bacteria from the gastrointestinal tract to the mesenteric lymph nodes and other organs. *Microecol. Ther.*, **15**, 85–91.
Ravin, H.A. and Fine, J. (1962) Biological implications of intestinal endotoxins. *Fed. Proc.*, **21**, 65–8.
Schatten, W.F. (1954) The role of intestinal bacteria in liver necrosis following experimental excision of the hepatic arterial supply. *Surgery*, **36**, 256–69.
Schweinburg, F.B., Frank, H.A., Frank, E.D., Heimberg, F. and Fine, J. (1949) Transmural migration of intestinal bacteria during peritoneal irrigation in uremic dogs. *Proc. Soc. Exp. Biol. Med.*, **71**, 150–3.
Schweinburg, F.B., Seligman, A.M. and Fine, J. (1950) Transmural migration of intestinal bacteria. A study based on the use of radioactive *Escherichia coli*. *New Engl. J. Med.*, **242**, 747–51.
Scott, M.T. (1972) Depression of delayed-type hypersensitivity by *Corynebacterium parvum*: Mandatory role of the spleen. *Cell. Immunol.*, **13**, 251–63.
Spaeth, G., Berg, R.D., Specian, R. and Deitch, E.A. (1990) Food without fiber promotes bacterial translocation from the gut. *Surgery*, **108**, 240–7.
Staley, T.E., Jones, E.W. and Marshall, A.E. (1968) The jejunal absorptive cell of the newborn pig: An electron microscopic study. *Anat. Rec.*, **161**, 497–516.
Staley, T.E., Jones, E.W. and Corly, L.D. (1969) Attachment and penetration of *E. coli* into epithelium of the ileum in newborn pigs. *Amer. J. Pathol.*, **56**, 371–92.
Steffen, E.K. and Berg, R.D. (1983) Relationship between cecal population levels of indigenous bacteria and translocation to the mesenteric lymph nodes. *Infect. Immun.*, **39**, 1252–9.
Steffen, E.K., Berg, R.D. and Deitch, E.A. (1988) Comparison of the translocation rates of various indigenous bacteria from the gastrointestinal tract to the mesenteric lymph nodes. *J. Infect. Dis.*, **157**, 1032–7.
Surawicz, C.M., Elmer, G.W., Speelman, P. *et al.* (1989) Prevention of antibiotic-associated diarrhea by *Saccharomyces boulardii*: A prospective study. *Gastroenterology*, **96**, 981–8.
Takeuchi, A. (1967) Electron microscope studies of experimental *Salmonella* infection. *Amer. J. Pathol.*, **50**, 109–36.
Tancrede, C.H. and Andremont, A.O. (1985) Bacterial translocation and gram-negative bacteremia in patients with hematological malignancies. *J. Infect. Dis.*, **152**, 99–103.
Trusler, H.M. and Reeves, J.R. (1934) Significance of anaerobic organisms in peritonitis due to liver autolysis. *Arch. Surg.*, **28**, 479–91.
Vincent, P., Colombel, J.F., Lescut, D. *et al.* (1988) Bacterial translocation in patients with colorectal cancer. *J. Infect. Dis.*, **158**, 1395–6.
van der Waaij, D., Berghuis de Vries, J.M. and Lekkerkerk van der Wees, J.E. C. (1972) Colonization resistance of the digestive tract and the spread of bacteria to the lymphatic organs in mice. *J. Hyg.* (Camb.), **70**, 335–42.
Volkheimer, G. (1974) Passage of particles through the wall of the gastrointestinal tract. *Environ. Health Perspect.*, **9**, 215–25.
Wells, C.L. and Balish, E. (1979) Immune response modulation by colonization of germfree rats with *Propionibacterium acnes*. *Infect. Immun.*, **26**, 473–8.

Wells, C.L., Jechorek, R.P. and Erlandsen, S.L. (1990) Evidence for the translocation of *Enterococcus faecalis* across the mouse intestinal tract. *J. Infect. Dis.*, **162**, 82–90.

Wells, C.L., Maddus, M.A. and Simmons, R.L. (1987) Role of the macrophage in the translocation of intestinal bacteria. *Arch. Surg.*, **122**, 48–53.

Wells, C.L., Maddus, M.A., Erlandsen, S.L. and Simmons, R.L. (1988) Evidence for the phagocytic transport of intestinal particles in dogs and rats. *Infect. Immun.*, **56**, 278–82.

Wilson, S.J. and Walzer, M. (1935) Absorption of undigested proteins in human beings. IV. Absorption of unaltered egg protein in infants and in children. *Amer. J. Dis. Child.*, **50**, 49–54.

Wolbach, S.B. and Saiki, T. (1909) A new anaerobic spore-bearing bacterium commonly present in the livers of healthy dogs, and believed to be responsible for many changes attributed to aseptic autolysis of liver tissue. *J. Med. Res.*, **21**, 267–78.

Wolochow, G., Hildebrand, G.J. and Lamanna, C. (1966) Translocation of microorganisms across the intestinal wall of the rat: Effects of microbial size and concentration. *J. Infect. Dis.*, **116**, 523–8.

Wooles, W.R. and DiLuzio, N.R. (1962) Reticuloendothelial function and the immune response. *Science*, **142**, 1078–80.

Chapter Five

Gut flora and disease resistance

DAVID J. HENTGES

5.1. INTRODUCTION

It is now recognized that the indigenous microflora of humans and animals provides protection against infections with pathogenic microorganisms. Evidence of a protective role for the intestinal flora was obtained largely from studies with either germ-free or antibiotic-treated experimental animals that are much more susceptible to infections with intestinal pathogens than are conventional animals with an intact flora.

Germ-free mice, for example, are easily colonized with *Salmonella enteritidis*, *Shigella flexneri* and *Campylobacter jejuni* whereas conventional mice are far more resistant. It was recently reported that an oral inoculum of 1.7×10^2 viable *Sal. enteritidis* resulted in a uniformly fatal infection of germ-free mice in about 7 days while all conventional control mice included in the experiment survived (Nardi *et al.*, 1990). Bacterial counts obtained from homogenates prepared from livers and spleens increased rapidly with time in the germ-free mice and reached lethal levels of approximately 10^9 bacteria per organ by day 6. By comparison, counts of only $10^{4.5}$ bacteria per organ were recovered from the conventional mice. In studies done a number of years ago, germ-free and conventional mice were inoculated orally with approximately 10^7 viable *Sh. flexneri* to determine the impact of the intestinal flora on resistance (Maier *et al.*, 1972). *Sh. flexneri* rapidly established in the caeca of the germ-free mice attaining populations of approximately 10^{10} viable organisms per gram content within 24h. The mice developed no symptoms of infection, however, because *Shigella* is not a natural pathogen for these animals. In conventional mice, by contrast, the population of *Sh. flexneri* in the caeca declined over a 2-day period to a level of approximately 10^3 organisms per gram content and persisted at this level for the duration of the experiment. In this case, the presence of flora resulted in a 10-million-fold decrease in the *Sh. flexneri* population. Similar results were obtained in studies with *Cam. jejuni* strains (Jesudason *et al.*, 1989). The intestines of conventional mice were irregularly colonized with small numbers of these organisms even when large oral inocula of 10^{11} organisms were given. Germ-free mice, on the other hand, were more susceptible to colonization. An oral challenge of 10^8 viable *Cam. jejuni* resulted in counts of between 10^4 and 10^9 organisms per gram of intestine content in all of the mice. Thus, the presence of a flora impedes colonization of the intestine with pathogenic bacteria.

The administration of oral antibiotics in high doses decreases the resistance of experimental animals to colonization with non-indigenous organisms, presumably by disrupting the intestinal flora. Some time ago, Bohnhoff *et al.* (1954) reported that oral administration of streptomycin to mice greatly increased the susceptibility of the animals to *Sal. enteritidis* infection. The antibiotic lowered the infectious dose (ID_{50}) of *Sal. enteritidis* from 1 million viable bacteria per mouse to fewer than 10 bacteria, representing a 10 000-fold decrease in resistance (Bohnhoff and Miller, 1962). In related studies, Freter (1955, 1956) demonstrated that oral administration of a combination of streptomycin, erythromycin and nystatin in large quantities to mice and guinea-pigs, to eliminate major flora components, rendered the animals susceptible to infection with antibiotic-resistant strains of *Sal. flexneri* and *Vibrio cholerae*. Simultaneous introduction of an antibiotic-resistant *Escherichia coli* strain with each pathogen into the intestinal tract resulted in elimination of the pathogen. These early studies lent strong support to the view that an intact flora is essential to prevent establishment of pathogenic bacteria in the intestinal tract.

5.2 COLONIZATION RESISTANCE

The protection against colonization of the intestinal tract with potentially pathogenic bacteria afforded by the intestinal microflora was termed *colonization resistance* by van der Waaij *et al.* (1971). These investigators studied colonization resistance in mice before, during and after oral administration of streptomycin and neomycin. The mice were infected perorally with *E. coli*, *Klebsiella pneumoniae* or *Pseudomonas aeruginosa* during each of the the three experimental periods. Colonization resistance was expressed quantitatively as the $\log_{10}$ of the oral dose of an organism required to colonize 50% of the mice for at least 2 weeks. Before antibiotic treatment, colonization resistance was high against all three organisms. During antibiotic administration, which virtually eliminated the intestinal flora, there was a precipitous decrease in resistance and all of the mice were easily colonized with the three organisms. Colonization resistance gradually returned to normal values after the cessation of antibiotic administration due to the repopulation of the intestinal tract with anaerobic organisms that had survived the antibiotic treatment. Gnotobiotic mice, contaminated with the surviving anaerobes, displayed a high level of colonization resistance against *E. coli*. Selective elimination of Enterobacteriaceae from the intestinal tract of the animals with nalidixic acid had no apparent effect on colonization resistance, as determined by population levels of enterococci (van der Waaij and Berghuis de Vries, 1974). The anaerobic components of the flora, which appeared to be responsible for colonization resistance in the digestive tract, were not affected by this treatment.

Systemic administration of antibiotics also decreased colonization resistance in the mice (van der Waaij *et al.*, 1972). Streptomycin-resistant strains of *E. coli* and enterococci developed into much larger populations and persisted for longer periods of time in the caeca of mice injected intraperitoneally with either streptomycin or ampicillin than in the caeca of mice injected with saline. Possibly, flora components responsible for colonization resistance are in close contact with the intestinal mucosa because of their sensitivity to antibiotics present in the peritoneal cavity.

Thijm and van der Waaij (1979) examined, in addition, the effect of commonly used absorbable antibiotics on colonization resistance in the digestive tract of mice. Ampicillin or epicillin administered orally in various doses strongly diminished colonization resistance against an antibiotic-resistant strain of *E. coli*, which attained population levels ranging between 10^8 and 10^{10} viable organisms per gram of faeces. On the other hand, oral administration of cephradine, which is also absorbed, had no effect on colonization resistance. The differences in the impact of the antibiotics on colonization resistance was explained by the fact that ampicillin and epicillin, which are excreted in the bile, presumably reach the intestine in concentrations sufficiently high to inhibit sensitive bacteria, whereas cephradine, which is not excreted in the bile, is not found in the intestinal tract.

5.3 SUPPRESSION OF THE MULTIPLICATION OF PATHOGENS BY THE INTESTINAL MICROFLORA

There is evidence that the intestinal flora provides protection against colonization with a broad range of enteric pathogens. These include the anaerobes *Clostridium difficile* and *C. botulinum*, the enteric bacteria

E. coli. *Salmonella*, *Shigella* and *Pseudomonas*, and the yeast *Candida albicans*.

5.3.1 *Candida albicans*

Patients who are compromised immunologically or who undergo prolonged antibiotic therapy frequently develop serious *Candida* infections that spread from the intestinal tract to the viscera. The overgrowth of *Candida* in the intestinal tract and its subsequent passage through the gut mucosa into the bloodstream is believed to be the proximate mechanism leading to systemic candidiasis (Stone *et al*., 1974).

A number of experimental animal models have been developed to examine the role of the indigenous intestinal flora in preventing *Candida* overgrowth. Clark (1971) found that *Can. albicans* administered to germ-free mice persisted in the gastrointestinal tract for a long period but was eliminated within a short time in pathogen-free mice habouring a complex flora. The addition of streptomycin (1 mg ml^{-1}) to the drinking water of the pathogen-free mice extended the time during which *Can. albicans* could be recovered from the faeces of the animals, although less than one-third of the animals continued to excrete *Can. albicans* in their faeces 1 month after suspension of treatment. Tetracycline administration also permitted colonization of mice with *Can. albicans* (Helstrom and Balish, 1979). Conventional mice that were not treated with antibiotics began excreting *Can. albicans* within 48 h after challenge and were devoid of detectable yeast by 16 days. Tetracycline-treated mice were also colonized with *Can. albicans* within 48 h but continued excreting the organisms in their faeces for 32 days, the duration of the experiments. More recently, Kennedy and Volz (1985a) demonstrated that a number of antibiotics decreased the colonization resistance of mice to *Can. albicans*. Mice were given either penicillin, clindamycin, vancomycin, erythromycin or gentamicin in drinking water for 3 days and were then challenged orogastrically with *Can. albicans*. Penicillin, clindamycin and vancomycin, but not gentamicin or erythromycin, decreased total anaerobic bacterial populations in the caeca and increased populations of enteric bacilli. These three antibiotics allowed *Can. albicans* to proliferate in the gut and to translocate extraintestinally. Similar results were obtained when the Syrian hamster was used as the experimental animal (Kennedy and Volz, 1985b).

Since antibiotic administration renders experimental animals susceptible to colonization with *Can. albicans* after oral challenge, it is believed that the indigenous intestinal flora suppresses growth of the yeast in the gut. Several studies have shown that intestinal flora components inhibit the *in vitro* multiplication of *Can. albicans* and in gnotobiotic mice associated with one or a few species of intestinal bacteria. However ecological conditions *in vitro* or in the environment in the intestinal tract of the gnotobiotic mice are far different from conditions in the intestinal tract of conventional animals harbouring a complete flora. Data obtained from *in vitro* studies or from experiments with gnotobiotic animals associated with a small number of bacterial species need to be interpreted with caution, therefore (Freter, 1983). Kennedy and Volz (1985a, b) found that antibiotics that diminished anaerobe populations in the intestinal tract of experimental animals most effectively compromised colonization resistance to *Can. albicans.* Colonization resistance could be restored, on the other hand, if the antibiotic-treated animals were reassociated with caecal homogenates from untreated animals (Kennedy *et al.*, 1988). The data suggest that anaerobes, which predominate in the intestinal ecosystem, are responsible for inhibition of the multiplication of *Can. albicans* and prevent its dissemination from the intestinal tract.

5.3.2 *Clostridium difficile*

A severe gastrointestinal disease of humans – pseudomembranous colitis – occurs as a consequence of the use of antimicrobial agents. It is widely believed that antibiotics suppress the indigenous intestinal flora allowing the causative agent, *C. difficile*, to survive and multiply. *C. difficile* produces potent toxins in the intestinal tract that are responsible for the disease symptoms.

There is evidence that intestinal flora components suppress the *in vitro* multiplication of *C. difficile*. This has been demonstrated on the surface of agar media (Barclay and Borriello, 1982; Malamou-Ladas and Tabaqchali, 1982; Rolfe *et al.*, 1981) in continuous-flow culture (Wilson and Freter, 1986) and in batch cultures consisting of human faecal or hamster caecal emulsions (Borriello and Barclay, 1986). Several faecal emulsions prepared from the stools of infants, who are frequently colonized by *C. difficile*, supported multiplication of the pathogen whereas faecal emulsions from the stools of adults, who are rarely colonized, were inhibitory to *C. difficile*.

Some time ago an animal model was developed in which symptoms of *C. difficile*-induced colitis could be reproduced (Bartlett *et al.*, 1977). Adult Syrian hamsters treated with antibiotics such as clindamycin, vancomycin or ampicillin and exposed to *C. difficile* developed ileocaecitis, presumably from the large quantities of toxins elaborated by the pathogen as it proliferated in the intestine (Rolfe and Iaconis, 1983). Attempts have been made to restore colonization resistance to *C. difficile* in these animals by inoculating them orally with faecal or ceacal homogenates from healthy animals. In general, these attempts have been successful. *C. difficile* populations were approximately 10 000 times smaller in the caeca of antibiotic-treated, flora-inoculated hamsters than in the caeca of uninoculated hamsters (Wilson *et al.*, 1981) The suppression of *C. difficile* by the flora prevented development of antibiotic-associated caecitis in the animals. Other evidence of a protective role of the flora is found in recent studies described by Borriello (1990). Gnotobiotic rats colonized with human faecal flora were compared with germ-free rats regarding susceptibility to infection with *C. difficile*. Thirty-three days after challenge, all the germ-free rats were colonized with *C. difficile* and excreted high levels of cytotoxin in their faeces, whereas none of the gnotobiotic rats harbouring human flora were colonized. Administration of clindamycin to human flora associated rats and environmental exposure of the animals to *C. difficile* resulted in colonization with the pathogen, indicating that the barrier provided by the flora was disrupted by the antibiotic.

The success of experiments employing whole flora for protection could not be reproduced when smaller collections of intestinal flora isolates were used. This is not surprising since normalization of other parameters in the intestinal tract requires the presence of whole flora. For example, Freter and Abrams (1972) measured the effects of a collection of 95 anaerobic bacteria isolated from mouse caecal contents on the normalization process in mice. Germ-free mice were first monoassociated with *E. coli* and then inoculated with the anaerobes. Caecal size was reduced and *E. Coli* populations suppressed 1000-fold, but the effects were not nearly as pronounced as when whole flora was used. Similar results were obtained by Wilson *et al.* (1986) in studies with *C. difficile*. One hundred and fifty isolates from the climax stage intestinal flora of hamsters were inoculated into gnotobiotic mice pre-colonized with *C. difficile* and *E. coli*. This complex flora suppressed the *E. coli* population 100-fold, and the *C. difficile* population 10-fold. However, the number of *C. difficile* remaining were theoretically sufficient to produce caecitis in the hamsters. In subsequent experiments designed to stimulate ecological succession, Wilson and Freter (1986) inoculated the *E. coli*, *C. difficile* harbouring gnotobiotic mice with 67 isolates obtained at various times from a continuous-flow culture of hamster caecal contents. The mice were also inoculated with 100 isolates obtained from hamster climax-stage caecal flora. *C. difficile* was suppressed nearly 1000-fold and *E. coli* nearly 10 000-fold by this extensive collection of isolates. Although the *C. difficile* population was reduced to non-pathogenic levels by the flora components, the degree of suppression was not as great as when whole flora was used. Thus, even moderate suppression of C. *difficile* requires vast numbers of flora components.

5.3.3 Clostridium botulinum

There is indirect evidence that resistance to infection of the gastrointestinal tract by *C. botulinum* is also due to the presence of a protective flora. In human infant botulism the large intestine is the site of colonization by *C. botulinum* (Mills and Arnon, 1987). It has been speculated that the flora of human infants, who are susceptible to infection, lacks essential components present in the adult flora that prevent colonization of the intestine by *C. botulinum*. Infection occurs as a consequence of ingestion of *C. botulinum* spores by infants who are between 3 and 26 weeks old (Johnson *et al.*, 1979). The spores, which are widely distributed in the soil and are acquired by infants *via* contaminated food and other vehicles, germinate in the intestinal tract and produce toxin. The toxin is responsible for symptoms of overt botulism, which include constipation, progressive muscle weakness, feeding difficulty and sometimes fever (Arnon *et al.*, 1977).

It has been known for a long time that adult laboratory animals resist colonization by *C. botulinum* spores given orally. By contrast, infant animals are susceptible to colonization. In a study carried out by Sugiyama and Mills (1978) at the University of Wisconsin, the presence of toxin could be demonstrated in the large intestine of infant mice after intragastric administration of spores of *C. botulinum* type A. Botulism was not observed, but 80% or more of 8–11-day-old mice challenged with 10^5 spores had toxin in the large intestine 3 days later. Mice younger than 7 days or older than 15 days were resistant to the challenge. The final resistant state correlated with the time that the infant mice began to sample solid food.

The possible role of the flora in preventing intestinal botulism was also studied using adult germ-free and antibiotic-treated mice. Intraintestinal botulism association was consistently produced when germ-free mice were fed ten *C. botulinum* type A spores (Moberg and Sugiyama, 1979). Control germ-free mice became enterically infected when placed in the same isolator with, but separated from, animals that had been fed spores. When transferred into a room holding a colony of conventional mice, the highly susceptible germ-free animals became resistant to challenge with 10^5 spores in about 3 days due to the establishment of a protective intestinal flora. The resistance of conventional mice to oral infection with *C. botulinum* was lost, however, when a mixture of erythromycin and kanamycin sulphate was administered perorally (Burr and Sugiyama, 1982). From 80 to 100% of the mice became infected when challenged 15–60 h after antibiotic administration. The mean infective dose of 1×10^4 spores per mouse for challenges given 23 h after treatment contrasted with the failure of 10^6 spores to infect control mice with an intact flora. Metronidazole also increases the susceptibility of conventional mice to colonization with *C. botulinum* (Wang and Sugiyama, 1984). In experiments designed to minimize coprophagy, antibiotic-treated mice began to develop symptoms of botulism fewer than 3 days after orogastric challenge with *C. botulinum* spores. Morbidity rates, which reached maximum levels within 5 days after challenge, were related to spore inoculum sizes. Mortality rates were also related to inoculum sizes. The antibiotics apparently eliminate organisms that ordinarily prevent the establishment of *C. botulinum* in the gut. Intestinal anaerobes of the genera *Bifidobacterium*, *Lactobacillus*, *Propionibacterium* and *Bacteroides* and *Clostridium*, as well as enterococci, are known to inhibit the *in vitro* multiplication of *C. botulinum* (Sullivan *et al.*, 1988). Susceptibility to botulism may possibly result from the absence of large numbers of these inhibitory organisms in the flora of the human infant intestine.

5.3.4 Gram-negative enteric organisms and enterococcus

Studies using antibiotics indicate that the intact flora hinders colonization of the intestinal tract of mice with Gram-negative enteric organisms and *Enterococcus*. The effects of administering individually to mice high concentrations of ampicillin, clindamycin, kanamycin, metronidazol or streptomycin on colonization resistance to *Ps. aeruginosa* was assessed (Hentges *et al.*, 1985). One week after the initiation of antibiotic administration, the treated mice and untreated control mice were challenged perorally with approximately 10^8 viable *Ps. aeruginosa* cells. Despite differences in spectra of activity, all of the antibiotics decreased the resistance of the animals to intestinal colonization with *Ps. aeruginosa* as reflected by an increased faecal carriage of the organism and an increase in the population levels of *Ps. aeruginosa* in the faeces as compared with untreated controls. Streptomycin, which was most effective in reducing colonization resistance was investigated further.

Streptomycin-treated mice and untreated control mice were challenged with either *Ps. aeruginosa*, representing an opportunistic organism, *Sal. typhimurium*, representing an enteric pathogen (Que and Hentges, 1985) or *Enterococcus faecalis*, representing a common intestinal flora organism which frequently overgrows during antibiotic therapy (Dougherty *et al.*, 1988). All of these organisms were resistant to high concentrations of streptomycin. The treated mice were given 5 mg ml^{-1} streptomycin sulphate in their drinking water for a period of 2 weeks. One week after initiation of antibiotic treatment, treated and untreated control animals were challenged perorally with either *Ps. aeruginosa*, *Sal. typhimurium* or *Ent. faecalis*. Seven days after challenge the animals were sacrificed and counts of the challenge organisms in homogenates of various organs were determined. The caecum and small intestine of all streptomycin-treated mice were colonized with *Ps. aeruginosa* and the organism translocated to either the mesentery, liver or spleen of the majority of these animals. None of the untreated mice was colonized. Although the *Sal. typhimurium* challenge dose was much smaller than the *Ps. aeruginosa* dose, the caecum, small intestine and extraintestinal organs of all streptomycin-treated mice were colonized with the organism. *Sal. typhimurium* failed to colonize untreated mice. Similarly, *Ent. faecalis* colonized the caecum and small intestine of all streptomycin-treated mice and the extraintestinal organs of a majority of the animals. The organs of 10% of untreated mice harboured *Ent. faecalis*, however. Streptomycin administration therefore greatly decreased the resistance of mice to intestinal colonization with *Ps. aeruginosa*, *Sal. typhimurium* and *Ent. faecalis* and facilitated translocation of the organisms to extraintestinal sites.

5.4 MECHANISMS RESPONSIBLE FOR SUPPRESSION OF PATHOGENS

There is considerable evidence, therefore, that when antibiotics are used to disrupt the intestinal flora, experimental animals become very susceptible to colonization with pathogens and other non-indigenous organisms, but when the flora is undisturbed they are quite resistant. A number of activities of the intestinal flora have been proposed as mechanisms responsible for the exclusion of non-indigenous organisms. They include competition between the flora and non-indigenous organisms for nutrients present in limited quantities, elaboration by the flora of metabolites that inhibit multiplication of non-indigenous organisms, establishment by the flora of environmental conditions that affect non-indigenous organisms adversely, and competition between the flora and non-indigenous organisms for intestinal mucosal association sites. Any one or all of these mechanisms may be operative.

5.4.1 Competition for nutrients

The hypothesis that nutrient competition is responsible for exclusion of non-indigenous organisms from the intestinal tract is based on a series of observations indicating that population control mechanisms in the intestine are consistent with the chemostat theory (Fredrickson, 1977). In a study described by Freter *et al.* (1983c), an anaerobic continuous-flow culture of mouse caecal flora reproduced a number of bacterial interactions that occur in the mouse large intestine. The numeric balance between 37 strictly anaerobic bacteria present in the intestinal tract of mice was maintained in a continuous-flow culture of mouse flora. *E. coli* populations were suppressed to similar levels in gnotobiotic mice and continuous-flow cultures inoculated with mixed populations of bacteria isolated from the caeca of mice. Contents from the continous-flow cultures when fed to germ-free mice redressed germ-free abnormalities, namely, a large caecal size and support of large *E. coli* populations after monocontamination. Furthermore, dense layers of bacterial growth formed on the walls of the continuous-flow culture vessels containing mouse caecal flora in a manner analogous to the colonization of the mouse large intestinal mucosa. The data of Kennedy *et al.* (1988) provide additional evidence that control mechanisms in the large intestine are reproduced in continuous-flow cultures. In these studies, both human faecal flora and mouse caecal flora were examined. The anaerobic continuous-flow culture reproduced total bacterial counts and levels of metabolic end-products present in human and mouse large intestines. Similar to the observations of Freter *et al.* (1983c), layers of bacterial growth formed on the walls of the continous-flow culture vessels and contents from the continuous-flow cultures, when fed to antibiotic-treated mice, rectified several abnormalities of the caecum and suppressed *Can. albicans* populations to a level observed in untreated mice. It is likely, therefore, that the major ecological control mechanisms operating in continuous-flow culture are the same, or at least similar to, those operating in the intestinal tract.

A fundamental concept of the chemostat theory is that bacteria in mixture in a continuous-flow culture compete for essential growth substances. Freter *et al.* (1983a) demonstrated that the multiplication of *E. coli*, *Fusobacterium* sp. and *Eubacterium* sp. was greatly suppressed when the organisms were inoculated individually into filtrates from a continuous-flow culture of mouse caecal flora, provided that care was taken to simulate the conditions of the caecal ecosystem. A major factor limiting multiplication of these organisms was lack of a source of utilizable carbohydrate. Examination of a similar continuous-flow culture system revealed that carbohydrate concentrations of the filtrate were also insufficient for *C. difficile* multiplication (Wilson, 1988). It was speculated that competition for utilizable carbohydrates is the activity of overriding importance in the regulation of populations of bacteria in the mouse intestinal tract. Kennedy *et al.* (1988) showed, however, that nutrient depletion was not responsible for inhibition of *Can. albicans* in continuous-flow cultures of mouse caecal flora or human faecal flora. Effluents from these cultures were filter sterilized and then inoculated with *Can. albicans*. The yeast was unable to multiply in the culture filtrates incubated either aerobically or anaerobically. To determine if the inhibition was due to depletion of nutrients, carbon and nitrogen sources, vitamins and trace elements were added to the filtrates. *Can. albicans* failed to multiply after these additions suggesting that some other inhibitory mechanism was operative, perhaps production of toxic metabolites by flora components.

5.4.2 Toxic metabolites and adverse environmental conditions

There is considerable evidence that toxic metabolites, including hydrogen sulphide, free bile acids and short-chain volatile fatty acids (VFA) are inhibitory for intestinal bacteria. Hydrogen sulphide contributes to the suppression of *E. coli* multiplication in the continuous-flow culture system (Freter *et al.*, 1983a), but its activity against enteric pathogens has not been investigated in recent years. Very limited data are available on the inhibitory activities of free bile acids but extensive data are available on the participation of VFA in excluding pathogens from the intestinal tract.

The primary bile acids, cholic acid and chenodeoxycholic acid, are synthesized by the liver and are then conjugated to either taurine or glycine. Human bile also contains conjugates of the secondary bile acid, deoxycholic acid, which is formed by the dehydroxylation of cholic acid. In the intestine, conjugates are hydrolysed to release free bile acids by a variety of bacteria, particularly anaerobes (Drasar and Hill, 1974). Only free acids are present in the faeces. The intestinal bacteria also convert some of the primary bile acids to secondary bile acids, deoxycholic and lithocholic acids. Flock *et al.* (1972) demonstrated that both Gram-positive and Gram-negative intestinal flora components were inhibited by free bile acids but not by either human whole bile or by conjugated bile acids. The dihydroxy acids, deoxycholic and chenodeoxycholic acids, were more toxic for bacteria than trihydroxy cholic acid, causing inhibition at 1 or 2 mM concentrations, which are physiological in humans. These studies did not examine the effects of the acids on enteric pathogens, however.

Apparently, toxic metabolites interfere with the ability of pathogens to associate with intestinal mucosal surfaces. Recent *in vitro* studies by Kennedy and Volz (1983, 1985a), and Kennedy *et al.* (1987) demonstrated that *Can. albicans* adhered in large numbers to excised intestinal tissue obtained from antibiotic treated hamsters. However, the numbers of the organisms adhering were greatly reduced when the tissue was suspended either in intestinal contents containing live flora or in a filtrate of these contents. The results suggest that filterable substances produced by the flora restrict the association of *Can. albicans* with intestinal mucosal tissue.

In an attempt to identify these substances, VFA and free bile acids were tested in a mucosal association assay. Both substances inhibited the association of *Can. albicans* with intestinal mucosa perhaps by modifying *Can. albicans* adhesins or mucosal receptors. It has been demonstrated that antibiotic treatment which predisposes hamsters to colonization with *Can. albicans* causes a decrease in the levels of free bile acids (Fekety *et al.*, 1978). These data, along with the reported suppressive activity of VFA and free bile acids on *Can. albicans* multiplication (Kennedy, 1989; Marshall *et al.*, 1987), support the concept that both classes of acids provide protection against colonization of the intestine with *Can. albicans*.

It has been known for years that VFA are toxic for Gram-negative bacteria. Evidence that the acids are involved with the exclusion of pathogens from the intestinal tract was first presented by Meynell (1963) and Bohnhoff *et al.* (1964a, b). These investigators demonstrated that the *in vitro* multiplication of *Sal. enteritidis* was inhibited by suspensions of intestinal contents from conventional mice. Most effective was material from the caecum and transverse colon. The inhibitory activity was not impaired by filtration or heat sterilization. Volatile acids were recovered from the intestinal contents of the mice in concentrations that inhibited *Sal. enteritidis* multiplication at the pH (6.1) and the oxidation-reduction potential (Eh, –200 mV) of the caecum. Under these conditions, the acids were weakly bactericidal for *Sal. enteritidis*. Oral administration of streptomycin eliminated an important segment of the intestinal flora of the mice. A decrease in total VFA concentration of intestinal contents was accompanied by an increase in oxidation-reduction potential and pH, producing conditions that favoured the multiplication of *Sal. enteritidis*. It was concluded that protection against *Sal. enteritidis* infection in mice is the result of VFA production by the intestinal flora and a concomitant decrease in the oxidation-reduction potential and pH of the intestinal contents.

Studies were also performed to determine whether VFA produced by the indigenous flora, are responsible for the inhibition of *Shs. flexneri* in the intestinal tract of mice (Maier *et al.*, 1972). When conventional mice were infected perorally with 10^7 viable *Sh. flexneri* organisms per mouse, the pathogen failed to multiply in the intestine and persisted at a population of approximately 10^3 organisms per gram caecal content. In germ-free animals, infected with the same number of organisms, *Sh. flexneri* multiplied rapidly, attaining populations of about 10^{10} organisms per gram content. Caecal contents from the two groups of animals were therefore compared to determine whether VFA concentrations, oxidation-reduction potential and pH were substantially different. The results showed that the oxidation-reduction potential and pH in caecal contents of conventional mice were lower and VFA concentrations were higher than in caecal contents of germ-free mice. *Sh. flexneri* multiplied in caecal contents taken from germ-free mice but failed to multiply in contents obtained from conventional mice (Figure 5.1). Therefore, caecal contents obtained from germ-free mice were adjusted to simulate the oxidation-reduction potential, pH and VFA concentrations found in the caecal contents from conventional animals. After adjustment, the bactericidal effect against *Sh. flexneri* was almost identical to the effect observed in the caecal contents from conventional animals. Essentially, the same results were obtained when strains of *Sh. sonnei* and *Sh. dysenteriae* were used (Arnold, 1974), indicating that the high concentrations of VFAs and the low pH and oxidation-reduction potential in the intestinal tract of conventional mice are major factors limiting the multiplication of *Shigella* species *in vivo*.

Several other studies support the theory that high concentrations of VFAs are of prime importance in regulating population levels of bacteria in the intestine. Levison (1973) demonstrated that pooled caecal contents of conventional mice, an anaerobic culture of mouse faeces, and several human intestinal bacterial isolates all suppressed the *in vitro* multiplication of *Ps. aeruginosa*. Activity was heat stable

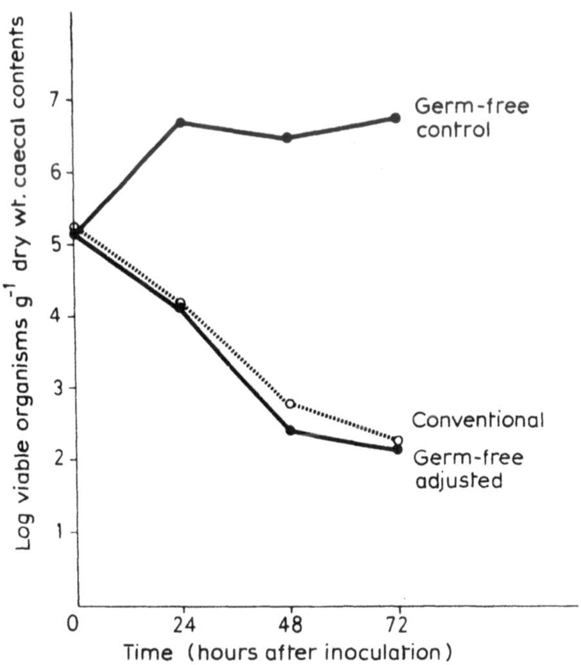

Figure 5.1 Viable counts of *Shigella flexneri* after inoculation into ceacal contents obtained from germ-free and conventional mice. From Maier *et al.* (1972).

and pH dependent. The lower the pH of the suspensions, the greater the antibacterial activity. Acetic and butyric acids were present in mouse colon contents and in the anaerobic culture of mouse faeces in concentrations that inhibited the *in vitro* multiplication of *Ps. aeruginosa* at the pH of the mouse caecum. There is additional evidence, moreover, that VFAs exert a repressive effect on Enterobacteriaceae in mice. Lee and Gemmell (1972) studied the development of the intestinal flora of young mice. Volatile fatty acids appeared to be responsible for changes in flora composition that occurred during weaning. The consumption of solid food by the animals was correlated with the appearance of anaerobic fusiform bacteria in the intestinal lumen and a 10000-fold decrease in the number of coliform organisms. Concomitant with these changes was the appearance of VFAs, especially butyric acid, in intestinal contents. Presumably, these acids were produced by the fusiform bacteria and were responsible for the decline in coliform populations. Byrne and Dankert (1979) also demonstrated an inverse relationship between VFA concentrations and Enterobacteriaceae population levels. The total VFA concentration in caecal contents of conventional mice fed *ad libitum* was 81.7 $\mu mol g^{-1}$ wet weight, which is antibacterial under *in vitro* conditions; in rectal samples it was 41.1 $\mu mol g^{-1}$ wet weight. The mean count of Enterobacteriaceae was only 10^2 g^{-1} in the caecum but was 10^5 g^{-1} in the rectum, with a lower total VFA concentration. Volatile fatty acid levels were influenced by food intake. When the mice were fasted for 4 days, caecal and rectal VFA concentrations fell to extremely low levels, and the mean count of Enterobacteriacae increased to 2×10^6 g^{-1} in the caecum and 1×10^7 g^{-1} in the rectum. The results indicate that VFAs are important factors controlling total Enterobacteriaceae population levels in the intestine.

Other evidence, however, indicates that *E. coli* populations are not influenced by VFA concentrations in the intestine. Freter and Abrams (1972) associated germ-free mice with either whole flora obtained from conventional mice or with various mixtures of facultative and anaerobic intestinal bacteria and then implanted an *E. coli* strain. The nature and concentrations of the VFAs associated with the different floras in the caeca of the mice did not correlate with the *E. coli* population levels, indicating that the acids are not the sole agents regulating *E. coli* populations in the intestine. Similar experiments of Koopman *et al.* (1981) confirmed these findings. No correlation was observed between the population levels of an implanted *E. coli* strain and total VFAs present in the caecal contents of mice. There was an inverse relationship, however, between the concentration of propionic acid and the *E. coli* population levels, indicating that this acid may be involved in the regulation of the *E. coli* population. Koopman *et al.* (1979) varied the total VFA concentrations in the caeca of mice by altering the animals' diets. Mice on diets resulting in the production of high concentrations of VFAs in the caecum harboured many more *E. coli* than did mice on diets resulting in the production of low concentrations of the acids. Clearly, volatile acids do not have a major influence on resident *E. coli* population levels in the intestines of mice.

More recent studies, comparing streptomycin-treated mice with untreated mice, indicate that VFAs suppress intestinal pathogens *in vivo*. The treated animals are far more susceptible to colonization with *Sal. typhimurium* (Que and Hentges, 1985), *Sh. sonnei* and enterotoxigenic *E. coli* (Pongpech and Hentges, 1989a) than are untreated mice. Streptomycin administration did not significantly alter the nutrient status of mouse intestinal contents, the Eh or the motility of the gut (Que *et al.*, 1986). However, the antibiotic caused an increase in the pH of caecal contents, from a mean value of 6.42 in untreated animals to a mean value of 6.73 in treated animals, and a significant decrease in the concentrations of acetic, propionic and butyric acids (Table 5.1). Valeric acid was detected in low concentration in contents of untreated mice but not streptomycin-treated mice. The increase in pH and decrease in VFA concentrations resulting from treatment theoretically provides a more hospitable environment for the multiplication of intestinal pathogens that are inhibited by the acids at low pH. This may explain why streptomycin treatment increases the susceptibility of the animals to colonization with the pathogens.

Table 5.1 Influence of streptomycin treatment on various ecological factors in caecal contents of mice.

Factor	Mean ± SD of results from Untreated mice	Treated mice
Eh (mV)	−128.90 ± 7.61	−118.57 ± 21.80
Protein (mg g^{-1})	6.17 ± 1.53	5.95 ± 1.79
Carbohydrate (mg g^{-1})	4.54 ± 3.22	6.90 ± 4.81
pH	6.42 ± 0.13	6.73 ± 0.28*
Acetic acid (μEq g^{-1})	74.80 ± 9.00	53.10 ± 7.90*
Propionic acid (μEq g^{-1})	19.60 ± 4.00	13.00 ± 4.40*
Butyric acid (μEq g^{-1})	60.80 ± 9.00	20.70 ± 4.90*
Valeric acid (μEq g^{-1})	2.50 ± 0.90	ND

* Value statistically significantly different ($p < 0.05$) from corresponding value in untreated group (From Hentges *et al.*, 1990)
ND = not detected.

To examine this question, caecal contents obtained from streptomycin-treated mice and from untreated mice were inoculated separately with

Sal. typhimurium (Que *et al.*, 1986), *Sh. sonnei* and enterotoxigenic *E. coli* (Pongpech and Hentges, 1989b). After inoculation of the pathogens into the two types of caecal contents, growth curves were plotted of the organisms multiplying under anaerobic conditions. The multiplication rates and population sizes of each of the pathogens were greater in contents from streptomycin-treated mice than untreated mice. When the pH level and the VFA concentrations of contents from streptomycin-treated mice were adjusted to simulate the values observed in contents from untreated mice, the multiplication rates and population sizes of the pathogens were lower in the adjusted than unadjusted contents. Growth curves obtained with adjusted contents were almost identical to curves obtained with contents from untreated mice. Conversely, neutralization of VFAs present in the caecal contents from untreated mice by increasing the pH level to 7.50, reversed inhibition and growth curves resembled those obtained with contents from streptomycin-treated mice. These results provide strong evidence that VFAs operating at the pH level of the intestinal tract represent formidable barriers to colonization of the region by Gram-negative enteric pathogens. Additional information supporting this conclusion is the observation that discontinuation of streptomycin treatment restores colonization resistance to enteric pathogens in mice within a period of 7 days (Pongpech and Hentges, 1989a). Restoration of resistance is accompanied by an increase in the concentration of total VFAs and a decrease in the pH of caecal contents to pre-treatment levels.

There is indirect evidence that VFAs also interfere with the multiplication of *C. difficile* and *Can. albicans* in the intestinal tract of experimental animals. Infant hamsters older than 4 days are readily colonized with *C. difficile* (Rolfe, 1984), whereas adult hamsters resist colonization. Volatile fatty acids, in concentrations present in the caecum of adult hamsters, inhibit the multiplication of *C. difficile* in broth culture at the pH level of caecal contents but the acids are non-inhibitory in concentrations observed in the caecum of infant hamsters. In similar studies, VFAs, in concentrations observed in continuous-flow cultures of mouse caecal flora, prevented multiplication of *Can. albicans* in broth culture (Kennedy *et al.*, 1988). All this information indicates that VFAs provide protection in the intestinal tract against a broad range of pathogens.

5.4.3 Competition for association sites

Disease processes depend on the persistence of the pathogens in the intestinal tract. As a consequence of the inhibitory activities of the flora, the doubling time of pathogens in intraluminal contents often exceeds the dilution rate, and under these conditions the pathogens are theoretically 'washed out' of the intestinal tract. To ensure survival in this ecosystem it becomes imperative for the pathogens to associate with the intestinal mucosa. Freter *et al.* (1983b) developed a mathematical model which predicts that two or more bacterial strains that compete in the gut for the same limiting nutrient can coexist if the metabolically less efficient strains have specific adhesion sites available. Ability to associate with the mucosa is therefore an important determinant for the successful colonization of the intestinal tract by pathogenic organisms.

Data from several experiments indicate that flora components compete with pathogens for mucosal association sites. The flora, firmly attached to the mucosa, blocks colonization by pathogenic organisms (Snoeyenbos, 1979).

In studies described by Kennedy and Volz (1985a, b) the degree of association of *Can. albicans* to mucosal surfaces was significantly different in antibiotic-treated and untreated experimental animals. In penicillin-treated hamsters, large numbers of *Can. albicans* were present both in intestinal contents and in association with intestinal mucosal surfaces. By contrast, significantly fewer *Can. albicans* were present in contents and on mucosal surfaces of untreated animals. Scanning electron microscopy revealed that large numbers of yeast cells colonized the villus surfaces and the mucous material adjacent to the villi in the penicillin-treated hamsters. In untreated hamsters, the same sites were colonized with fusiform-shaped organisms and few yeast cells were present (Kennedy *et al.*, 1987). The ability of *Can. albicans* to associate with intestinal mucosal surfaces was also tested in an *in vitro* adhesion assay (Kennedy and Volz, 1985b). Caecum and small bowel segments removed from both penicillin-treated and untreated hamsters were washed gently, cut into squares and suspended in phosphate-buffered saline. *Can. albicans* was added to the suspension and counts of the yeast associated with the tissues were determined. Large numbers of *Can. albicans* associated with intestinal slices obtained from penicillin-treated hamsters. Conversely, slices obtained from untreated hamsters (containing mucosa associated flora components) were associated with small numbers of *Can. albicans*. Apparently the dense layer of flora in the mucus gel blocks association of *Can. albicans* with the intestinal mucosa.

This phenomenon was also demonstrated in studies on the association of *Sh. sonnei* and enterotoxigenic *E. coli* with caecal tissues of streptomycin-treated and untreated mice (Pongpech *et al.*, 1989). After peroral challenge, both pathogens associated in significantly greater numbers with tissues of streptomycin-treated than untreated mice. Populations of the two pathogens were also greater in caecal contents of the antibiotic-treated animals. When excised caecal tissue from the two groups of mice was exposed to the pathogens under *in vitro* conditions the extent of the associations was greater with tissue slices from steptomycin-treated than untreated mice.

Streptomycin administration diminished the populations of fusiform-shaped bacteria associated with the mucus layer of the mouse caecum by a factor of 100 (Pongpech *et al.*, 1989). Perhaps the partial elimination of these bacteria by the antibiotic liberated association sites that were subsequently occupied by enteric pathogens. Fusiform bacteria are also an important source of intestinal VFAs (especially butyric acid) that suppress the multiplication of pathogens. In one study, Voravuthikunchai and Lee (1987) showed that caecectomy of mice caused a decrease in the number of fusiform bacteria and in the levels of VFAs in the intestinal tract, and an increase in the susceptibility of the animals to peroral challenge with *Sal. enteritidis*. Fusiform bacteria in the mouse intestine appear to act as major deterrents to colonization of the region with pathogenic bacteria by blocking association sites and producing toxic VFAs.

5.5 CONCLUSIONS

The intact intestinal flora resists implantation with non-indigenous microorganisms because it consists of a community in its climax stage. As such, its members have been selected for over a period of time and represent those microorganisms that can best cope with the biological and non-biological restrictions imposed by the ecosystem. Non-indigenous organisms, such as pathogens, that are less fit, cannot easily colonize the ecosystem. Colonization is prevented primarily by the activities of the established flora. The mechanisms involved, commonly referred to as 'colonization resistance', represent a powerful deterrent to infection of the region by enteric pathogens. Colonization resistance is apparent in two major regions of the intestinal habitat: the luminal contents and the mucosal surfaces. In the luminal contents, the most important resistance mechanism is the production by flora components, of toxic metabolites, such as VFAs and free bile acids, that suppress multiplication of pathogens. Competition between flora components and pathogens for nutrients present in limited quantities appears to function primarily to regulate populations of established indigenous components, such as *E. coli*. At the mucosal surfaces the resistance mechanism of prime importance is the occupation by flora components of association sites utilized by pathogenic organisms. In normal circumstances, all the above mechanisms effectively impede colonization by pathogens. They become impaired, however, when the intestinal flora is disturbed by stresses such as prolonged antibiotic therapy or drastic dietary changes.

5.6 THE PROBIOTIC CONCEPT

Efforts have been made to restore colonization resistance when it is diminished, by implanting viable bacteria. Partial restoration of resistance has been achieved with mixtures containing very large numbers of intestinal bacteria. Suppression of pathogens also occurs in gnotobiotic mice associated with one or a few flora components but it is unlikely that the ecosystem of the intestinal tract of these animals resembles that of conventional animals with a complete flora. Interactions occurring in the intestines of the gnotobiotic animals are probably not representative of interactions occurring in the complex ecosystems of the intestines of conventional animals.

Attempts to restore colonization resistance have also been made by implanting single strains of bacteria in compromised hosts. Lactobacilli, or dairy products containing these organisms, have been used for this purpose for years but reports of their efficacy are contradictory (Gorbach, 1990). It is highly unlikely that these bacteria alter the ecosystem of the intestinal lumen in any significant way because even very large numbers of intestinal bacteria administered to mice are not completely successful in this regard. It has been shown, however, that the non-pathogenic yeast, *Saccharomyces boulardii* (Surawicz *et al.*, 1989) and a strain of *Lactobacillus*, designated GG (Gorbach, 1990), decrease the incidence of antibiotic-induced diarrhoea in hospitalized patients. *Lactobacillus* GG colonizes the human intestinal tract and adheres more tightly to human intestinal and buccal cells than other lactobacillus strains. Perhaps it colonizes the intestinal mucosa and non-specifically blocks mucosal association by *C. difficile*.

The use of avirulent variants of pathogenic bacteria to increase colonization resistance to specific intestinal pathogens has met with some success. Borriello and Barclay (1985) examined the efficacy of administering non-toxigenic strains of *C. difficile* to hamsters and found that these strains protected the animals against lethal, antibiotic-associated colitis. Protection was species specific. Non-toxigenic *C. difficile*, but not toxigenic strains, were found associated with the caecal wall of the protected hamsters. Similar results were reported when an avirulent mutant of *Sal. typhimurium* was administered to newly hatched chickens in order to prevent infection (Berchieri and Barrow, 1990). The mutant provided species specific protection against challenge with very large numbers of virulent *Sal. typhimurium*. In both of these studies it is likely that the avirulent variants, associated with the colonic mucosa or mucus gel, occupied specific sites usually available to the pathogens.

In the restoration of colonization resistance, it is improbable, therefore, that probiotic measures designed to alter the ecology of intestinal luminal contents will be successful. Efforts focused on the identification of bacterial strains that colonize the mucosa and effectively prevent mucosal association by enteric pathogens appear to be more promising.

REFERENCES

Arnold, W.J. (1974) Inhibition of *Shigella* species by environmental factors in the mouse intestine. M.S. Thesis, University of Missouri School of Medicine, Columbia.

Arnon, S.S., Midura, T.F., Clay, S.A., Wood, R.M. and Chin, J. (1977) Infant botulism: epidemiological, clinical and laboratory aspects. *J. Amer. Med. Assoc.*, **237**, 1946–51.

Barclay, F.E. and Borriello, S.P. (1982) *In vitro* inhibition of C. *difficile*. Eur. *J. Chemother. Antibiot.*, **2**, 155–6.

Bartlett, J.G., Onderdonk, A.B., Cisneros, A.B. and Kasper, D.L. (1977) Clindamycin-associated colitis due to toxin producing species of clostridium in hamsters. *J. Infect. Dis.*, **136**, 701–5.

Berchieri, A. Jr, and Borrow, P.A. (1990) Further studies on the inhibition of colonization of the chicken alimentary tract with *Salmonella typhimurium* by pre-colonization with an avirulent mutant. *Epidemiol. Infect.* **104**, 427–41.

Bohnhoff, M. and Miller, C.P. (1962) Enhanced susceptibility to *Salmonella* infection in streptomycin-treated mice. *J. Infect. Dis.*, **111**, 117–27.

Bohnhoff, M., Drake, B.L. and Miller, C.P. (1954) Effect of streptomycin in susceptibility of intestinal tract to experimental *Salmonella* infection. *Proc. Soc. Exp. Biol. Med.*, **86**, 132–7.

Bohnhoff, M., Miller, P.C. and Martin, W.R. (1964a) Resistance of the mouse's intestinal tract to experimental *Salmonella* infection. I. Factors which interfere with initiation of infection by oral inoculation. *J. Exp. Med.*, **120**, 805–16.

Bohnhoff, M., Miller, P.C. and Martin, W.R. (1964b) Resistance of the mouse's intestinal tract to experimental *Salmonella* infection. II. Factors responsible for its loss following streptomycin treatment. *J. Exp. Med.*, **120**, 817–28.

Borriello, S.P. (1990) The influence of the normal flora on Clostridium *difficile* colonisation of the gut. *Ann. Med.*, **22**, 61–7.

Borriello, S.P. and Barclay, F.E. (1985) Protection of hamsters against *Clostridium difficile* ileocaecitis by prior colonisation with non-pathogenic strains. *J. Med. Microbiol.*, **19**, 339–49.

Borriello, S.P. and Barclay, F.E. (1986) An *in vitro* model of colonisation resistance to *Clostridium difficile* infection. *J. Med. Microbiol.*, **21**, 299–309.

Burr, D.H. and Sugiyama, H. (1982) Susceptibility to enteric botulinum colonization of antibiotic-treated adult mice. *Infect. Immun.*, **36**, 103–6.

Byrne, B.M. and Dankert, J. (1979) Volatile fatty acids and aerobic flora in the gastrointestinal tract of mice under various conditions. *Infect. Immun.*, **23**, 559–63.

Clark, J.D. (1971) Influence of antibiotics or certain intestinal bacteria on orally administered *Candida albicans* in germ-free and conventional mice. *Infect. Immun.*, **4**, 731–7.

Dougherty, S.H., Hentges, D.J., Casey, S.W. and Thal, W.R. (1988) Impact of LY 146032 on *Streptococcus* (*Enterococcus*) *faecalis* translocation in mice. *Antimicrob. Agents Chemother.*, **32**, 337–40.

Drasar, B.S. and Hill, M.J. (1974) *Human Intestinal Flora*, Academic Press, New York.

Fekety, R., Brown, R., Silva, J. and Hoffman, A.F. (1978) Fecal bile acids and cholestyramine in hamsters with clindamycin-associated colitis. *ICCAC Abstracts*, **129**, 150.

Flock, M.H., Binder, J.J., Filburn, B. and Gershengoren, W. (1972) The effect of bile acids on intestinal microflora. *Amer. J. Clin. Nutr.*, **25**, 1418–26.

Fredrickson, A.G. (1977) Behavior of mixed cultures of organisms. *Ann. Rev. Med. Microbiol.*, **31**, 63–87.

Freter, R. (1955) The fatal enteric cholera infection in the guinea pig, achieved by inhibition of normal enteric flora. *J. Infect. Dis.*, **97**, 57–65.

Freter, R. (1956) Experimental enteric *Shigella* and *Vibrio* infections in mice and guinea pigs. *J. Exp. Med.*, **104**, 411–18.

Freter, R. (1983) Mechanisms that control the microflora in the large intestine, in *Human Intestinal Microflora in Health and Disease* (ed. Hentges D.J), Academic Press, New York, pp. 33–54.

Freter, R. and Abrams, G.D. (1972) Function of various intestinal bacteria in converting germfree mice to the normal state. *Infect. Immun.*, **6**, 119–26.

Freter, R., Brickner, H., Botney, M. *et al.* (1983a) Mechanisms that control bacterial populations in continuous-flow culture models of mouse large intestinal flora. *Infect. Immun.*, **39**, 676–85.

Freter, R., Brickner, H., Fekete, J. *et al.* (1983b) Survival and implantation of *Escherichia coli* in the intestinal tract. *Infect. Immun.*, **39**, 686–703.

Freter, R., Stauffer, E., Cleven, D. *et al.* (1983c) Continuous-flow cultures as *in vitro* models of the ecology of large intestinal flora. *Infect. Immun.*, **39**, 666–75.

Gorbach, S.L. (1990) Lactic acid bacteria and human health. *Ann. Med.*, **22**, 37–41.

Helstrom, P.B. and Balish, E. (1979) Effect of oral tetracycline, the microbial flora, and the athymic state on gastrointestinal colonization and infection of BALB/c mice with *Candida albicans*. *Infect. Immun.*, **23**, 764–74.

Hentges, D.J., Pongpech, P. and Que, J.U. (1990) How streptomycin treatment compromises colonisation resistance against enteric pathogens in mice. *Microb. Ecol. Health Dis.*, **3** 105–11.

Hentges, D.J., Stein, A.J., Casey, S.W. and Que, J.U. (1985) Protective role of intestinal flora against infection with *Pseudomonas aeruginosa* in mice: influence of antibiotics on colonization resistance. *Infect. Immun.*, **47**, 118–22.

Johnson, R.O., Clay, S.A. and Arnon, S.S. (1979) Diagnosis and management of infant botulism. *Amer. J. Dis. Child.*, **133**, 586–93.

Jesudason, M.V., Hentges, D.J. and Pongpech, P. (1989) Colonization of mice by *Campylobacter jejuni*. *Infect. Immun.*, **57**, 2279–82.

Kennedy, M.J. (1989) Regulation of *Candida albicans* populations in the gastrointestinal tract: mechanisms and significance in GI and systemic candidiasis, in *Current Topics in Medical Mycology*, vol. 3 (eds M.R. McGinnis and M. Borgers) Springer-Verlag, Berlin, pp. 315–402.

Kennedy, M.J. and Volz, P.A. (1983). Dissemination of yeasts after gastrointestinal inoculation in antibiotic-treated mice. *Sabouraudia*, **21**, 27–33.

Kennedy, M.J. and Volz, P.A. (1985a) Effect of various antibiotics on gastrointestinal colonization and dissemination by *Candida albicans*. *Sabouraudia*, **23**, 265–73.

Kennedy, M.J., Rogers, A.L. and Yancey, R.J. (1988) An anaerobic continuous-flow culture model of interactions between intestinal microflora and *Candida albicans*. *Mycopathologia*, **103**, 125–34.

Kennedy, M.J. and Volz, P.A. (1985b) Ecology of *Candida albicans* gut colonization: inhibition of *Candida* adhesion, colonization, and dissemination from the gastrointestinal tract by bacterial antagonism. *Infect. Immun.*, **49**, 654–63.

Kennedy, M.J., Volz, P.A., Edwards, C.A. and Yancey, R.J. (1987) Mechanisms of association of *Candida albicans* with intestinal mucosa. *J. Med. Microbiol.*, **24**, 333–41.

Koopman, J.P., Kennis, H.M., Lankhorst, A. and Prins, R.A. (1979) Influence of the composition of the diet on some parameters connected with gasto-intestinal colonization resistance in mice. *Z. Versuchstierkd.*, **21**, 21–6.

Koopman, J.P., Welling, G.W., Huybregts, A.W.M. *et al.* (1981) Association of germ-free mice with intestinal microfloras. *Z. Versuchstierkd.*, **23**, 145–54.

Lee, A. and Gemmell, E. (1972) Changes in the mouse intestinal microflora during weaning: role of volatile fatty acids. *Infect. Immun.*, **5**, 1–7.

Levison, M.E. (1973) Effect of colon flora and short-chain fatty acids on growth in vitro of *Pseudomonas aeruginosa* and Enterobacteriaceae. *Infect. Immun.*, **8**, 30–5.

Maier, B.R., Onderdonk, A.B., Baskett, R.C. and Hentges, D.J. (1972) *Shigella*, indigenous flora interactions in mice. *Amer. J. Clin. Nutr.*, **25**, 1433–40.

Malamou-Ladas, H. and Tabaqchali, S. (1982) Inhibition of *Clostridium difficile* by faecal streptococci. *J. Med. Microbiol.*, **15**, 569–74.

Marshall, S.E., Marples, B.A., Salt, W.G. and Stretton, R.J. (1987) Aspects of the effect of bile salts on *Candida albicans*. *J. Med. Vet. Mycol.*, **25**, 307–18.

Meynell, G.G. (1963) Antibacterial mechanisms of the mouse gut. II. The role of Eh and volatile fatty acids in the normal gut. *Brit. J. Exp. Pathol.*, **44**, 209–19.

Mills, D.C. and Arnon, S.S. (1987) The large intestine as the site of *Clostridium botulinum* colonization in human infant botulism. *J. Inf. Dis.*, **156**, 997–8.

Moberg, L.J. and Sugiyama, H. (1979) Microbial ecological basis of infant botulism as studied with germfree mice. *Infect. Immun.*, **25**, 653–7.

Nardi, R.M., Vieira, E.C., Crocco-Afonso, L.C. *et al.* (1990) Experimental salmonellosis in conventional and germfree mice: bacteriological and immunological aspects. *Program and Abstracts, 10th Int. Symp. on Gnotobiology*, P-23.

Pongpech, P. and Hentges, D.J. (1989a) Inhibition of *Shigella sonnei* and enterotoxigenic *Escherichia coli* by volatile fatty acids in mice. *Microb. Ecol. Health Dis.*, **2**, 153–61.

Pongpech, P. and Hentges, D.J. (1989b) Inhibitory effects of volatile fatty acids and low pH on the multiplication of enteric pathogens *in vitro* in caecal contents of mice. *Microb. Ecol. Health Dis.*, **2**, 247–53.

Pongpech, P., Hentges, D.J., Marsh, W.W. and Eberle, M.E. (1989) Effect of streptomycin administration on association of enteric pathogens with cecal tissue of mice. *Infect. Immun.*, **57**, 2092–7.

Que, J.U. and Hentges, D.J. (1985) Effect of streptomycin administration on colonization resistance to *Salmonella typhimurium* in mice. *Infect. Immun.*, **48**, 169–74.

Que, J.U., Casey, S.W. and Hentges, D.J. (1986) Factors responsible for increased susceptibility of mice to intestinal colonization after treatment with streptomycin. *Infect. Immun.*, **53**, 116–23.

Rolfe, R.D. (1984) Role of volatile fatty acids in colonization resistance to *Clostridium difficile*. *Infect. Immun.*, **45**, 185–91.

Rolfe, R.D and Iaconis, J.P. (1983) Intestinal colonization of infant hamsters with *Clostridium difficile*. *Infect. Immun.*, **42**, 480–6.

Rolfe, R.D., Helebian, S. and Finegold, S.M. (1981) Bacterial interference between *Clostridium difficile* and normal fecal flora. *J. Infect. Dis.*, **143**, 470–5.

Snoeyenbos, G.H. (1979) Role of native intestinal microflora in protection against pathogens. *Proc. Ann. Meeting U.S. Animal Health Assoc.*, **83**, 388–93.

Stone, H.H., Kolb, L.D., Currie, C.H. *et al.* (1974) *Candida* sepsis: pathogenesis and principles of treatment. *Ann. Surg.*, **179**, 697–711.

Sugiyama, H. and Mills, D.C. (1978) Intraintestinal toxin in infant mice challenged intragastrically with *Clostridium botulinum* spores. *Infect. Immun.*, **21**, 59–63.

Sullivan, N.M., Mills, D.C., Riemann, H.P. and Arnon, S.S. (1988) Inhibition of growth of *Clostridium botulinum* by intestinal microflora isolated from healthy infants. *Microb. Ecol. Health Dis.*, **1**, 179–92.

Surawicz, C.M., Elmer, G.W., Speelman, P., McFarland, L.V., Chinn, J. and van Belle, G. (1989) Prevention of antibiotic associated diarrhea by *Saccharomyces boulardii*: a prospective study. *Gastroenterology*, **96**, 981–8.

Thijm, H.A. and van der Waaij, D. (1979) The effect of three frequently applied antibiotics on the colonization resistance of the digestive tract of mice. *J. Hyg. (Camb.)*, **82**, 397–405.

van der Waaij, D. and Berghuis de Vries, J.M. (1974) Selective elimination of Enterbacteriaceae species from the digestive tract of mice and monkeys. *J. Hyg. (Camb.)*, **72**, 205–11.

van der Waaij, D., Berghuis de Vries, J.M. and Lekkerkerk van der Wees, J.E.C. (1971) Colonization resistance of the digestive tract in conventional and antibiotic-treated mice. *J. Hyg.* (Camb.), **69**, 405–11.
van der Waaij, D., Berghuis, J.M. and Lekkerkerk, J.E.C. (1972) Colonization resistance of the digestive tract of mice during systemic antibiotic treatment. *J. Hyg.* (Camb.), **70**, 605–10.
Voravuthikunchai, S.P. and Lee, A. (1987) Cecectomy causes long-term reduction of colonization resistance in the mouse gastrointestinal tract. *Infect. Immun.*, **55**, 995–9.
Wang. Y. and Sugiyama, H. (1984) Botulism in metronidazole-treated conventional adult mice challenged orogastrically with spores of *Clostridium botulinum* type A or B. *Infect. Immun.*, **46**, 715–19.
Wilson, K.H. (1988) Microbial ecology of *Clostridium difficile*, in Clostridium difficile: *Its Role in Intestinal Disease* (eds R.D. Rolfe and S.M. Finegold), Academic Press, New York, pp. 183–200.
Wilson, K.H. and Freter, R. (1986) Interactions of *Clostridium difficile* and *Escherichia coli* with microfloras in continuous flow cultures and gnotobiotic mice. *Infect. Immun.*, **54**, 354–8.
Wilson, K.H., Silva, J. and Fekety, F.R. (1981) Suppression of *Clostridium difficile* by normal hamster cecal flora and prevention of antibiotic-associated colitis. *Infect. Immun.*, **34**, 626–8.
Wilson, K.H., Sheagren, J.N., Freter, R. *et al.* (1986) Gnotobiotic models for study of *Clostridium difficile* and *E. coli*. *J. Infect. Dis.*, **153**, 547–51.

Chapter Six

Factors affecting the microecology of the gut

ROLF FRETER

6.1 INTRODUCTION

This author began his career in medical bacteriology in the early 1950s, i.e. shortly after the introduction of antibiotics into clinical practice. In those years the often similar side-effects of these chemically different drugs were usually (and correctly) attributed to population shifts among the indigenous microflora with a resulting overgrowth or superinfection by undesirable or outright pathogenic microorganisms. Early experimental data included the demonstration that oral administration of streptomycin rendered guinea-pigs susceptible to enteric infection with cholera vibrios (Freter, 1954, 1955) and made mice susceptible to colonization by cholera vibrios or shigellae (Freter, 1956a). The most exciting event at the time seemed to be the discovery that colonization of streptomycin-treated animals by these human pathogens could be completely prevented by the feeding of a streptomycin-resistant strain of *Escherichia coli* (Freter, 1956a). This raised the (then) justifiable expectation that the oral administration of a properly selected strain of *E. coli* or other bacterial species would allow it to colonize the gut and take over some or all of the normal functions of an indigenous microflora, when the latter had been disturbed by antibiotics or by other forms of stress. This strategy, described as probiotics in recent years, would have been the realization of the dreams of Carre (1887) and Metchnikoff (1907) and his followers, and this author expected to become rich and famous very quickly by simply pursuing this line of applied research. Unfortunately, none of these expectations has been realized to date.

Subsequent research by the author and others revealed that the mechanisms governing homeostasis and other functions of the indigenous microflora are exceedingly complex and are not amenable to seemingly straightforward manipulation. Moreover, data obtained with seemingly straightforward experimental techniques designed to explore these mechanisms, often led to oversimplified and misleading conclusions, because the ecological complexities (many of them still unknown) were not, or could not be, considered in the interpretation. It is for these reasons that most of the diverse practices currently combined under the heading 'probiotics' are still highly controversial. These controversies are a consequence of inconsistent results reported in the literature by various authors [cf. relevant chapters in this volume and recent reviews by Fuller (1989), Gorbach (1990), Vanbelle *et al.* (1990)]. Most reviewers list a number of possible explanations for such inconsistencies. These usually include factors such as the use of different strains, variations in the stability and the consequent variations in viability at the time of oral administration of different culture preparations, and variations in dosage and dosage schedules employed by different investigators. While it is obvious that these variables need to be controlled, there are more important considerations based on basic ecological principles which suggest that the currently favoured approaches to probiotics are imperfect and must be expected to give inconsistent results in practice. This chapter, therefore, will discuss the current status of the ecology of gastrointestinal flora as it relates to the aims and practices of probiotics. Finally, suggestions will be made as to the kinds of information that must become available through future research, before significant progress in probiotics can be anticipated.

6.2 DEFINITIONS

Some inconsistencies have arisen in the literature because of the lack of precise definitions. For this reason, definitions of important ecological terms used in this chapter are given below.

'Colonization' and 'to colonize' will describe a bacterial population in the gastrointestinal tract that is stable in size over time, without the need for periodic re-introduction of the bacteria by repeated oral doses or other means. Colonization implies, therefore, that the colonizing bacteria multiply in at least some part of the gastrointestinal tract at a rate that exactly equals their rate of physical elimination. This definition of colonization conforms to common usage in microbial ecology (Ellwood *et al.*, 1980).

'Indigenous' microflora will refer to the usually complex mixture of bacterial populations that colonize a given area of the gastrointestinal tract in individual human or animal hosts that have not been affected by medical or experimental intervention, or by disease.

The terms 'to invade', 'invasion' and 'invader' will refer to a bacterial strain that is newly introduced into the gastrointestinal tract by oral (or sometimes rectal) administration. 'Implantation' and 'to implant' will be used to describe one of the possible outcomes of invasion, namely, that the invader has been able to achieve colonization. It is important to distinguish this outcome from a superficially similar situation that occurs when an invader is introduced repeatedly and frequently (e.g. three times a day) in large numbers, such that it can be cultured regularly from various regions of the gastrointestinal tract and from the faeces, but where the invader disappears again from the gut once the repeated oral doses are discontinued. This situation is often found in the practice of probiotics. It is indicative of the fact that the invader did not implant, i.e. that it did not multiply at all *in vivo* or, at best, multiplied at a rate insufficient to compensate for its physical removal. Such populations of invaders which are unable to implant, will be referred to as 'transients' or as being in a 'sustained transient state'. The distinction between implantation and the establishment of a sustained transient state is potentially of great practical importance, because the attributes that bacteria in a probiotic preparation must possess for optimal clinical effectiveness are likely to differ in each case.

The ability of the indigenous microflora to prevent the implantation of invaders is well known. In fact, it is this activity that most probiotics are intended to augment or restore. This phenomenon has been studied by many authors, and various terms have been used to describe this effect and/or its underlying mechanisms, e.g. 'bacterial antagonism' (Freter, 1956a), 'bacterial interference' (Dubos, 1963), 'barrier effect' (Ducluzeau *et al.*, 1970) and 'competitive exclusion' (LLoyd *et al.*, 1977). To avoid monotony, three of these terms, 'antagonism', 'interference' and 'barrier effect' will be used in this chapter more or less synonymously. Another widely used term, 'colonization resistance' (van der Waaij *et al.*, 1971) is nicely descriptive and, no doubt for this reason, has been used by some writers synonymously with the other terms mentioned. It must be realized, however, that the original definition of 'colonization resistance' was considerably more specific, as will be detailed later. This distinction is important, and 'colonization resistance' will be used in this chapter in accordance with the original authors' definition.

The term 'association' of bacteria, either with an inert surface, with the mucosa or with the mucus gel, will be used to denote the fact that the bacteria have reacted with these structures in such a way that they can resist to some degree the physical removal by such forces as washing or peristalsis. This term is intended to be neutral with respect to the mechanisms involved, i.e. it does not specify whether the interactions involve specific or non-specific adhesion or simple trapping in the mucus gel. Conforming to standard usage, this chapter will also describe germ-free animals that have been deliberately exposed to or 'associated' with one ('monoassociated'), two ('diassociated') or several known kinds of bacteria that then proceed to colonize the animals.

This term is preferred, by workers in that field, to 'contamination' which implies accidental exposure to unwanted bacteria due to faulty technique. The appropriate connotation of 'associate' or 'association' should be obvious from the context.

The term 'conventional' microflora refers to the indigenous flora of healthy humans who have not recently been exposed to antibacterial agents, of healthy animals caught in the wild or of laboratory animals that had not been derived from germ-free stock. This definition excludes the flora of so-called pathogen-free animals. The latter, derived from germ-free stock, have been fed a curious mixture of a few bacterial strains and acquire the remainder of their flora by chance contamination. Typically, such animals lack most of the strict anaerobes that are characteristic of the conventional microflora of the large intestine. For this reason, pathogen-free animals are unsuitable for most experimental studies concerned with intestinal microecology or probiotics (Freter *et al.*, 1983a; Lee, 1985). Unfortunately, as far as this author is aware, most commercial suppliers of laboratory animals in the USA and western Europe cannot supply conventional animals, but carry only pathogen-free stock.

6.3 USE OF ONE OR A LIMITED NUMBER OF BACTERIAL STRAINS IN PROBIOTIC PREPARATIONS

6.3.1 Choice of bacterial species

If two bacterial species or strains share the same environment, e.g. a continuous-flow broth culture or the large intestine, they will fail to interact only in very unusual circumstances. The most likely form of interaction will be antagonism, i.e. the presence of one kind of bacterium will reduce the population size of the other, or eliminate it entirely. This intuitively obvious principle represents a major rationale of the probiotic concept. In view of the near universality of this phenomenon, and the fact that the aim of most probiotics is to eliminate undesirable or pathogenic bacteria from the host, it is rather surprising that only a few bacterial species have been used for this purpose. After the original suggestions of Metchnikoff (1907) promoting the suitability of lactobacilli (including the later differentiated bifidobacteria) for probiotics, there followed a period during which attempts at implantation of *E. coli* were the preferred techniques (Nissle, 1916). *E. coli* is still being considered by some contemporary workers (Herget and Weinrauch, 1979; Rusch, 1980; Rusch *et al.*, 1980; Barrow and Tucker, 1986; Fukata *et al.*, 1989). More recently, enterococci have been added to this repertoire (Lewenstein *et al.*, 1979; Rusch, 1980). Few workers have used bacterial species other than the above. This narrow selection is surprising because neither of those species is predominant among the indigenous microflora of any region of the gastrointestinal tract (with the possible exception of lactobacilli in the chicken crop and in the mouse and rat stomach).

One possible explanation for the frequent choice of lactobacilli as probiotics may be found in the opinion of some early workers (e.g. Shahani and Ayebo, 1980; Pollmann *et al.*, 1980), that a 'healthy' flora is characterized by a high ratio of lactobacilli to *E. coli* in the faeces. However, unless the larger *Lactobacillus* populations can be shown to be causally related to (rather than being the indicator of) the proper functioning of the indigenous microflora, attempts to improve such a balance by the feeding of lactobacilli would be the equivalent of trying to cure a fever by shaking the thermometer to a lower reading.

A priori, one would have to assume that the predominant members of the indigenous microflora would be the most promising choices for inclusion in probiotic preparations. Lee (1985) discussed this paradox and suggested that the bacterial species included in probiotic preparations are chosen mainly for historical reasons and because they are easy to culture. An equally important consideration may have been safety, because the species currently singled out for use in probiotics were at one time regarded as harmless symbionts among the indigenous microflora. We now know, of course, that *E. coli* and enterococci can cause a variety of serious infections. Even the venerable lactobacilli have recently been reported in septicaemia of a compromised patient (Andriessen *et al.*, 1991). Moreover, lactobacilli have been identified as a major source of intestinal bile salt hydrolases (Tannock *et al.*, 1989) which, in turn, have been implicated in growth depression of livestock of the type that in current practice is relieved by the addition of subtherapeutic doses of antibiotics to the feed (Feighner and Dashkevicz, 1987). There is, consequently, no compelling ecological or other scientific rationale for the current narrow choice of bacterial species utilized in probiotic preparations. Strong arguments can be made for the inclusion of many other species, as will be discussed below.

6.3.2 Ecological considerations in the use of single strains

As mentioned in the introduction to this chapter, the implantation of an *E. coli* strain into streptomycin-treated mice could restore the animals' resistance to implantation of some human pathogens. Much of the rationale underlying probiotics is based on the work of a large number of workers who expected to see similar effects on intestinal microecology with single strains that either were able to implant or were maintained *in vivo* in a sustained transient state. It is important to realize that, in most of these situations, the newly introduced probiotic strains are present in the gastrointestinal tract in numbers much higher than the populations these same bacterial species can attain in an undisturbed conventional microflora. The consequences of such numerical differences have been illustrated in an early experiment from this author's laboratory: two *E. coli* strains, designated C25 and 40T, respectively, were used. When introduced as monoassociates into germ-free mice, each strain attained populations of approximately 5×10^9 per caecum. When both strains were fed simultaneously to germ-free mice, strain 40T reached populations similar to those in monoassociated animals, but the presence of strain 40T reduced the populations of strain C25 to 8.4×10^7 per caecum of the diassociated animals, i.e. its population was reduced by a factor of 60 by the presence of strain 40T. In conventionalized mice harbouring a synthetic indigenous microflora plus one of these *E. coli* strains, each strain reached populations in the order of 1×10^7 per caecum. When both *E. coli* strains were introduced simultaneously into conventionalized mice, each reached similar population sizes, i.e. in the order of 1×10^7 per caecum (Freter and Abrams, 1972). It appears, therefore, that strain 40T was able to antagonize strain C25 and reduce its population size in diassociated gnotobiotic mice when the populations of 40T were abnormally high. In contrast, the much lower populations of 40T in conventionalized mice no longer had any effect on C25. Rather, the populations of both *E. coli* strains were controlled in conventionalized mice by other elements of the indigenous microflora. It is therefore misleading to conclude that a bacterial species that has a demonstrable biological effect when used as a probiotic, is *ipso facto* also responsible for that same effect in an undisturbed ('healthy') indigenous microflora. This caveat must be considered in attempts at selecting bacterial strains with optimal characteristics for use as probiotics: such characteristics may be very different from the attributes of a strain of the same species that successfully colonizes a host as part of a 'healthy' conventional microflora.

Another basic problem with the use of single strains lies in the specificity of bacterial interactions. In the experiment described in the preceding paragraph, *E. coli* C25, which did not exert any effect on the populations of strain 40T in any experimental setting, was able to eliminate a strain of *Shigella flexneri* not only from diassociated gnotobiotic mice, but also when both C25 and the shigellae were colonizing mice at low population densities in the presence of an indigenous microflora (Freter and Abrams, 1972). In recent, more extensive studies, Barrow and Tucker (1986) demonstrated a very high degree of specificity of bacterial antagonism in newly hatched chicks: it required the implantation of a mixture of three *E. coli* strains to protect the animals against a given strain of *Salmonella typhimurium*, but, significantly, these *E. coli* strains were not equally effective against other *Salmonella* strains. Additional examples of a high degree of specificity with respect to *in vivo* inhibition were reported in a subsequent paper (Barrow *et al.*, 1987).

A further limitation on the use of single-strain probiotic preparations must be considered when these strains are intended to implant in the host (rather than to act in a sustained transient state). Almost by definition, probiotics are used when the indigenous microflora is incomplete, as in newborns, or disturbed by stress such as the administration of antibiotics. Such conditions are characterized by the colonization ('overgrowth') of the gut by one or a few types of bacteria that reach abnormally high population densities. The latter bacteria are often the cause of the harmful effects that one wishes to counteract by the use of probiotics, but their very presence must also be expected to antagonize certain strains of invaders, i.e. to exert a barrier effect. If this antagonism happens to be effective against the constituents of a probiotic preparation, the latter cannot be expected to be beneficial. For example, while streptomycin fed to conventional mice rendered these animals susceptible to colonization with streptomycin-resistant shigellae (Freter, 1956a), conventional mice which happened to incorporate the streptomycin-resistant *E. coli* strain C25 as part of their indigenous microflora, remained resistant to implantation of shigellae even after administration of streptomycin. This resistance was caused by the explosive expansion of the *E. coli* C25 population in the presence of streptomycin which, in turn, prevented the implantation of the drug-resistant shigellae (unpublished data). Barrow and Tucker (1986) also suggested that interference by *E. coli* strains already in the intestinal tract of chicks, accounted for inconsistencies in the effect of three *E. coli* strains that were fed to protect the animals against subsequent challenge with *Sal. typhimurium*.

The barrier effect described above, caused by the overgrowth of 'undesirable' bacteria, is not always considered by workers concerned with the effect of various stresses on the intestinal flora. For example, Gorbach and co-workers (Barza *et al.*, 1987; Giuliano *et al.*, 1987; Gorbach *et al.*, 1988) noted that the administration of antibiotics to human volunteers and the resulting changes in total anaerobe populations had no apparent correlation with the ability of marked strains of *E. coli* and *Pseudomonas aeruginosa* to appear, after ingestion, in the stools of the subjects. This finding does not suggest, as the authors assume, that the indigenous anaerobes may have no role in creating the barrier effect ('colonization resistance') in the *undisturbed* human gut. The ecologically most plausible interpretation of their data would be that the interference with implantation of those particular invading strains had shifted from certain strains of anaerobes (i.e. from the normal situation) to other, antibiotic-resistant strains of anaerobes and/or to other bacteria (Enterobacteriaceae, enterococci, yeasts), the populations of which increased after the administration of antibiotics.

Finally, we must consider another potentially major restriction on the efficacy of probiotics containing single strains of bacteria. This is based on the ecological principle that bacterial interactions differ in different habitats. In a study relevant to the use of probiotics (Itoh and Freter, 1989), lactobacilli were found to drastically reduce (i.e. by a factor of 10^3) the populations of *E. coli* in the stomach of gnotobiotic mice. In contrast, the same lactobacilli had relatively little effect on populations of the same *E. coli* in the large intestine of these animals, even though viable counts of the lactobacilli were slightly higher in the latter organ than in the stomach. The lactobacilli obviously colonized the stomach of these animals, but the bacterial counts suggested that they were merely in a sustained transient state in the large intestine, i.e. the *Lactobacillus* populations in the latter organ appeared to derive from the accumulation of lactobacilli originating in the proximal regions of the gastrointestinal tract, rather than from local multiplication. This finding also illustrates the intuitively obvious principle that the antagonism against other bacteria exerted by a given bacterial strain is likely to differ, depending on whether the latter is colonizing and therefore actively multiplying and metabolizing, or whether it is in a sustained transient state with a residual metabolism that is probably much reduced compared to that of actively multiplying cells.

6.3.3 Recapitulation

Evidence presented in this section indicates that probiotics containing one (or only a few) bacterial strains will face severe theoretical limitations to their broad and predictable effectiveness. The distinction between colonization and the sustained transient state is important. Strains that are intended to protect by colonization of the host will encounter unpredictable interference from the very bacteria whose overgrowth they are supposed to offset. Bacteria in probiotic preparations that are used to create a sustained transient state are not as likely to be affected by this form of interference. Nevertheless, like colonizing bacteria, they must still be expected to give highly variable results, because bacterial antagonism is specific both with respect to the bacterial species and strains that are susceptible, and with respect to the area in the gastrointestinal tract in which a given type of antagonism can be manifested. In view of these considerations it is rather surprising that only a limited number of bacterial species have been utilized for use in probiotics.

The mechanisms by which single strains interact with other bacteria in the intestine have not been widely studied. Some of the available evidence will be discussed in section 6.4.3, in the context of mechanisms controlling the indigenous microflora.

6.4 ECOLOGICAL CONSIDERATIONS

This section discusses the use in probiotic preparations of complex mixtures of bacteria resembling the indigenous microflora.

6.4.1 Habitats and microhabitats

Earlier workers often disregarded the fact that the gastrointestinal tract is not a single homogeneous habitat for the indigenous microflora, and thus considered only the composition of the faecal flora as a sufficient indicator of the functioning of the indigenous microflora. Today, even introductory texts of medical microbiology may present tables listing the different types of flora found in various regions of the gut.

Some habitat preferences may differ in various host species. For example, lactobacilli are among the characteristic indigenous flora colonizing the chicken crop, the stomach of mice and rats, and the lower ileum of man. Most of these habitats are distinguished by a high rate of transit, i.e. the intestinal contents are replaced at a rate that is much higher than that of even optimal bacterial multiplication. The early work of investigators such as Fuller (1973) and Savage (1972) demonstrated that indigenous bacteria colonizing such sites must firmly adhere to the mucosal epithelium. An understanding of the need for bacterial pathogens to associate with the wall of the small intestine, and the consequent protective effect by antiadhesive local antibody (Freter, 1956b) is now well established (Beachey, 1980). The ability to adhere to mucosal or tooth surfaces is also an important prerequisite for bacterial colonization of the mouth (Gibbons and van Houte, 1975). It is important to realize, of course, that the ability to associate with surfaces *in vivo* is not the only prerequisite that a bacterium must possess in order to colonize areas of the body that experience a high rate of physical removal. For example, in early work that contributed much to the recognition of the role of adhesion in colonization, Gibbons and co-workers introduced marked strains of various streptococcal species into the mouth of human volunteers and showed that the adhesion patterns paralleled those of the wild-type parent species (reviewed in Gibbons and van Houte, 1975). In spite of their ability to adhere in a normal pattern, however, the marked strains apparently did not colonize, most likely because they failed to multiply in a timely fashion, i.e. they failed to multiply before they eluted from the body surfaces.

In contrast to the above, some areas of the gastrointestinal tract are characterized by a relatively slow rate of transit, such that bacteria which can sustain multiplication at a doubling time as slow as 3h, have the theoretical potential to form constant populations in the lumen, i.e. they could colonize without having to rely on association with the mucosa (Freter, 1983). It is somewhat surprising, however, that this potential is not realized among the flora of the caecum and colon, where colonization by the indigenous microflora also requires some form of bacterial association with the mucosa, as will be detailed below. A common feature of such habitats is a dense and complex indigenous microflora.

Equally as important as the above-described distinct macroscopic habitats along the gastrointestinal tract, is the differentiation of at least four microhabitats within each of these areas. The first of these is the surface of epithelial cells. As mentioned above, a number of indigenous and pathogenic bacteria colonize this site. A common feature is the specificity of these adhesive reactions, often mediated by special organelles, such as fimbriae (Gibbons and van Houte, 1975; Beachey, 1980).

A second microhabitat is the deep layer of the mucus gel of the crypts of the ileum, caecum and colon. Typically, the organisms colonizing these sites are motile, spiral-shaped bacteria of the genera *Borrelia*, *Treponema*, *Spirillum* and others. Some of these may be attached to the underlying epithelial cells, whereas others appear to reside free in the mucus gel. Lee and co-workers have published extensive studies of these populations (reviewed in Lee, 1980, 1985). A major consequence of spiral morphology appears to be the ability to traverse viscous media such as the mucus gel. By virtue of this special aptitude, spiral-shaped bacteria are thought to resist removal with the mucus flow by active motility directed, perhaps, by chemotactic stimuli towards the bottom of the crypts. In such a situation, special means of attachment to the epithelial surface would not be required for successful colonization (Lee, 1985). Spiral bacteria, including a newly described pathogen, *Helicobacter pyloris*, also inhabit the mucus layer adjacent to the mucosa of the stomach of humans (Blaser, 1990). Even though spiral bacteria are a part of the indigenous microflora of the gastrointestinal tract of man and animals, little is known about their possible effect on other bacteria, i.e. whether they contribute to a barrier effect.

The third microhabitat is the mucus gel that overlays the epithelium of the entire gastrointestinal tract. Its contributions both to colonization by indigenous microflora and to the pathogenesis of enteric infection have been the subject of speculation for many years. A major reason for the tentative nature of our current understanding lies in the difficulty of demonstrating the mucus gel in microscopic preparations without incurring artefacts during drying and fixation. A number of techniques for overcoming this difficulty have been reported. Frozen sections of intestinal tissue have been used by Davis (1976) and Freter *et al.* (1981). Garland *et al.* (1982) employed acrolein vapours. A technique involving the stabilization of mucus gel in histological specimens with antimucus antibody has been described by Rozee *et al.* (1982), and its usefulness confirmed by Cornish *et al.* (1987) and Bollard *et al.* (1986).

Not least because of the above-described difficulties in demonstrating mucus gel in its native configuration, the types of evidence for many putative functions of the mucus gel often seem contradictory. Edwards (1978) thought of the mucus gel as an impermeable barrier to macromolecules and colloidal size particles such as bacteria. Florey (1933) first described the expulsion of extraneous colloidal particles by intestinal mucus gel and, consequently, viewed the flow of mucus gel as a protective mechanism that keeps mucosal surfaces free of contamination. Secretory antibodies are present in the intestinal mucus gel and can prevent the adhesion of bacteria to the surface of the underlying cells (McSweegan *et al.*, 1987). Walker and co-workers (1977) showed that mucus flow is increased when antigen–antibody complexes are present on the lumenal side of intestinal tissue. This phenomenon, which might represent an additional mechanism of local immunity, may result in a more rapid removal of microorganisms to which antibodies are present.

In addition to being a protective structure, the mucus gel can also serve as a habitat for indigenous and pathogenic bacteria. It appears that association with the mucus gel, even in the absence of adhesion to the epithelial surface, may suffice to make colonization possible. Bacterial chemotaxis, i.e. the attraction of motile bacteria into the mucus gel, has been shown to be an important mechanism that allowed bacteria to rapidly enter the mucus (Freter *et al.*, 1981). In typical experiments of this series, mixed suspensions of *Vibrio cholerae* and polystyrene microspheres (1.1 μm diameter) were injected *in vivo* into ileal loops of rabbits. Alternatively, slices of rabbit or mouse ileum were incubated in these suspensions. Fifteen minutes after contact with the suspension the tissues were removed, washed, frozen and sectioned in a cryostat. The locations of particles were then determined microscopically. The initial vibrio/particle ratio in the suspensions had been adjusted to unity. When a motile, chemotactic wild-type strain of vibrios was used, this ratio became successively larger as areas along the intervillous spaces were viewed, often reaching factors of increase greater than 10-fold near the bases of the villi. This finding indicates that the vibrios actively penetrated the mucus gel and proceeded towards the bases of the villi with much greater efficiency than the inert particles. In contrast, the vibrio/particle ratios deep in the intervillous spaces remained unchanged or even decreased slightly when motile but non-chemotactic or non-motile vibrios were used. Motile, chemotactic vibrios were also better colonizers of the gut of gnotobiotic mice than motile but non-chemotactic or non-motile mutants (*loc. cit.*).

More recently, this work has been extended to a member of the strictly anaerobic flora of the mouse large intestine. When a conventional mouse is dissected inside an anaerobic chamber and suspensions of the caecal contents viewed under a phase contrast microscope located inside the chamber, most bacteria are seen to be actively motile. Most motility is lost when the specimens are exposed to air. A highly motile, strictly anaerobic, Gram-positive bacterium, probably of the genus *Clostridium*, was isolated from such a specimen, and a non-chemotactic but normally motile, smooth swimming mutant selected. Interestingly, this mutant showed mainly negative chemotaxis, responding to short-chain fatty acids (the major metabolic end-products of the predominant anaerobes in the large intestine). The chemotactic parent strain showed superior association with caecal mucosa and was a much better colonizer in gnotobiotic mice than the non-chemotactic mutant (Freter, 1988). It is possible, therefore, that the metabolic end-products of the anaerobes, which presumably accumulate in the lumen, caused the chemotactic strain to move into the mucus gel towards the epithelium. Stanton and Savage (1984) also described a non-motile mutant of an anaerobic bacterium isolated from the mouse caecum that was less efficient than the motile parent in colonizing gnotobiotic mice, but the role of chemotaxis was not explored in that study. This means of traversing the mucus gel apparently is not a universal feature of all motile bacteria, however, and strains of *Sal. typhimurium* and *E. coli* have been described which exhibited motility in laboratory media but lost their motility when grown in intestinal mucus (McCormick *et al.*, 1988, 1990).

Cohen and co-workers in an interesting series of reports have extensively explored the role of mucus gel in colonization of the gut of streptomycin-treated mice. Their early studies showed a correlation between the ability of enteric bacteria to colonize the mouse intestine and the ability of the bacterial strains to adhere to mucus gel that had been immobilized in polystyrene wells (Wadolkowsky *et al.*, 1988). In contrast, it has been postulated on theoretical grounds that receptors in mucus gel may competitively inhibit bacterial adhesion to analogous receptor sequences on the surface of the underlying epithelial cells (Freter, 1980). Indeed, mucosal extracts can inhibit bacterial adhesion to brush border membranes of gut epithelial cells (Freter and Jones, 1976; Drumm *et al.*, 1988; Mantle *et al.*, 1989; Conway *et al.*, 1990), and salivary mucin competes for streptococcal adhesins with receptors on buccal epithelium (Williams and Gibbons, 1975). In a later report from Cohen's group (McCormick *et al.*, 1988), these workers showed that the ability of a strain of *Sal. typhimurium* to bind to components of the mucus gel may also impair, rather than promote colonizing ability, possibly by preventing the efficient penetration of the mucus gel. It is thus apparent that the ecological consequences of bacterial binding to components of the mucus gel are variable, and the presently available data do not permit the formulation of simple, generally applicable 'rules'. Nevertheless, the likely possibility must be considered that strains selected for probiotic preparations on the basis of their ability to adhere to mucosal epithelial cells, may actually be impaired in their ability to colonize, if the overlaying mucus gel carries receptors similar to those of the epithelial cells. Consequently, the only valid test for the ability of bacterial strains to associate with the mucosa of the gastrointestinal tract must be carried out *in vivo* or, at least, with fresh autopsy specimens that retain the mucus blanket.

Dubos and co-workers (1965) first drew attention to the extensive layers of bacterial populations covering the walls of the large intestine – a finding that has been confirmed by many workers (reviewed by Savage, 1983). They considered these to be embedded in the 'mucus layer'. Mucus is degraded by several of the indigenous bacteria in the large intestine (Miller and Hoskins, 1981), and may also serve as a nutritional substrate for the flora. The latter idea is supported by data suggesting that bacteria capable of degrading the host's blood group specific glycoproteins may have an ecological advantage in the human gut (Hoskins and Boulding, 1976). The finding of van de Merwe *et al.* (1983) that the human faecal flora may be determined by genetic differences, may conceivably relate to differences in glycoproteins as well. Unfortunately, the differences between monozygotic and dizygotic twins noted by these authors did not extend to the predominant anaerobes, possibly because only total counts of anaerobes were determined. In view of the mucin-degrading activity of the flora, it is not clear whether the mucosa-associated bacterial populations of the large intestine are indeed embedded in host-derived mucin, or whether the host's mucin is completely degraded and the matrix in which the bacteria are embedded is actually of bacterial origin.

The data summarized in the preceding paragraphs indicate that at least some of the seemingly contradictory data in the literature are not necessarily incompatible, and that the mucus gel can protect the host against bacterial colonization in some circumstances, and can form a bacterial habitat in others. It is important to realize that the gel can be traversed, though inefficiently, by non-motile bacteria and by inert particles the size of bacteria. Most likely, this penetration is *via* channels that develop in the mucus along planes of stress (Gibbons and Sellwood, 1973).

The fourth microhabitat to be considered is the lumen of the intestine.

It is certainly of minor importance in areas of high physical removal. The classical studies of Dixon (1960) sought to explain the common observation that the jejunum and upper ileum of most species harbour mostly transient bacteria and have little or no indigenous microflora. He injected a mixture of non-absorbable tracers and live bacteria into the jejunal lumen and, after various intervals of time, determined whether the original ratio of tracers *vs* bacteria had changed in the more distal regions of the small intestine. The results showed little change in these ratios, indicating that there was no bactericidal mechanism in the lumen. In fact there was some indication of bacterial multiplication which, however, was insufficient to compensate for the rapid peristaltic removal of lumenal contents. His conclusion was that mechanical forces alone are responsible for the paucity of flora in the small intestine, in spite of the presence of ample supplies of nutrients and in the absence of antibacterial activity. This notion is strongly supported by the events occurring in intestinal obstruction: in the absence of flow through the lumen, large bacterial populations develop rapidly proximally to the obstruction, leading to a surgical emergency. The lumen of the large intestine contains very large bacterial populations, but it is not clear whether these are actually multiplying significantly in the lumen or whether they represent merely the accumulated inactive daughter cells of the mucosa associated populations. As will be detailed later, the latter populations are certainly of prime importance in the ecology of the large intestine.

6.4.2 Models and tests

The kinds of experimentation that are ethical and compatible with the survival of an experimental animal are very limited. For this reason, research concerned with the mechanisms underlying intestinal microecology must rely extensively on *in vitro* models of the gut environment. The failure of many investigators to identify and use a proper *in vitro* model that can reproduce bacterial interactions as they actually occur *in vivo* has been a major shortcoming in this field of investigation. It seems preposterous to imagine that an ecologist interested in the interactions between desert cactus and other desert plants would go about this project by transplanting and observing the study objects in a tropical rain forest. The latter environment is so different from a desert, that whatever ecological interactions were to be observed, they would give no clue as to the interactions among the same species in their natural desert environment. Unfortunately, many workers in intestinal microecology have used a similarly uncritical approach in the choice of *in vitro* models. The subject has been reviewed in detail several times by this author (Freter, 1976, 1983, 1988). Hentges and Freter

(1962) first demonstrated that the interaction between *Sh. flexneri* and a number of other enteric bacteria differed considerably depending on the *in vitro* culture system used for testing these interactions. Thus most of the time honoured *in vitro* methods for testing the interactions between pairs (or a few) microorganisms cannot be expected to yield data that correlate with other methods or with the *in vivo* interactions among the tested strains. This includes experimental techniques such as mixed liquid cultures, inhibitory substances produced in liquid media (detected in the filtrates or across cellophane barriers), and diffusible inhibitory substances detected on solid media by 'cross streaks' or as inhibition zones around macrocolonies. Inhibitory reactions in most of these systems also can be a consequence of exhaustion of nutrients, rather than indicate the presence of inhibitory substances.

A reasonably well-established exception to the inadequacy of most *in vitro* models appears to be the anaerobic continuous-flow (CF) culture, which can duplicate the numerical relationships among the complex flora of the large intestine, as well as reproduce bacterial interactions as they occur in the large intestine (Zubrzycki and Spaulding, 1957; Hentges and Freter, 1962; Freter *et al.*, 1973, 1983a; Veilleux and Rowland, 1981; Edwards *et al.*, 1985; Wilson and Freter, 1986; Bernhardt *et al.*, 1987, 1988). The mere fact that CF cultures are able to maintain a natural balance among the numerous species populating the large intestine, is a strong argument supporting the conclusion that the ecological control mechanisms in CF cultures are similar to those operating *in vivo*. It is difficult to imagine two different sets of mechanisms which fortuitously would bring about similar equilibria in populations as complex as those of the indigenous microflora of the large intestine.

This somewhat surprising distinction of anaerobic CF cultures appears to be due to a considerable number of features that this culture device shares with the mammalian large intestine. The most obvious of these is the physical feature of continuous flow. In addition, the CF culture shares with the large intestine the large and complex bacterial populations associated with the wall. As will be detailed below, these adherent populations are critical for the ability of a CF culture to simulate the intestinal ecosystem. Also critical are the strictly anaerobic conditions that prevail in the large intestine, and which must be established in the CF culture for proper functioning. The metabolic activities of the conventional flora appear to be somewhat sensitive to changes in diet (Rowland and Wise, 1985). On the other hand, the species composition of the indigenous microflora appears to be less affected by changes in the diet, at least with respect to the populations of the major genera, even though more subtle influences cannot be excluded (Savage, 1977; Aries *et al.*, 1971; Finegold *et al.*, 1974, 1975; Bounous and Devroede, 1974; Moore and Holdeman, 1975; Bornside and Cohn, 1975; Hentges *et al.*., 1977; Simon and Gorbach, 1984). Consequently, it does not seem unreasonable to accept the possibility that a CF culture can also maintain the equilibrium of intestinal populations, even though the mix of nutrients in the growth medium may not closely resemble that available to those populations *in vivo*. Certainly, the major metabolic end-products (short-chain fatty acids and hydrogen sulphide) in a CF culture of mouse caecal flora resembled in quality and concentration those found *in vivo* (Freter *et al.*, 1983b).

It seems likely, however, that discrepancies between CF cultures and the large intestine will become obvious, as increasingly fine details of microbial interactions are studied. For example, Wilson and Freter (1986) determined that the interactions between mouse indigenous flora and *Clostridium difficile* are best reproduced in CF cultures using a medium based on faecal extracts from germ-free mice. Also, the rate constants of adhesion and elution of *E. coli*, as determined by the interpretation of experimental data with a mathematical model, were slightly different in the caecum of conventional mice, as compared to CF cultures of mouse caecal flora. Adhesion in both systems, however, was similar in other respects: both were non-specific, in the sense that they were non-saturable (Freter, unpublished). Obviously, then, a CF culture is not a promising system in which to study the biochemical and physicochemical mechanisms of *E. coli* association with the mucosa. Nevertheless, the ecological consequences of bacterial association with the wall of CF cultures appear to be sufficiently similar to the *in vivo* situation to simulate bacterial interactions as they occur *in vivo*.

Because of the difficulties with most *in vitro* systems discussed above, many workers (including this author in his early studies) have turned to an *in vivo* system – the gnotobiotic animal – in the hope that this would be a more relevant model for the study of microbial interactions. A number of laboratories have used streptomycin-treated conventional animals as an inexpensive, though imperfect substitute for germ-free animals. Unfortunately, the germ-free animal differs profoundly from its conventional counterpart. For example, the intestinal contents differ, peristalsis is slowed, local and systemic immune mechanisms are absent or reduced, and glycoproteins are not degraded by the indigenous microflora, and therefore are of different composition. As has been described by a number of workers, colonization of germ-free animals with a single or a few bacterial strains rarely redresses the germ-free abnormalities. Moreover, as discussed earlier in this chapter, when only a limited number of bacteria colonize a gnotobiotic animal, their population sizes are increased by several orders of magnitude over those present in the conventional host. Consequently, the ecological impact of these unnaturally high populations is greatly exaggerated, and may have no counterpart in the conventional intestine. One must conclude, then, that research with gnotobiotic animals harbouring a limited flora is useful as a preliminary, serving to identify the types of microbial interactions that potentially can occur in the intestine. Nevertheless, this kind of research cannot determine whether such reactions do indeed take place in a conventional microflora, and cannot identify the bacterial species that are normally involved in these reactions. The subject of relevance and interpretation of gnotobiotic and germ-free animal research has been reviewed in detail elsewhere (Freter, 1986).

A part of the problem of modelling the association of bacteria with the mucosal surface has already been discussed above in connection with the role of the mucus gel. A more detailed review of the subject has been published elsewhere (Freter and Jones, 1983). In addition to the complicating role of the mucus gel, there is the difficulty of obtaining mucosal specimens of human origin. This problem has prompted some investigators to use substitutes, such as buccal mucosal cells. Because of the specificities of adhesive reactions for different tissues from the same host, studies employing such substitutes must be interpreted with caution. Moreover, adhesive reactions depend very much on the nature of the suspending medium (pH, ionic composition, presence of surface active agents), and on the conditions under which the bacteria to be tested have been grown. For these reasons, *in vitro* tests of bacterial adhesion to isolated cell suspensions can be regarded only as screening devices, and must be confirmed by *in vivo* observations. Indeed, Pedersen and Tannock (1989) showed that *in vitro* tests of various *Lactobacillus* strains for adhesion to gut epithelial cells from piglets, did not predict whether a given strain would associate *in vivo* with the gut epithelium of these animals. Another relevant illustration is the number of recent studies involving *Lactobacillus* GG, a strain which had been selected for its ability to adhere to human buccal mucosal cells and perhaps, for that reason, looked most promising to a number of investigators. There is interesting but anecdotal evidence that this strain may be useful in the treatment of recurrent *C. difficile* diarrhoea (Gorbach, 1990) but, as far as this author is aware, no data are available comparing the efficacy of this strain to other lactobacilli, nor is there evidence that this strain does indeed associate with mucosa when given orally to humans. This strain gave variable results in the prevention of travellers' diarrhoea (Oksanen *et al.*, 1990), and its therapeutic effect in antibiotic associated diarrhoea (Siitonen *et al.*, 1990) was no greater than that of other, non-selected lactobacilli that had been used earlier by other groups (e.g. in studies cited by Siitonen *et al.*, 1990; and by Tankanow *et al.*, 1990). Consequently, the studies of *Lactobacillus* GG by various groups lend no support to the idea that *in vitro* adhesion to epithelial cells may parallel either *in vivo* association with the mucosa, colonization potential, or therapeutic effectiveness.

A major obstacle to understanding how the populations of the intestinal flora are controlled, is the complication that each species is affected, at any given time, by a large number of factors. Mechanisms such as the rate of flow through the gut; rates of association with, and elution from, the mucosa; competition for nutrients and for adhesion sites; the lag phase of growth – all act *simultaneously* to determine the fate of invaders and of indigenous bacteria. Consequently, the study of any single factor in isolation, no matter how thorough, cannot lead to an understanding of the whole system. Mathematical models, however, can simulate all of these interactions and, in conjunction with appropriate experimental studies, are an indispensable tool for the study of intestinal microecology (reviewed in Freter *et al.*, 1986).

6.4.3 Mechanisms that control the indigenous microflora

(a) Habitats with a high rate of transit

Surprisingly little is known about mechanisms that control bacterial populations in areas with high rates of physical removal. The crucial role of association with mucosal surfaces has already been discussed. In areas where the wall-associated bacterial populations are very dense, e.g. lactobacilli in the chicken crop or the rodent stomach, they exert a barrier effect. Some authors assume that this phenomenon is mediated by competition for adhesion sites. Proof of this theory would require the demonstration that *in vivo* adhesion in these areas is indeed saturated, i.e. that exposure of the surfaces to dense suspensions of bacteria does not result in additional adhesion. Alternatively, one can show microscopically that the surfaces of epithelial cells lining the wall are so densely covered with the indigenous flora that invading bacteria are physically prevented from approaching the cell surfaces. Less often, authors have considered the opposite possibility – namely, that the dense masses of adhering bacteria are themselves the substratum for the adhesion of other bacteria. Pedersen and Tannock (1989) cited this phenomenon as a possible explanation for the observed discrepancies between *in vitro* adhesion of lactobacilli to epithelial cells and their *in vivo* association with the mucosa. They suggested that strains which did not adhere *in vitro* might have associated with the mucosal surface *in vivo* by adhering to other *Lactobacillus* strains that were already attached, rather than by attaching directly to the epithelial surfaces. Adhesive reactions among different species of the indigenous microflora have received little attention in intestinal microecology, even though such phenomena are well known to oral microbiologists (Eifuku *et al.*, 1990; and earlier references and reviews cited therein).

Changes in pH or the production of inhibitory substances have frequently been proposed as mechanisms of bacterial antagonism that create a barrier effect. Unfortunately, most of these interference reactions have been studied *in vitro* by culture methods which, as discussed earlier, cannot be relied upon to simulate *in vivo* microbial interactions. For example, the control of *E. coli* populations in the mouse stomach and upper small intestine observed by Itoh and Freter (1989), was not correlated with an increase in acidity over that found in the same areas of the gut of germ-free mice.

It is well known that local immune mechanisms protect against colonization of the small intestine by enteric pathogens. There is little evidence, however, that local immunity has a significant effect on indigenous bacterial populations. In fact, it is hard to imagine how stable indigenous populations could exist at all, if they were indeed affected by immune mechanisms at the mucosal surface. Shedlofsky and Freter (1974) reported that local immunity had relatively little effect on the large populations of *V. cholerae* that colonized monoassociated gnotobiotic mice (without causing disease in this host species). In contrast, when the *Vibrio* populations in the gnotobiotes were reduced by associating the animals with several additional enteric species, then local immunity directed against the cholera vibrios had a much greater effect in still further reducing the *Vibrio* populations. The authors concluded that there was a synergistic effect between local immunity and bacterial antagonism in controlling the *Vibrio* populations. They speculated further, that those indigenous bacteria that form large populations might be little affected by local immune mechanisms because they are in a situation analogous to that of the vibrios in monoassociated mice, i.e. they are not significantly constrained by antagonisms exerted by other bacteria. Other authors who tried to reconcile the existence of a stable indigenous microflora with the pronounced effect of local immunity against pathogens, have presented evidence suggesting that the immune response to some indigenous bacterial species is impaired (Berg, 1983). It is not clear, however, whether this is a general phenomenon, because the evidence is not entirely consistent (Berg, 1983; Wold *et al.*, 1989).

(b) Habitats with a slow rate of transit

The indigenous flora of the large intestine consists of several hundred different kinds of bacteria, and it seemed unlikely to most students of intestinal microecology, that every one of these should be contributing to homeostasis or to the creation of the barrier effect. For this reason, many investigators have attempted to identify certain groups of indigenous bacteria that are 'responsible' for antagonizing invading bacteria. Meynell and Subbaiah (1963) showed that streptomycin-treated mice which had vastly increased susceptibility to oral *Salmonella* infection, lacked short-chain fatty acids in the contents of the large intestine. Because these acids were inhibitory to salmonellae *in vitro*, and because they are primarily the metabolic end-products of the indigenous strict anaerobes, the latter populations were implicated in the barrier effect. Van der Waaij *et al.*, (1971) reported that mice recovering from antibiotic treatment showed a barrier effect against *E. coli* at a time when their caecal flora consisted predominantly of clostridia, and when other bacterial species including *Bacteroides* and *Lactobacillus* were absent. A major, though by no means exclusive role of indigenous clostridia in controlling the populations of *E. coli* and other enteric bacteria has been demonstrated by several workers, most recently by Itoh and Freter (1989).

It is important to note that 'colonization resistance', as originally defined by van der Waaij and co-workers (1971), is a quantitative indicator that is measured in terms of the number of invading bacteria that must be fed in order to achieve colonization in 50% of the animals. As the figures in their paper suggest, antagonism exerted by the clostridia appears to affect the lag phase of growth that the invading bacteria undergo *in vivo*. If there are few invaders, they are already eliminated from the gut before their lag phase ends and they begin to multiply. With larger numbers of the same invaders there are some residual invader populations left at the time they start multiplication, and colonization may result. Colonization resistance, as determined in the original narrowly defined sense, may therefore be regarded as an indirect measure of the length of the *in vivo* lag phase for a particular invader that has to contend with a particular intestinal microflora. It is most important to realize that this indicator is quite different from another parameter that is often used to describe the normal homeostatic mechanisms of the intestinal flora, namely the population size that a certain species or strain (e.g. *E. coli*) can achieve during colonization. As will be detailed shortly, the currently available evidence suggests that metabolic competition is the most important factor governing the size of colonizing bacterial populations in an indigenous microflora at equilibrium, whereas the length of the lag phase of invaders is, in addition, determined by the rates of bacterial adhesion to the wall and by the presence of certain inhibitors, such as short-chain fatty acids. Relevant evidence will be pointed out below, as the discussion progresses. Much of today's uncertainties concerning the mechanisms that control intestinal populations may be traced to the failure to distinguish between these two equally important but different parameters.

Syed *et al.* (1970) and Freter and Abrams (1972) attempted to reduce the size of the *E. coli* population in gnotobiotic mice to the levels found in conventional mice, by associating the animals with increasing numbers of anaerobic strains isolated from conventional mice. It required 95 of these strains to bring the *E. coli* population within the conventional range, and later (unpublished) attempts by this author to further reduce that number below 95 were unsuccessful. This effect was diet dependent. Fewer strains were required for mice fed a refined diet based on polished rice and casein (another suggestion of the involvement of nutrient competition). Koopman *et al.* (1981) also reported that as many as 110 anaerobes were required to reverse a variety of germ-free abnormalities, including the barrier effect, in the mouse. Some abnormalities could only be partially reversed. Interestingly, the different kinds of abnormalities tested by these investigators required different numbers of bacterial species for reversal. These and later reports from other laboratories allow the following conclusions: (1) the strict anaerobes are absolutely necessary for the proper functioning of the large intestinal flora and (2) it is unlikely that one or a few of these anaerobes are of predominant importance. Rather, each strain has to fill its own distinct ecological niche (another observation compatible with metabolic competition). This is not to say that facultative anaerobes, such as *E. coli*, do not contribute to bacterial antagonism exerted by the indigenous flora. As discussed in detail in section 6.3.2, Freter and Abrams (1972) reported that an *E. coli* strain contributed to the control of *Shigella* populations in conventional mice, even though the *E. coli* population had been reduced to conventional levels by the anaerobes. However, such functions of facultative anaerobes are secondary and, in the absence of anaerobes, their populations expand by orders of magnitude and a balanced conventional flora ceases to exist.

It is important to realize that the above description refers only to a normally functioning indigenous microflora. As discussed in some detail in section 6.3.2, when the microflora is immature or disturbed by antibiotics or other stresses, one or a few bacterial species will expand their populations in unpredictable ways. Experimental evidence, practical experience and ecological theory indicate that these abnormally large populations will also exert a barrier effect against invading bacteria. Unlike the barrier effect exerted by an undisturbed indigenous flora, however, the antagonism generated by such abnormally large populations of one or a few species is necessarily of narrow specificity. It is effective only against certain invader strains, whereas other invaders may antagonize the original expanded populations and, unpredictably, may become predominant in turn.

Little is known about the mechanisms underlying interactions among members of an abnormal intestinal flora. Early work with CF cultures and gnotobiotic mice *monoassociated* with a single *E. coli* strain indicated that this population could exert a barrier effect against another invading *E. coli*. This was shown conclusively to be due to metabolic competition between resident and invader strains: the feeding to mice or the addition to CF cultures of a carbon source that could be utilized by the invader but not by the resident strain mitigated or abolished the barrier effect (Freter, 1962; Ozawa and Freter, 1964). These phenomena conform to the strict definition of colonization resistance given by van der Waaij and co-workers (1971): the phenomenon was mediated by a prolonged lag phase of the invader, which caused it to be eliminated from the CF culture before it was able to multiply. With large inocula of the invader, however, a sufficient number of cells remained in the culture at the end of the lag. It is noteworthy that, after the lag, the remaining invaders could colonize, albeit at population sizes that were orders of magnitude below those of the resident. At that point the newly colonizing invader populations obviously must have multiplied at the same rate as the residents, because constant populations of any size in CF cultures indicate a growth rate that equals the rate of elimination from the system. In other words, beyond the extended lag phase, the invaders behaved like any other member of the indigenous flora.

Extensive later studies (Freter, 1976; Freter *et al.*, 1983a) of the ecology of a *complete conventional* mouse caecal flora in CF cultures showed considerable similarities to the interactions described in the preceding paragraph. When an invading *E. coli* strain, harvested from an agar plate, was introduced into such a culture, it failed to multiply and was washed out. The inhibitory principle could be isolated across cellophane membranes in a diffusion chamber that was immersed into the CF culture. The invading *E. coli* would not multiply in such a diffusion chamber unless one of two changes was made: either hydrogen sulphide was removed from the chamber, or a carbon/energy source was added. The conclusion drawn from these experiments was that the lag phase of this invader was caused by the lack of a carbon/energy source that could be utilized under the prevailing conditions of strict anaerobiosis and in the presence of hydrogen sulphide. Both anaerobiosis and hydrogen sulphide appeared to restrict the range of substrates that can support bacterial growth. Moreover, the addition of a carbon/energy source or removal of hydrogen sulphide also caused the invading *E. coli* to multiply in the diffusion chamber at a rate that was much higher than that of the indigenous bacteria in the parent CF culture, indicating that these factors were also involved in controlling the size of colonizing bacterial populations. Similar data were also obtained with a strain each of a *Fusobacterium* sp. and *Eubacterium* sp., indicating that these mechanisms were also involved in the control of at least some of the predominant anaerobes. Wilson and Perini (1988) reported that the barrier effect against *C. difficile* in CF cultures of mouse caecal flora was mediated by competition for nutrients.

Guiot (1982) used an ingenious method involving a 2-hour pre-incubation of agar slices containing *E. coli* organisms in the contents of conventional rat caecum or human faeces. After subsequent immersion of the slices in small amounts of saline the authors noted relatively poor growth of the *E. coli* in the slices, especially when the experiments were conducted under anaerobic conditions. This growth could be increased dramatically by supplying nutrients from brain heart infusion broth. Guiot concluded that these data can be interpreted to demonstrate that bacterial antagonism in the large intestine is caused by a high degree of competition for substrate that can be utilized under anaerobic conditions. Nevertheless, the multiplication rate of *E. coli* in the pre-incubated slices, even under anaerobic conditions, was sufficient to permit colonization if it had occurred *in vivo* (cf. figure 4 in the quoted paper). This is not consistent with Guiot's finding that the *E. coli* strain used did not colonize when fed to conventional animals. The discrepancy might be explained by the fact that substances such as hydrogen sulphide or fatty acids, which have also been implicated in intestinal ecology, were lost or diluted, respectively, under the experimental conditions. Nevertheless, the relative scarcity in large intestinal contents of nutrients that support anaerobic growth, was clearly demonstrated, a finding which supports the data obtained with CF cultures reviewed above.

Earlier work (Freter *et al.*, 1983c) had shown that an inoculum of *E. coli* taken from another CF culture did multiply to some extent in a CF culture of mouse caecal flora, whereas a plate-grown inoculum failed to multiply and was washed out with the flow rate of the culture. While these data support the idea of the importance of a lag phase in the barrier effect, they were marred by the fact that even the CF culture-adapted inoculum was eventually eliminated, i.e. its growth rate was insufficient to support colonization. More recent (unpublished) studies from this author's laboratory have shown that an *E. coli* strain taken from a CF culture of conventional mouse caecal flora would implant when transferred to another, similar, CF culture. The later experiments differed from those reported earlier in that the CF-adapted inoculum was transferred *via* glass slides that had been immersed in the primary CF culture for 2 weeks and were subsequently immersed in the secondary culture for 2 days. The *E. coli* inoculum thus consisted of cells that eluted from the slides during the 2 days of immersion. This resulted in constant populations of the inoculum for 2 days which, presumably, allowed the *E. coli* sufficient time to associate with the walls of the CF culture. This again is consistent with the theory that failure to implant (i.e. colonization resistance in the strict sense) is due to a prolonged lag phase of growth caused by transplanting the invader into the new and rather hostile environment of the indigenous microflora. It also is consistent with the theory that association with the wall is important in intestinal ecology, because it allows a bacterium to colonize at a lower rate of growth than would be required if colonization were restricted to the lumen. The latter phenomenon will be discussed in the next paragraph.

The importance of bacterial association with the wall of a CF culture of conventional mouse caecal flora was investigated by transferring the contents of the culture vessel to a new one. The culture in the new vessel lost its barrier effect and allowed an invading *E. coli* to colonize (Freter *et al.*, 1983a). Consequently, association with the wall by at least some of the indigenous flora appears to be important for the proper functioning of the intestinal ecosystem. This problem has been explored further by means of a mathematical model. The model (Freter, 1983) demonstrated that association with the wall is not absolutely necessary for colonization of the large intestine or CF culture, but that it is necessary for the establishment of a barrier effect. This seeming paradox is due to the fact that large wall-associated bacterial populations will reduce the limiting nutrients below the level necessary for the multiplication rate required to maintain a constant population in the lumen, where physical removal is more rapid than on the wall and where, consequently, more rapid multiplication is required. For this reason, an invader will be washed out of the lumen, even if it were physiologically adapted to wall-associated growth in the intestinal environment and, therefore, were able instantly to begin the relatively slow rate of multiplication necessary to maintain constant wall-associated populations. Whether or not an invader will be able to colonize depends, therefore, on its ability to associate quickly with the wall before it is washed out of the lumen (Freter, 1983). Consequently, not only the mere ability to adhere, but also more precise parameters such as the rate constants of adhesion and elution, must be known if one wants to predict the colonizing ability of a given invader strain. Another mathematical analysis showed that the ability to adhere strongly to the wall can compensate for a less than optimal rate of multiplication, and thereby allow an organism to colonize, even if it multiplies somewhat more slowly than other members of the indigenous flora (Freter *et al.*, 1983c).

The possible role of inhibitory substances in the ecology of the intestinal flora is controversial. As reviewed elsewhere (Freter, 1983), most studies of this nature were done with *in vitro* culture models that do not simulate the intestinal environment. Confirmatory tests *in vivo* were seldom done, and the few reported in the literature are contradictory. Even on theoretical grounds one must postulate that inhibitory substances cannot account for relatively low but constant populations of species of the indigenous microflora (Freter, 1983). Short-chain fatty acids have received by far the most attention from investigators as possible inhibitors active in intestinal microecology. As reviewed by Hentges (1983) the evidence is controversial. Freter *et al.* (1983c) suggested that much of the controversy might be resolved by interpreting the data available in the literature to show that fatty acids prolong the lag phase of bacterial multiplication, but have little or no effect on the subsequent growth rate exhibited during colonization. Most workers who studied the inhibitory effects of fatty acids *in vitro* carried out short-term experiments which did not allow a distinction to be made between effects on the length of the lag phase *vs* effects on logarithmic growth. In other studies (e.g. Que *et al.*, 1986) the *in vivo* antibacterial mechanisms did allow *in vivo* multiplication to constant populations after an initial lag phase. Moreover, in broth containing fatty acids (their Figure 4), there was a prolonged lag phase followed by logarithmic growth; the latter was not affected by the added fatty acids. In tests that did show a decrease in growth rate (their Figure 2), the decreased growth rate was still ample to support colonization if the tested bacteria had been able to maintain that rate *in vivo*. This suggests that other inhibitory mechanisms that had been operating *in vivo* may have been lost during the experimental manipulations. The data of Rolfe (1984) do show a rapid death of *C. difficile* in broth containing volatile fatty acids. Anaerobes are known, however, to show rapid death in culture once multiplication is prevented or has stopped, possibly due to the action of endogenous bacteriolytic systems (e.g. Freter *et al.*, 1983b). If this were true for *C. difficile* as well, data showing rapid bacterial death would not be incompatible with the assumption of a prolonged lag phase caused by the presence of fatty acids, probably acting in conjunction with other mechanisms. Most *in vivo* data show that the rate of bacterial multiplication in the intestine is not affected by varying levels of fatty acids, once the lag phase has passed (e.g. Maier *et al.*, 1972; review by Hentges, 1983). Confusion has resulted from the failure of some writers to realize that a finding of constant populations *in vivo* indicates similar rates of growth, regardless of the size of the population. Thus, as reviewed by Hentges (1983), some reports in the literature show that *in vivo* population levels correlate negatively with the concentrations of fatty acids present in the gut lumen, whereas others failed to show such a correlation. One must realize that the volatile fatty acids are end-products of the metabolism of many of the predominant anaerobes indigenous to the large intestine. Consequently, high levels of fatty acids may simply indicate the presence of such anaerobes that may affect the population sizes of other bacteria by mechanisms such as metabolic competition. In such cases, the fatty acids would be mere indicators of the presence of anaerobes that happen to be particularly active in producing these metabolites, and no consistent correlations could be expected. Be this as it may, the only consistent effect of fatty acids that may be related to colonization resistance (in the strict sense) is their ability to prolong the lag phase. A demonstration of this latter effect, or the lack of it, however, is significant only if the data were obtained *in vivo*, or under experimental conditions that closely simulate the intestinal environment.

6.4.4 Recapitulation

In considering homeostatic mechanisms among the flora of the large intestine, much confusion can be avoided by distinguishing between observations of the size of indigenous intestinal populations and evidence based on resistance to implantation (i.e. colonization resistance in the strict sense). Most of the evidence available today is consistent with the theory that the sizes of populations in the large intestine are controlled by metabolic competition under the peculiar environmental conditions prevailing. By definition, indigenous populations do not change much in size and, therefore, all constant bacterial population at a given site must multiply at similar rates, regardless of their actual size. This relation is modified by bacterial association with the wall, which reduces the rates of elimination and allows the bacteria to form constant populations at lower rates of multiplication and, consequently, at lower concentrations of a limiting nutrient.

The implantation of invaders, whether they be pathogens or indigenous species, is strongly influenced by the length of the lag phase of growth that bacteria commonly undergo when entering a new environment. Unless invaders are able to start multiplication before being eliminated from the gut, they obviously cannot implant. The length of the lag phase is determined by a variety of factors, including the low concentration of nutrients that can be utilized in the gut environment, i.e. by the same mechanisms that control the population sizes. However, even if the invader begins multiplication at the same rate as the indigenous species, it cannot colonize unless it is able to associate with the wall. Thus, while the mere ability to associate with the wall may be sufficient for colonization, the *rate* at which this association occurs is peculiarly important to invaders. Short-chain fatty acids are the best studied substances that appear to affect the length of the lag phase more that the size of colonizing populations.

In all areas of the gastrointestinal tract, competition for adhesion sites is a possible mechanism for bacterial interference, but direct evidence of this effect is not convincing. Metabolic competition may be an alternative explanation for the antagonism exerted by large bacterial populations that colonize areas with high rates of transit, such as the rodent stomach.

A possible role for inhibitory substances in the barrier effect of the indigenous microflora has not been established.

6.5. Recommendations for future developments

Preparations containing a single or few types of bacteria are limited by ecological necessity. From a practical standpoint, the situation is made worse by an almost complete lack of understanding of whether, and how, the organisms in probiotic preparations are required to associate with the mucosa. While it may seem reasonable to assume that mucosal association would be helpful, one needs to know the mechanisms involved, and the macroenvironment (stomach, small intestine, large intestine) and microenvironment (epithelial surface, mucus gel) in which association takes place. In addition, little is known about the mechanisms by which bacteria in probiotic preparations are expected to bring about beneficial physiological effects or antagonize unwanted microorganisms in the gut. A rational selection of promising species or strains for probiotic use is not possible without this information, all of which must be obtained *in vivo*, or by means of assays that can be relied upon to test parameters that are actually operative in the gut. It is especially important to know whether probiotics do indeed colonize, or whether they remain in a sustained transient state, i.e. in a prolonged lag phase. In the latter case, little or no bacterial multiplication takes place, and *in vivo* assays for desirable physiological activities must test non-multiplying bacteria, rather than bacteria during logarithmic growth.

The only reliable means to restore a broadly specific barrier effect after it has been diminished or lost due to stress, is to restore the complete indigenous flora. This has been done in animals by a number of workers, as well as in humans (by enemas of human faecal suspensions; Bowden *et al.*, 1981; Schwan *et al.*, 1984). A more practical approach might be to maintain a human flora in CF cultures, and to use the washed effluent as an oral probiotic. CF cultures also would be free of protozoa (unpublished data) and human viruses.

Effective implantation of a complete indigenous microflora would probably require the prior administration of antibiotics to eliminate competing organisms that may have overgrown during the period of stress. The only exception would be newborns or newly hatched chicks. Alternatively, it may be possible to establish in CF cultures a variety of human conventional floras, each of which is resistant to a certain antibiotic. Such preparations may then be given prophylactically when therapy with the respective antibiotic is begun. Future research might well establish that not every single species of the indigenous flora is required for a broadly specific barrier effect. A more limited number of bacterial species, e.g. the indigenous clostridia, might suffice for practical purposes. This would also simplify distribution, because clostridial spores could be stored easily.

Finally, well-controlled field trials, as suggested by Fuller (1989) and others, are an absolute necessity to establish the practical and scientific soundness of what today is in many ways still the probiotic *concept*.

The above suggestions for further research are obviously difficult to execute. It should be considered, however, that the one hundredth anniversary of the probiotic concept occurred four years ago (cf. Carre, 1887), without this idea having emerged from controversy. Unfortunately, without future critical studies of basic mechanisms, as well as field trials, much of the commercial practice of probiotics will remain to be viewed by many as an exploitation of oversimplified ecological theories.

REFERENCES

Andriessen, M. P., Mulder, J. G. and Sleijfer, D. T. (1991) Lactobacillus septicemia, an unusual complication during the treatment of metastatic choriocarcinoma. *Gynecol. Oncol.*, **40**, 87–9.

Aries, V. C., Crowther, J. S., Drasar, J. S., Hill, M. J. and Ellis, F. R. (1971) The effect of a strict vegetarian diet on the faecal flora and faecal steroid concentrations. *J. Pathol.*, **103**, 54–6.

Barrow, P. A. and Tucker, J. F. (1986) Inhibition of colonization of the chicken cecum by pre-treatment with strains of *Escherichia coli*. *J. Hyg. (Camb.)* **96**, 161–9.

Barrow, P. A., Tucker, J. F. and Simpson, J. M. (1987) Inhibition of colonization of the chicken alimentary tract with *Salmonella typhimurium* gram-negative facultatively anaerobic bacteria. *Epidemiol. Infect.*, **98**, 311–22

Barza, M., Giuliano, M., Jacobus, N. V. and Gorbach, S. L. (1987) Effect of broad-spectrum parenteral antibiotics on 'colonisation resistance' of intestinal microflora of humans. *Antimicrob. Agents Chemother.*, **31**, 723–7.

Beachey, E. H. (1980) *Bacterial Adherence*, Chapman and Hall, London.

Berg, R. D. (1983) Host immune response to antigens of the indigenous intestinal flora, in *Human Intestinal Microflora in Health and Disease* (ed. D. J. Hentges), Academic Press, New York, pp. 101–26.

Bernhardt, H., Knoke, M., Bootz, T. and Zschiesche, M. (1987) Simulierung des intestinalen mikrobiellen Overgrowth durch kontinuierliche Kultur. *Dtsch. Z. Verdau. – Stoffwechs.krankh*, **47**, 261–7.

Bernhardt, H., Knoke, M. and Bootz, T. (1988) Simulation of the intestinal microflora in a continuous flow culture, in *Bengt E. Gustafsson Symp.: The*

Regulatory and Protective Role of the Normal Microflora, Stockton Press, Stockholm.
Blaser, M. J. (1990) Epidemiology and pathophysiology of *Campylobacter pylori* infections. *Rev. Infect. Dis.*, **12**(S1), S99–S106.
Bollard, J. E., Vanderwee, M. A., Smith, G. W. *et al.* (1986) Location of bacteria in the mid-colon of the rat. *Appl. Environ. Microbiol.*, **51**, 604–8.
Bornside, G. H. and Cohn, I. (1975) Stability of normal human fecal flora during a chemically defined, low residue liquid diet. *Ann. Surg.*, **181**, 58–60.
Bounous, G. and Devroede, G. J. (1974) Effect of an elemental diet on human fecal flora. *Gastroenterol.*, **66**, 210–14.
Bowden, T. A., Mansberger, A. R. Jr and Lykins, L. E. (1981) Pseudomembranous Enterocolitis: Mechanisms of restoring floral homeostasis. *Amer. Surg.*, **47**, 178–83.
Carre, C. (1887) Ueber Antagonisten unter den Bacterien. *Correspondenz-Blatt fuer Schweizer Aerzte*, **17**, 385–92.
Conway, P. L., Welin, A. and Cohen, P. S. (1990) Presence of K88-specific receptors in porcine ileal mucus is age dependent. *Infect. Immun.*, **58**, 3178–82.
Cornish, J. M., Vanderwee, M. and Miller, T. (1987) Mucus stabilization in the urinary bladder. *Brit. J. Exp. Pathol.*, **68**, 369–75.
Davis, C. P. (1976) Preservation of gastrointestinal bacteria and their environmental association in rats by freezing. *Appl. Environ. Microbiol.*, **31**, 304–12.
Dixon, (1960) The fate of bacteria in the small intestine. *J. Path. Bacteriol.*, **79**, 131–40.
Drumm, B., Robertson, A. M. and Sherman, P. M. (1988) Inhibition of attachment of *Escherichia coli* RDEC-1 to intestinal microvillus membranes by rabbit ileal mucus and mucin *in vitro*. *Infect. Immun.*, **56**, 2437–42.
Dubos, R. (1963) Staphylococci and infection immunity, *Amer. J. Dis. Child.*, **105**, 643–5.
Dubos, R., Schaedler, R. W., Costello, R. and Hoet, P. (1965) Indigenous, normal and autochthonous flora of the gastrointestinal tract. *J. Exp. Med.*, **122**, 67–76.
Ducluzeau, R., Bellier, M. and Raibaud, P. (1970) Transit digestif de divers inoculums bactériens introduits 'per os' chez des souris axéniques ou 'holoxéniques' (conventionelles): Effet antagoniste de la microflore du tractus gastro-intestinal. *Zbl. Bacteriol. I. Abt. Orig.*, **213**, 533–48.
Edwards, C. A., Duerden, B. I. and Read, N. W. (1985) Metabolism of mixed human colonic bacteria in a continous culture mimicking the human cecal contents. *Gastroenterology*, **88**, 1903–9.
Edwards, P. A. W. (1978) Is mucus a selective barrier to macromolecules? *Brit. Med. Bull.*, **34**, 55–6.
Eifuku, H., Yakushiji, T., Mizunu, J. *et al.* (1990) Cellular coaggregation of oral *Streptococcus milleri* with actinomyces. *Infect. Immun.*, **58**, 163–8.
Ellwood, D. C., Hedger, J. N., Latham, M. J. *et al.* (1980) *Contemporary Microbial Ecology*, Academic Press, London, p. 254.
Feighner, S. D. and Dashkevicz, M. P. (1987) Subtherapeutic levels of antibiotics in poultry feeds and their effects on weight gain, feed efficiency,

and bacterial cholytaurine hydrolase activity. *Appl. Environ. Microbiol.*, **53**, 331–6.

Finegold, S. M. Attebery, H. R. and Sutter, V. L. (1974) Effect of diet on human fecal flora: Comparison of Japanese and American diets. *Amer. J. Clin. Nutr.*, **27**, 1456–69.

Finegold, S. M., Attebery, H. R. and Sutter, V. L. (1975) Fecal bacteriology of colonic polyp patients and control patients. *Cancer Res.*, **35**, 3407–17.

Florey, H. W. (1933) Observations on the functions of mucus and the early stages of bacterial invasion of the intestinal mucosa. *J. Pathol. Bacteriol.*, **37**, 283–9.

Freter, R. (1954) The fatal enteric cholera infection in the guinea pig. *Bacteriol. Proc. 1954*, p. 56.

Freter, R. (1955) The fatal enteric cholera infection in the guinea pig achieved by inhibition of normal enteric flora. *J. Infect. Dis.*, **97**, 57–64.

Freter, R. (1956a) Experimental enteric *Shigella* and *Vibrio* infections in mice and guinea pigs. *J. Exp. Med.*, **104**, 411–18.

Freter, R. (1956b) Coproantibody and bacterial antagonism as protective factors in experimental enteric cholera. *J. Exp. Med.*, **104**, 419–26.

Freter, R. (1962) *In vivo* and *in vitro* antagonism of intestinal bacteria against *Shigella flexneri*. II. The inhibitory mechanism. *J. Infect. Dis.*, **110**, 38–46.

Freter, R. (1976) Factors controlling the composition of the intestinal microflora, in *Microbial Aspects of Dental Caries. (eds H. M. Stiles. et al.)* Special Suppl. Microbiology Abstracts, vol. 1, pp. 109–20.

Freter, R. (1980) Prospects for preventing the association of harmful bacteria with host mucosal surfaces, in *Bacterial Adherence* (ed E.H. Beachey) Chapman & Hall, London, pp. 441–58.

Freter, R. (1983) Mechanisms that control the microflora in the large intestine, in *Human Intestinal Microflora in Health and Disease*, (ed. D. J. Hentges., Academic Press, London.

Freter, R. (1986) Gnotobiotic and germfree animal systems, in *Bacteria in Nature* (eds J. S. Poindexter, and E. R. Leadbetter), vol. 2, Plenum Publ. Co., New York, pp. 205–27.

Freter, R. (1988) Mechanism of bacterial colonization of the mucosal surfaces of the gut, un *Virulence Mechanisms of Bacterial Pathogens*, (ed. J. A. Roth,) American Society for Microbiology, Washington, DC, pp. 45–60.

Freter, R. and Abrams, G. D. (1972) Function of various intestinal bacteria in converting germfree mice to the normal state. *Infect. Immun.*, **6**, 119–26.

Freter, R. and Jones, G. W. (1976) Adhesive properties of *Vibrio cholerae*: Nature of the interaction with intact mucosal surfaces. *Infect. Immun.*, **14**, 241–9.

Freter, R. and Jones, G. W. (1983) Models for studying the role of bacterial attachment in virulence and pathogenesis. *Rev. Infect. Dis.*, **5** (Suppl. 4), S647–58.

Freter, R., Abrams, G. D. and Aranki, A. (1973) Patterns of interaction in gnotobiotic mice among bacteria of a synthetic 'normal' intestinal flora, in *Germfree Research. Biological Effects of Gnotobiotic Environments* (ed. J. B. Henegan.), Academic Press, London pp. 429–33.

Freter, R., O'Brien, P. C. M. and Macsai, M. S. (1981) Role of chemotaxis in the association of motile bacteria with intestinal mucosa: *In vivo* studies. *Infect. Immun.*, **34**, 234–40.

Freter, R., Stauffer, E., Cleven, D. *et al.* (1983a) Continuous flow cultures as *in vitro* models of the ecology of large intestinal flora. *Infect. Immun.*, **39**, 666–75.

Freter, R., Brickner, H. and Temme, S. (1986) An understanding of colonization resistance of the mammalian large intestine requires mathematical analysis. *Microecol. Ther.*, **16**, 147–55.

Freter, R., Brickner, H., Botney, M. *et al.* (1983b) Mechanisms that control bacterial populations in continuous-flow culture models of mouse large intestinal flora. *Infect. Immun.*, **39**, 676–85.

Freter, R., Brickner, H., Fekete, J. *et al.* (1983c) Survival and implantation of *Escherichia coli* in the intestinal tract. *Infect. Immun.*, **39**, 686–703.

Fuller, R. (1973) Ecological studies on the lactobacillus flora associated with the crop epithelium of the fowl. *J. Appl. Bacteriol.*, **36**, 131–9.

Fuller, R. (1989) Probiotics in man and animals. *J. Appl. Bacteriol.*, **66**, 365–78.

Fukata, T., Baba, E. and Arakawa, A. (1989) Population of *Salmonella typhimurium* in the cecum of gnotobiotic chickens. *Poultry Sci.*, **68**, 311–14.

Garland, C. D., Nash, G. V. and McMeekin, T. A. (1982) The preservation of mucus and surface-associated microorganisms using acrolein vapour fixation. *J. Microsc.*, **128**, 307–12.

Gibbons, R. A. and Sellwood, R. (1973) The macromolecular biochemistry of cervical secretions, in *The Biology of the Cervix* (eds R. J. Blanden and K. Moghissi) Univ. of Chicago Press, Chicago, pp. 251–65.

Gibbons, R. J. and van Houte, J. (1975) Bacterial adherence in oral microbial ecology. *Ann. Rev. Microbiol.*, **29**, 19–44.

Giuliano, M., Barza, M., Jacobus, N. V. and Gorbach, S. L. (1987) Effect of broad-spectrum parenteral antibiotics on components of intestinal microflora of humans. *Antimicrob. Agents Chemother.*, **31**, 202–6.

Gorbach, S. L. (1990) Lactic acid bacteria and human health. *Ann. Med.*, **22**, 37–41.

Gorbach, S. L., Barza, M., Giuliano, M. and Jacobus, N. V. (1988) Colonization resistance of the human intestinal microflora: testing the hypothesis in normal volunteers. *Eur. J. Clin. Microbiol. Infect. Dis.*, **7**, 98–102.

Gorbach, S. L. and Goldin, B. R. (1990) The intestinal microflora and the colon cancer connection. *Rev. Infect. Dis.*, **12** (Suppl. 2), S252–61.

Guiot, H. F. L. (1982) Role of competition for substrate in bacterial antagonism in the gut. *Infect. Immun.*, **38**, 887–92.

Hentges, D. J. (1983) Role of intestinal flora in host defense against infection, in *Human Intestinal Flora in Health and Disease* (ed. D. J. Hentges), Academic Press, New York, pp. 311–31.

Hentges, D. W. and Freter, R. (1962) *In vivo* and *in vitro* antagonism of intestinal bacteria against *Shigella flexneri*. I. Correlation between various tests. *J. Infect. Dis.*, **110**, 30–7.

Hentges, D. J., Maier, B. R., Burton, G. C. *et al.* (1977) Effects of a high beef diet on the fecal bacterial flora of humans. *Cancer Res.*, **37**, 568–71.

Herget, H. F. and Weinrauch, E. (1979) Criteria and application of microbiological therapy of the gastrointestinal tract in the department of anaesthesia and intensive medicine of Justus-Liebig-University. *Microecol. Ther.*, **9**, 63–73.

Hoskins, L.C. and Boulding, E. T. (1976) Degradation of blood group antigens in human colon ecosystems. II. A gene interaction in man that affects the fecal population density of certain enteric bacteria. *J. Clin. Invest.*, **57**, 74–82.

Itoh, K. and Freter, R, (1989) Control of *Escherichia coli* populations by a combination of indigenous clostridia and lactobacilli in gnotobiotic mice and continuous flow cultures. *Infect. Immun.*, **57**, 559–65.

Koopman, J. P., Welling, G. W., Huysregts, A. W. M. *et al.* (1981) Association of germ-free mice with intestinal microfloras. *Z. Versuchstierk.*, **23**, 145–154.

Lee, A. (1980) Normal flora of animal intestinal surfaces, in *Adsorption of Microorganisms to Surfaces* (eds G. Bitton and K. C. Marshall), Wiley, New York, pp. 145–73.

Lee, A. (1985) Neglected niches, the microbial ecology of the gastrointestinal tract. *Adv. Microb. Ecol.*, **8**, 115–62.

Lewenstein, A., Frigerio, G. and Moroni, M. (1979) Biological properties of SF 68, a new approach for the treatment of diarrheal diseases. *Curr. Ther. Res.*, **26**, 967–81.

Lloyd, A. B., Cumming, R. B. and Kent, R. D. (1977) Prevention of *Salmonella typhimurium* infection in poultry by pretreatment of chickens and poults with intestinal extracts. *Austral. Vet. J.*, **53**, 82–7.

Maier, B. R., Onderdonk, A. B., Baskett, A. C. and Hentges, D. J. (1972) Shigella, indigenous flora interactions in mice. *Amer. J. Clin. Nutr.*, **25**, 1433–40.

Mantle, M., Basaraba, L., Peacock, S. C. and Gall, D. G. (1989) Binding of *Yersinia enterocolitica* to rabbit intestinal brush border membranes, mucus, and mucin. *Infect. Immun.*, **57**, 3292–9.

McCormick, B. A., Stocker, B. A., Laux, D. C. and Cohen, P. S. (1988) Roles of motility, chemotaxis, and penetration through and growth in intestinal mucus in the ability of an avirulent strain of *Salmonella typhimurium* to colonize the large intestine of streptomycin-treated mice. *Infect. Immun.*, **56**, 2209–17.

McCormick, B. A., Laux, D. C. and Cohen, P. S. (1990) Neither motility nor chemotaxis plays a role in the ability of *Escherichia coli* to colonize the streptomycin-treated mouse large intestine. *Infect. Immun.*, **58**, 2957–61.

McSweegan, E., Burr, D. H. and Walker, R. A. (1987) Intestinal mucus gel and secretory antibodies are barriers to *Campylobacter jejuni* adherence to INT 407 cells. *Infect. Immun.*, **55**, 1431–5.

Metchnikoff, E. (1907) *The Prolongation of Life. Optimistic Studies*, Heinemann, London.

Meynell, G. G. and Subbaiah, T. V. (1963) Antibacterial mechanisms of the mouse gut. *Brit. J. Exp. Pathol.*, **44**, 197–219.

Miller, R. S. and Hoskins, L. C. (1981) Mucin degradation in human colon ecosystems. Fecal population densities of mucin-degrading bacteria estimated by a 'most probable number' method. *Gastroenterology*, **81**, 759–65.

Moore, W. E. C. and Holdeman, L. V. (1975) Discussion of current bacteriological investigations of the relationship between intestinal flora, diet, and colon cancer. *Cancer Res.*, **35**, 3418–20.

Nissle, A. (1916) Ueber die Grundlagen einer neuen ursaechlichen Bekaempfung der pathologischen Darmflora. *Dtsch. Med. Wochenschr.*, **42**, 1181–4.
Oksanen, P. J., Salminen, S., Saxelin, M. *et al.* (1990) Prevention of travellers' diarrhea by *Lactobacillus* GG. *Ann. Med.*, **22**, 53–6.
Ozawa, A. and Freter, R. (1964) Ecological mechanism controlling growth of *Escherichia coli* in continuous flow cultures and in the mouse intestine. *J. Infect. Dis.*, **114**, 235–42.
Pedersen, K. and Tannock, G. W. (1989) Colonization of the gastrointestinal tract by lactobacilli. *Appl. Environ. Microbiol.*, **55**, 279–83.
Pollman, D. S., Danielson, D. M. and Peo, E. R. (1980), Effect of *Lactobacillus acidophilus* on starter pigs fed a diet supplemented with lactose. *J. Animal Sci.*, **51**, 638–44.
Que, J. U., Casey, S. W. and Hentges, D. J. (1986) Factors responsible for increased susceptibility of mice to intestinal colonization after treatment with streptomycin. *Infect. Immun.*, **53**, 116–23.
Rolfe, R. D. (1984) Role of volatile fatty acids in colonization resistance to *Clostridium difficile*. *Infect. Immun.*, **45**, 185–91.
Rowland, I. R. and Wise, A (1985) The effect of diet on the mammalian gut flora and its metabolic activities. *CRC Crit. Rev. Toxicol.*, **16**, 31–103.
Rozee, K. R., Cooper, D., Lam, K. and Costerton, J. W. (1982) Microbial flora of the mouse ileum mucus layer and epithelial surface. *Appl. Environ. Microbiol.*, **43**, 1451–63.
Rusch, V. (1980) Medicine and the microbial world. *Microecol. Ther.*, **10**, 163–72.
Rusch, V., Hyde, R. M. and Luckey, T. D. (1980) Documentation on the identity, manufacture, safety, efficacy, and stability of SYMBIOFLOR products. *Microecol. Ther.*, **10**, 173–203.
Savage, D. C. (1972) Associations and physiological interactions of indigenous microorganisms and gastrointestinal epithelia. *Amer. J. Clin. Nutr.*, **25**, 1372–9.
Savage, D. C. (1977) Microbial ecology of the gastrointestinal tract. *Ann. Rev. Microbiol.*, **31**, 107–33.
Savage, D. C. (1983) Associations of indigenous microorganisms with gastrointestinal epithelial surfaces, in *Human Intestinal Microflora in Health and Disease* (ed. D. J. Hentges), Academic Press, New York, pp. 55–78.
Schwan, A., Sjolin, S., Trottestam, U. and Aronsson, B. (1984) Relapsing *Clostridium difficile* enterocolitis cured by rectal infusion of normal faeces. *Scand. J. Infect. Dis.*, **16**, 211–15.
Shahani, K. M. and Ayebo, A. D. (1980) Role of dietary lactobacilli in gastrointestinal microecology. *Amer. J. Clin. Nutr.*, **33**, 2448–57.
Shedlofsky, S. and Freter, R. (1974) Synergism between ecologic and immunologic control mechanisms of intestinal flora. *J. Infect. Dis.*, **129**, 296–303.
Siitonen, S., Vapaatalo, H., Salminen, S. *et al.* (1990) Effect of *Lactobacillus* GG yoghurt in prevention of antibiotic-associated diarrhoea. *Ann. Med.*, **22**, 57–9.
Simon, G. L. and Gorbach, S. L. (1984) Intestinal flora in health and disease. *Gastroenterology*, **86**, 174–93.
Stanton, T. B. and Savage, D. C. (1984) Motility as a factor in bowel colonization

by *Roseburia cecicola*, an obligately anaerobic bacterium from the mouse cecum. *J. Gen. Microbiol.*, **130**, 173–83.

Syed, S. A., Abrams, G. D. and Freter, R. (1970) Efficiency of various bacteria in assuming normal functions of enteric flora after association with germ-free mice. *Infect. Immun.*, **2**, 376–86.

Tankanow, R. M., Ross, M. B., Ertel, I. J. *et al.* (1990) A double blind, placebo controlled study of the efficacy of Lactinex in the prophylaxis of amoxicillin-induced diarrhea. *DICP, Ann. Pharmacother.*, **24**, 382–4.

Tannock, G. W., Dashkevicz, M. P. and Feighner, S. D. (1989) Lactobacilli and bile salt hydrolase in the murine intestinal tract. *Appl. Environ. Microbiol.*, **55**, 1848–51.

Vanbelle, M., Teller, E. and Focant, M. (1990) Probiotics in animal nutrition: a review. *Arch. Tiernahr.*, **40**, 543–67.

van de Merwe, J. P., Stegeman, J. H. and Hazenberg, M. P. (1983) The resident faecal flora is determined by genetic characteristics of the host. Implications for Crohn's disease? *Antonie van Leeuwenhoek*, **49**, 119–24.

van der Waaij, D., Berghuis de Vries, J. M. and Lekkerkerk, J. E. C. (1971) Colonization resistance of the digestive tract in conventional and antibiotic-treated mice. *J. Hyg. (Camb.)*, **69**, 405–41.

Veilleux, B. G. and Rowland, I. (1981) Simulation of the rat intestinal ecosystem using a two-stage continuous culture system. *J. Gen. Microbiol.*, **123**, 103–15.

Wadolkowsky, E. A., Laux, D. C. and Cohen, P. S. (1988) Colonization of the streptomycin-treated mouse large intestine by a human fecal *Escherichia coli* strain: role of adhesion to mucosal receptors. *Infect. Immun.*, **56**, 1036–43.

Walker, W., Wu, M. and Bloch, K. J. (1977) Stimulation by immune complexes of mucus release from goblet cells of the rat small intestine. *Science*, **197**, 370–2.

Williams, R. C. and Gibbons, R. J. (1975) Inhibition of streptococcal attachment to receptors on human buccal epithelial cells by antigenically similar salivary glycoproteins. *Infect. Immun.*, **11**, 711–18.

Wilson, K. H. and Freter, R. (1986) Interactions of *Clostridium difficile* and *E. coli* with microfloras in continuous flow cultures and gnotobiotic mice. *Infect. Immun.*, **54**, 354–8.

Wilson, K. H. and Perini, F. (1988) Role of competition for nutrients in suppression of *Clostridium difficile* by colonic microflora. *Infect. Immun.*, **56**, 2610–14.

Wold, A. E., Dahlgren, U. I. H., Hanson, L. A., *et al.* (1989) Difference between bacterial and food antigens in mucosal immunogenicity. *Infect. Immun.*, **57**, 2666–73.

Zubrzycki, L. and Spaulding, E. H. (1957) Studies on the stability of the normal human fecal flora. *Bacteriol. Proc. 1957*, p. 101.

Chapter Seven

Probiotics and the immune state

GABRIELA PERDIGÓN AND SUSANA ALVAREZ

7.1 INTRODUCTION

The role played by lactic acid bacteria in various biological functions of the host has been extensively reported. During the last two decades, numerous studies have demonstrated the anticarcinogenic properties of lactic acid bacteria and great emphasis has been laid on the antitumour activity exerted by yoghurt and by milks fermented with *Lactobacillus acidophilus*. Shahani *et al.* (1983) has demonstrated that, in mice fed with fermented colostrum, the growth of experimentally induced tumours was inhibited, but only in animals dosed before the onset of tumour growth. Reddy *et al.* (1983) and Ayebo *et al.* (1982) studied whether the antitumour effect was exerted by the presence of lactic acid bacteria, by components of their cell wall, or by products formed as a consequence of the fermentation process. Ayebo *et al.* (1982) were able to isolate a dialysable antitumour component from yoghurt. Goldin and Gorbach (1980) demonstrated that in mice dosed with *L. acidophilus* there was a decrease in the incidence of the colon cancer induced by 1,2-dimethylhydrazine dihydrochloride. Kato *et al.* (1981, 1985) demonstrated that the intraperitoneal administration of *L. casei* inhibited tumour growth in both syngeneic and allogeneic mice. The effect depended on the dose and on the time of administration of *L. casei*; the antitumour activity was effective only in pre-treated animals. However, Yasutake *et al.* (1984a) observed that the administration of a mixture of *L. casei* to tumour cells from a methylcolanthrene-induced tumour produced a total inhibition in its growth, while a simultaneous injection of *L. casei* and tumour cells at different body sites had no effect on tumour growth. The antitumour activity of *L. casei* administered intravenously was also observed in syngeneic mice and guinea-pigs with carcinoma of the lung and liver, respectively (Matsuzaki *et al.*, 1985).

Although the tumour-growth suppressing capacity of the above microorganisms might be related to the integrity of the cells, whether viable or non-viable, or to fractions of the cell wall (Bogdanov *et al.*, 1975; Sekine *et al.*, 1985), this capacity is also undoubtedly associated with the immune system of the host; that is to say, lactic acid bacteria or their components would act by activating or suppressing the cells involved in the immune response. It has been demonstrated that viable *L. plantarum* bacteria (Bloksma *et al.*, 1979) administered intraperitoneally stimulate only the delayed type hypersensitivity (DTH) reaction, while non-viable bacteria act as adjuvants in the production of antibodies.

The previous studies sought to establish which of the cells involved in the immune response were modified or activated by lactic acid bacteria. Kato *et al.* (1983) demonstrated that *L. casei* inoculated intraperitoneally activated the peritoneal macrophages, increasing both their phagocytic capacity and the activity of the enzymes involved in the phagocytic process. They also observed an enhanced activity of the mononuclear phagocytic system as reflected by an increase in the colloidal carbon clearance index rate. This would mean that *L. casei* possesses immunopotentiator properties similar to those of *Mycobacterium bovis*, Bacillus Calmette–Guerin (BCG), *Corynebacterium parvum* and *Streptococcus pyogenes*.

It has been demonstrated (Saito *et al.*, 1983) that the subcutaneous administration of *L. casei* induces an increase in the production of circulating antibodies for the antigens *Pseudomonas aeruginosa* and sheep red blood cells (SRBC), as well as an increase in the levels of production of IgM as reflected in an augmentation in the number of plaque-forming cells (PFC).

The inoculation of *L. casei* by the intravenous or intraperitoneal (IP) route induced the activation of natural killer cells (Kato *et al.*, 1984), which plays an important role in tumoural processes. When *L. casei* was compared other bacteria used as immunumudulators such as *Cor. parvum* and BCG, it proved to be just as effective as the other microorganisms, with the added advantage that it did not produce hepatomegaly or splenomegaly, a very common effect of immunomodulators, but only caused a local transitory cellular infiltration (Yasutake *et al.*, 1984b).

In vitro assays demonstrated that Kupffer cells, spleen, lung and peritoneal macrophages, when they are activated by *L. casei*, produced a cytotoxic factor (Hashimoto *et al.*, 1985). De Simone *et al.* (1986) observed that feeding with yoghurt induced the production of interferon by circulating blood lymphocytes stimulated with Concanavalin A.

On the other hand, lactic acid bacteria may not always produce beneficial effects on the host. *L. casei* subsp. *rhamnosus* can produce endocarditis or abscesses (Sharpe *et al.*, 1973). Some strains of *L. acidophilus* and *L. plantarum*, under special conditions, may possess undesirable properties. Iwasaki *et al.* (1983) observed that the administration of *L. arabinosus* to mice with intestinal tumours induced with methylazoximethane strengthened the tumoricidal effect.

Although previous works have suggested the use of certain lactobacilli as immunopotentiators, for their use as such to be possible and also for their utilization for therapeutic purposes, the elucidation of many unknown factors is required. For example, it is essential to know (a) the most active strains, and (b) the dose required for maximum effect and when it should be administered. Further work is required that would take account not only of their possible use for industrial purposes, but also their effect on the host, since their ingestion occurs through foods and dairy products. But even in the present state of our knowledge, the possibility of using lactic acid bacteria as immunomodulators seems very attractive. Although other substances (oil substances, synthetic peptides) and microorganisms have been shown to be effective, they have undesirable side-effects and are, therefore, less attractive than the lactic acid bacteria for use as supplements for human beings and animals.

On the whole, it has been admitted that the substances that enhance or stimulate the immune response (adjuvants or immunomodulators) can enhance the non-specific defence mechanisms of the host, activate the cells involved in the specific response or produce a systemic increase, that is, activate the whole immune system by stimulating the two types of response. Before a new substance can be used as an immunomodulator, it will have to be checked not only as regards its efficiency in the enhancement of the immune response, but also as regards the absence of harmful effects for the host, that is, the absence of side-effects that might occur as a result of long-term administration.

Since lactic acid bacteria are usually ingested as part of the normal daily diet, it is important to find out the action of these bacteria on the secretory and systemic immune systems, that is, the systemic and local response in the intestinal mucosae.

It is well known that the oral administration of antigens may induce or inhibit an immune response. Whether one or the other effect occurs depends on various factors such as the nature of the antigen (soluble or particulate) and number of administrations.

Soluble antigens in general have been shown to produce oral tolerance (Miller and Hanson, 1979). As regards particulate antigens, although they may induce an immune response, under certain conditions they may inhibit it, as in the case of SRBC, which may provoke a tolerance effect (Enders *et al.*, 1986).

7.2 EFFECT OF ORALLY ADMINISTERED LACTIC ACID BACTERIA ON IMMUNITY: NON-SPECIFIC AND SPECIFIC IMMUNE RESPONSE

Four species of lactic acid bacteria have been used: *L. casei* CRL 431, *L. acidophilus* ATCC 4356, *L. delbrueckii* subsp. *bulgaricus* CRL 423 and *S. salivarius* subsp. *thermophilus* CRL 412. The experimental model consisted of Swiss albino mice from a closed colony.

7.2.1 Effect of the Cells Involved in the Non-Specific Defence Mechanisms of the Host

The non-specific defence mechanisms of the host include phagocytosis, which is effected by macrophages, polymorphonuclear leucocytes, histiocytes and monocytic cells that are part of the mononuclear phagocytic system, which removes foreign antigens. In this system, the most important cells are macrophages. The state of activation of these cells is a measure of the non-specific immune response of the host, and can be measured by assessing the activity of enzymes released as a consequence of cellular activation, or by a functional test.

The above microorganisms (viable and non-viable) were compared for ability to activate macrophages by the oral and the intraperitoneal routes. The cell concentration used was 1.2×10^9 cells day^{-1} $mouse^{-1}$, and the administration time for both routes was 2, 5 and 7 consecutive days. In the case of oral administration, the viable and non-viable lactic cultures were suspended in sterile non-fat milk at 10% concentration and included in the drinking water at 20% concentration (v/v) and fed *ad libitum*.

The state of macrophage activation was determined *in vitro*, using peritoneal macrophages (attached cells and total population) by means of biochemical tests to measure the release of the lysosomal enzymes, β-glucuronidase and β-galactosidase (Stossel, 1980; Conchie *et al.*, 1959). A non-lysosomal enzyme, lactate dehydrogenase (LDH), was determined by the method of Gasser and Gasser (1971) by measuring the NADH oxidation percentage. The *in vitro* functional test of peritoneal macrophage activation was assessed by measuring the phagocytic capacity for the *Salmonella typhimurium* antigen either non-opsonized or opsonized with anti-*Salmonella*-specific antibody; the percentage of macrophages with ingested bacteria was determined.

The *in vivo* assay was measured by the colloidal carbon clearance test; the phagocytic index K (Tolone *et al.*, 1970), and the mean time $t_{1/2}$ (Kato *et al.*, 1984) were determined.

Lactic acid bacteria administered orally or intraperitoneally activated the peritoneal macrophages; the optimum dose for the oral route was 6×10^9 cells, and 2.4×10^9 cells for the intraperitoneal route. No differences in the activation of macrophages were observed when using either viable or non-viable cells in the animals treated with *L. delbrueckii* subsp. *bulgaricus* and *S. salivarius* subsp. *thermophilus*. Although in the case of mice treated with *L. casei* and *L. acidophilus* the response was slightly higher when viable bacteria were used, no significant differences occurred (Perdigón *et al.*, 1986a, d; 1987).

Table 7.1 Comparative effect of viable culture of *Lactobacillus delbrueckii* subsp. *bulgaricus* and *Streprococcus salivarius* subsp. *thermophilus* on the release of lysosomal enzymes from peritoneal macrophages of mice.

Enzymes	Days of Treatment	Mean enzymatic activity ± SD *L. delbrueckii* subsp. *bulgaricus* Oral route	IP route	*S. salivarius* subsp *thermophilus* Oral route	IP route
	2	10.13 ± 3.7	67.18 ± 3.8	13.4 ± 3.7	20.3 ± 3.5
β-Glucuronidase	5	46.61 ± 7.5	43.18 ± 5.5	10.33 ± 3.2	29.9 ± 4.2
	7	52.43 ± 4.8	60.67 ± 2.6	10.00 ± 3.2	28.6 ± 3.5
	2	22.03 ± 6.1	22.74 ± 4.2	25.7 ± 5.7	25.5 ± 4.8
β-Galactosidase	5	64.28 ± 4.0	56.4 ± 4.2	21.3 ± 2.0	36.8 ± 3.7
	7	55.91 ± 3.9	25.84 ± 5.2	18.8 ± 2.1	19.5 ± 2.0

(Normal Values: β-Glucuronidase= 10.09 ± 3.4 nmol of PNP h^{-1} per 10^6 cells; β-Galactosidase= 18.8 ± 2.0 nmol of ONP h^{-1} per 10^6 cells.
Effect of *L. delbrueckii* subsp. *bulgaricus* and *S. salivarius* subsp. *thermophilus* on the release of β-glucuronidase and β-galactosidase enzymes from peritoneal macrophages of mice orally or intraperitoneally treated with 1.2×10^9 cells day^{-1} $mouse^{-1}$ for 2, 5 and 7 consecutive days. Values represent means of the 5 mice ± standard deviation.

When analysing the production of lysosomal enzymes (β-glucuronidase and β-galactosidase) (Tables 7.1 and 7.2), we observed that intraperitoneal stimulation was very effective in releasing enzymes, probably due to the fact that the *in situ* inoculation of lactic acid bacteria might be more effective because of the direct contact between antigen and macrophage. This effect does not occur with oral administration, in which macrophages would be stimulated by lymphokines. In all cases the LDH values obtained, when lactic acid bacteria were administered by the oral route, were very low whereas by the intraperitoneal route significant LDH values were obtained.

This enzyme is released as a consequence of cellular damage. We believe that the higher production of LDH when using intraperitoneal stimulation is due to the aggressiveness of the antigen, which would damage peritoneal cells, an effect that would not occur when administered by the oral route.

Table 7.2 Comparative effect of viable culture of *Lactobacillus acidophilus* and *L. casei* on the release of lysosomal enzymes from peritoneal macrophages of mice.

Mean enzymatic activity ± SD					
Enzymes	Days of Treatment	*L. acidophilus*		*L. casei*	
		Oral route	IP route	Oral route	IP route
β-Glucuronidase	2	9.8 ± 2.46	23.6 ± 2.5	49.55 ± 6.8	43.31 ± 1.0
	5	14.5 ± 2.16	25.8 ± 2.7	65.51 ± 7.8	41.84 ± 3.2
	7	10.5 ± 3.20	28.5 ± 3.2	65.83 ±6.8	33.37 ± 1
β-Galactosidase	2	51.3 ± 7.95	55.7 ± 2.9	37.91±2.67	35.75 ±2.30
	5	29.4 ± 5.63	32.9 ± 4.3	100.71 ± 3.2	29.27±4.05
	7	18.8 ± 2.01	35.9 ± 3.2	121.83 ± 4.8	26.90±1.2

Normal values: β-Glucuronidase = 10.09 ± 3.4 nmol of PNP h^{-1} per 10^6 cells; β-Galactosidase = 18.8 ± 2.0 nmol of ONP h^{-1} per 10^6 cells; Effect of *L. acidophilus* and *L. casei* on the release of β-glucuronidase and]-galactosidase enzymes from peritoneal macrophages of mice treated orally or intraperitoneally with 1.2 × 10^9 cells day^{-1} $mouse^{-1}$ for 2, 5 and 7 consecutive days. Values represent means of the 5 mice ± standard deviation.

Table 7.3 Percentage phagocytosis of peritoneal macrophages in mice.

		Percentage phagocytosis			
Route of administration	Days of treatment	*L. delbrueckii* subsp. *bulgaricus*	*S. salivarius* subsp. *thermophilus*	*L. acidophilus*	*L.casei*
Oral	2	45.0 ± 5	49 ± 2.5	36 ± 5.2	61 ± 6
	5	51.5 ± 7.2	49 ± 2.2	33 ±2.5	47 ± 5.5
	7	44.0 ± 6	45 ± 3	35 ± 3.2	43.5 ± 2.5
IP	2	80 ± 2	86 ± 5.1	63 ± 6.1	62.5 ± 5
	5	76 ± 1	64 ± 3.5	75 ± 5.2	62.0 ± 5
	7	76 ± 2.5	63 ± 2	70 ± 5	64.0 ± 2

Normal value: 33% Peritoneal macrophages isolated from the mice treated orally or intraperitoneally with different lactic acid bacteria, were incubated with opsonised or non-opzonized *Salmonella typhimurium* at 37°C for 15 min. The macrophages phagocytosing bacteria were counted microscopically after incubation. Values represent mean of the 5 mice ± standard deviation for each groupp of mice. Differences between phagocytosis with opsonized and non-opsonized systems were not observed.

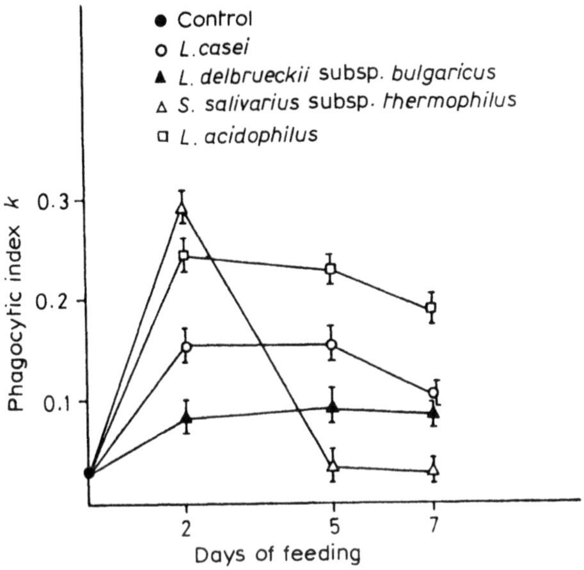

Figure 7.1 Kinetics of phagocytosis of colloidal carbon in mice fed with different lactic acid bacteria during 2, 5 and 7 consecutive days at a dose of 1.2×10^9 cells $day^{-1}mouse^{-1}$. K (phagocytic index) was calculated by the equation: $K = (\log C_2 - \log C_1) / (T_2 - T_1)$ (Tolone *et al.* 1970), where C_1 and C_2 represent the carbon concentration in the blood at times T_1 and T_2, respectively. Points and bars represent mean of the 5 mice ± standard deviation. Normal value = 0.025.

The *in vitro* phagocytosis assays (Table 7.3) agreed in large measure with the data obtained from the determination of macrophage activation, using biochemical criteria. The percentage of phagocytosis was higher in all cases when the intraperitoneal route was used. *L. acidophilus* was the least effective by the oral route, and *L. casei* induced a moderate activation of the macrophages by both administration routes. No significant differences were observed between viable and non-viable cells, or between opsonized or non-opsonized systems.

The results of the activation of peritoneal macrophages using lactic acid bacteria orally administered agree with the works of other investigators (Ianello *et al.*, 1984; Gemsa *et al.*, 1984), who reported that bacterial antigens administered by the oral route can activate peritoneal macrophages by lymphokines produced by T-cells. Namba *et al.* (1981) also demonstrated an increase in cellular and humoral immune response, induced by antigens of the cell wall of bacteria orally administered.

When the activation of the mononuclear phagocytic system was determined by the colloidal carbon clearance test (Figures 7.1–7.3), the values found for the clearance, phagocytic indices K and $t_{1/2}$

proved to be very different from those of the control when *L. casei*, *L. acidophilus* and *L. delbrueckii* subsp. *bulgaricus* were administered by either route. The values of $t_{1/2}$ for these bacteria were between 1 and 3 min, as against 10 min for the control. *L. casei* and *L. acidophilus* proved to be more effective by the oral route, probably because of their capacity for survival and colonization in the intestinal tract. By either route *S. salivarius* subsp. *thermophilus* was poorly effective in the activation of the mononuclear phagocytic system. This was probably due to structural differences in the cell walls, which may possess a lower activation capacity.

Although we are as yet ignorant of the mechanisms by which lactic acid bacteria stimulate macrophages, from the analysis of the above results we may conclude that: (a) *L. casei*, *L. acidophilus*, *L. delbrueckii* subsp. *bulgaricus* and certain doses of *S. salivarius* subsp. *thermophilus* are capable of activating the cells involved in the non-specific immune response; (b) the lactic acid bacteria capable of survival and growth in the intestinal tract, such as *L. casei* and *L. acidophilus*, are more efficient in the activation; and (c) the oral administration of lactic acid bacteria is as effective, or even more so, than intraperitoneal administration.

The above results are of great interest because of the importance of lactic acid bacteria in human feeding, as well as the significance of antigenic stimulation with bacteria in general, in the maturation of the immune system. This latter fact has been demonstrated when comparing the state of the immune system of germ-free and conventional animals (Bauer *et al.*, 1966; Moreau *et al.*, 1986).

7.2.2 Effect on the Cells Involved in the Specific Immune Response

Antigenic stimulation in an adult vertebrate animal leads to the proliferation of clones of T-lymphocytes (TL) and B-lymphocytes (BL); the latter become differentiated into plasma cells which secrete immunoglobulin molecules. The role played by the lymphokines released by the TL is of fundamental importance in the proliferation and differentiation of the BL, with the consequent enhancement of the immune response. The cells of the immune system communicate by means of chemical substances such as interleukins. The substances used as adjuvants or immunomdulators may induce the synthesis of interleukines.

In order to determine the state of activation of TL and BL, the increase in the cellular immune response due to a stronger activation of the TL, and the increase in the humoral response due to the increase in the activity of BL, should be measured.

The influence of the oral administration of lactic acid bacteria, on the humoral immune response (B function) by means of the PFC

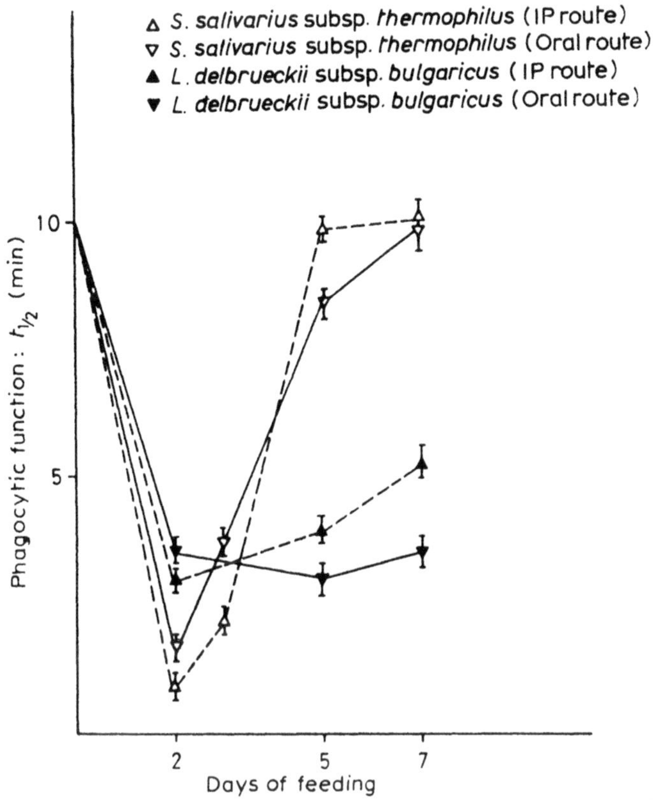

Figure 7.2 Effect of *Streptococcus salivarius* subsp. *thermophilus* and *Lactobacillus delbrueckii* subsp. *bulgaricus* on the phagocytic function of the reticuloendothelial system of mice. Lactobacilli were administered orally (—) or intraperitoneally (---) during 2, 5 and 7 consecutive days at a dose of 1.2×10^9 cells $day^{-1}mouse^{-1}$. Points and bars represent mean of the 5 mice ± standard deviation. The clearance rate of carbon ($t_{1/2}$) was calculated by the formula of Kato *et al.* (1984). Control value $T_{1/2}$ = 10 min.

assay by the direct method for the SRBC antigen, was studied. This assay allows the detection of IgM-producing cells. The increase in the levels of IgG-type serum antibodies for the SRBC antigen, was also investigated. In the case of the PFC assay, animals were fed with the lactic acid bacteria under study for 2, 5, 7 and 10 consecutive days. In order to study the levels of anti-SRBC circulating antibodies, mice were fed for only 7 consecutive days. The serum levels of antilactic acid bacteria antibodies were also analysed. T function was determined by the delayed type hypersensitivity assay, for the SRBC antigen and for each of the lactic acid bacteria studied, in mice fed for 5 and 7 consecutive days.

Taking into account the fact that *L. acidophilus* and *L. casei* are capable of surviving in the intestinal tract, the effect on the non-specific and on the specific immune response of the mixture of

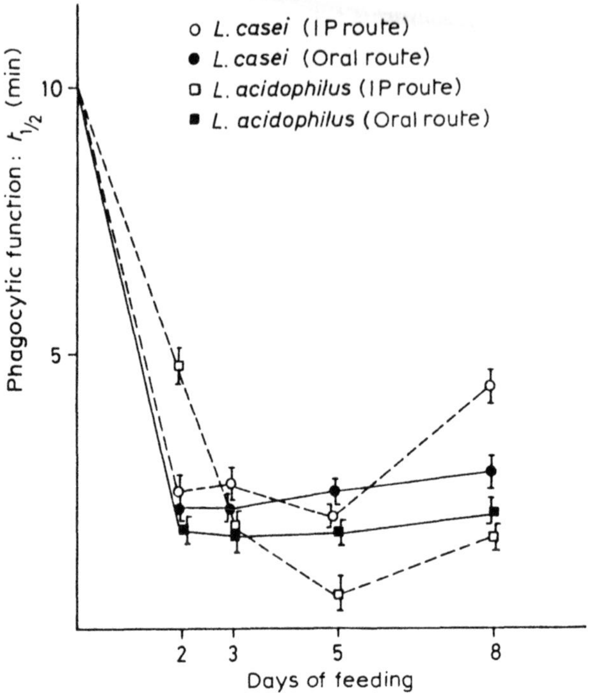

Figure 7.3 Effect of *Lactobacillus casei* and *L. acidophilus* on the phagocytic function of the reticuloendothelial system in mice. Lactobacilli were administered orally (—) or intraperitoneally (---) for 2, 5 and 7 consecutive days at a dose of 1.2 $\times 10^{10}$ cels day^{-1}mouse-1. Points and bars represent mean of the 5 mice ± standard deviation. The clearance rate of carbon ($t_{1/2}$) was calculated by the formula of Kato *et al.* (1984). Control value $T_{1/2}$ = 10 min.

these microorganisms in fermented and non-fermented milk, was also studied. Results showed (Perdigón *et al.*, 1986b, c, 1988a, b) that the oral administration of *L. casei*, *L. acidophilus* and *L. delbrueckii* subsp. *bulgaricus* enhanced both the cellular and the humoral immune response.

It was found that IgM production for the SRBC antigen, determined by the number of PFC, was increased in relation to normal values (Figure 7.4). The species capable of survival in the intestinal tract, such as *L. casei* and *L. acidophilus*, proved to require only 5 days of stimulation in order to reach a maximum number of PFC; this immunopotentiator effect remained higher throughout the assay. *L. delbrueckii* subsp. *bulgaricus* and *S. salivarius* subsp. *thermophilus* required a longer administration period (7 days) to become effective, probably because of their inability to survive in the gut. Among the microorganisms studied, *S. salivarius* subsp. *thermophilus* showed the lowest capacity for lymphocyte activation, since the PFC values decreased on the tenth day of feeding, coming close to normal values. This result might indicate that a higher administration of S. *salivarius* subsp. *thermophilus* would not increase its capacity to activate lymphocytes; on the contrary, such an administration would exert a negative effect, probably because this microorganism may induce early oral tolerance against its own epitopes, being unable to activate immunocompetent cells, an effect that was not observed in any of the other lactic acid bacteria under study.

When analysing the influences of a 7 consecutive days feeding trial with lactic acid bacteria on the production of circulating antibodies for the SRBC antigen, it was observed that the lactobacilli assayed (Figure 7.5) produce from a two-fold to a six-fold increase in the levels of serum antibodies compared with the controls not previously fed. However, S. *salivarius* subsp. *thermophilus* did not significantly increase the production of antibodies and this result would agree with that obtained in the PFC assay. The different behaviour of the genera *Lactobacillus* and *Streptococcus* on the production of antibodies was probably due to antigenic differences or to oral tolerance induction by S. *salivarius* subsp. *thermophilus*, as indicated above.

The lactic acid bacteria under study, when orally administered, did not induce serum antibodies against their own epitopes. This is an important finding because it means that the consumption of foods that include lactic acid bacteria could have a beneficial effect by activating the immune system without producing antilactic acid bacteria serum antibodies. This effect might be caused not by the lack

Table 7.4 *In vivo* phagocytosis assays.: determination of the clearance rate of carbon ($t_{1/2}$) in mice.

	$t_{1/2}$(min) : Mean ± SD	
Days of feeding	Mixture of *L.casei* and *L. acidophilus*	Milk fermented with *L. casei* and *L. acidophilus*
2	3.72 ± 0.50	1.95 ± 0.50
3	2.91 ± 0.25	1.00 ± 0.30
5	3.92 ± 0.52	0.55 ± 0.25
7	4.56 ± 0.60	1.10 ± 0.30

Control value: 10.0 ± 0.50 min.

The mixture was orally administered at a dose of 2.4 cells day^{-1}. Values represent means of the 6 mice ± standard deviation of each group of animals. The clearance rate of carbon ($t_{1/2}$) was calculated by the formula of Kato *et al.* (1984).

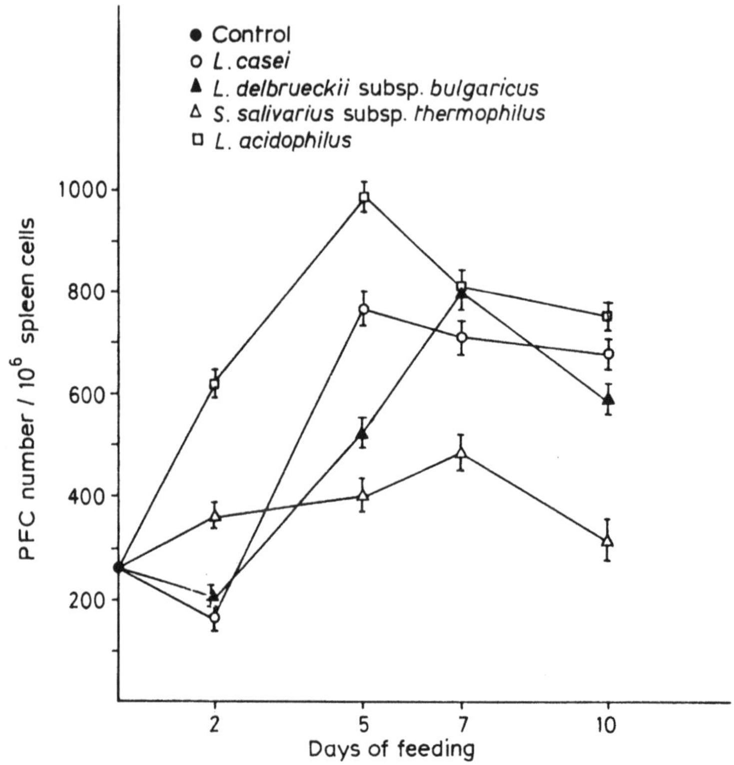

Figure 7.4 Plaque-forming cell (PFC) response against sheep red blood cells (SRBC) in mice fed with different lactic acid bacteria 2, 5 and 7 consecutive days at a dose of 1.2×10^9 cells $day^{-1}mouse^{-1}$. The SRBC were inoculated at the end of each feeding period, and the mice were sacrificed on the 5th day post-SRBC inoculation. Points and bars represent mean of the 6 mice ± standard deviation of each group of animals. Control value = 260 PFC/10^6 spleen cells.

of immunogenicity of such organisms but because of the presence of the muramyl dipeptide (MDP) structure in the cell wall. This is a common factor in Gram-positive bacteria and induces the activation of lymphoid cells without immunogenicity (Lowy *et al.*, 1980). Other studies indicate that the MDP acts directly on the stimulation of macrophages and lymphocytes (Tanaka *et al.*, 1980) and that it selectively suppresses the response to IgE (Kishimoto *et al.*, 1979).

Figure 7.6 shows the T-lymphocyte activation values obtained by DTH assay for the SRBC antigen and for lactic acid bacteria. There is a remarkable increase in the inflammation percentage observed for the SRBC antigen in mice fed for 5 and 7 days with the lactobacilli, but not with *S. salivarius* subsp. *thermophilus*. Similar results were obtained for the DTH assay against lactic acid bacteria, although inflammation percentages were lower than those obtained for SRBC. These results

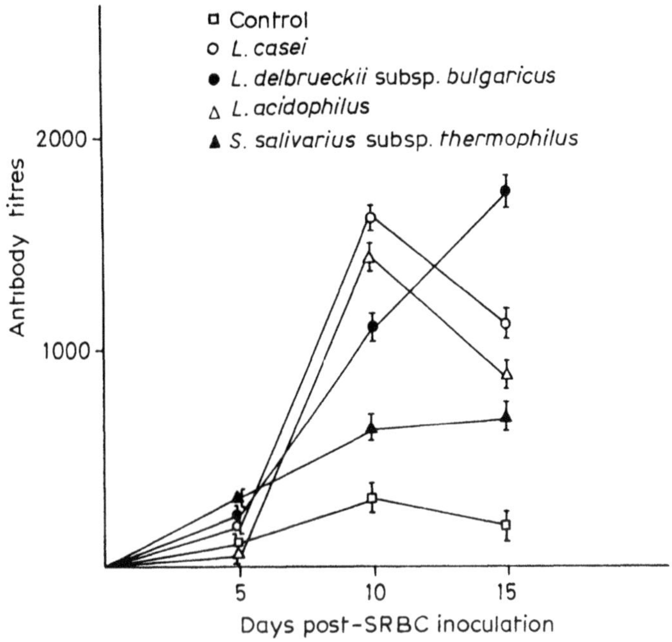

Figure 7.5 Anti-sheep red blood cell (SRBC) circulating antibodies from mice fed with different lactic acid bacteria at a dose of 1.2×10^9 cells $day^{-1}mouse^{-1}$ for 7 consecutive days. SRBC were inoculated on the 8th, 9th and 10th day and mice bled on the 5th, 10th and 15th day post-SRBC inoculation. Antibody titres were determined by the haemagglutination reaction. The control group was without *Lactobacillus* administration. Points and bars represent mean ± standard deviation for each group of 6 animals.

would indicate the effectiveness of lactobacilli feeding in the induction of the cell immune response, since the effect observed after 7 days of oral administration was the same as that obtained after only 5 days.

The increase in the systemic immune response was also determined in fermented and non-fermented mixtures of *L. casei* and *L. acidophilus*. A remarkable increase was observed in both responses in the case of the fermented milk (Table 7.4, Figure 7.7), probably because, during the fermentation process, the peptides produced are immunologically active and add to the effect of lactic acid bacteria, thus enhancing both types of response. It is known that casein and its peptides can locally activate the immune response at the level of the intestinal mucosae (Slobodianik *et al.*, 1984). This stimulation is probably enhanced by the presence of lactic acid bacteria, strengthening the systemic immune response.

Although further investigation is required for the elucidation of the mechanisms through which lactobacilli produce a systemic increase in the immune response, it is an undeniable fact that these bacteria exert an adjuvant effect on the immune system, enhancing the cellular and

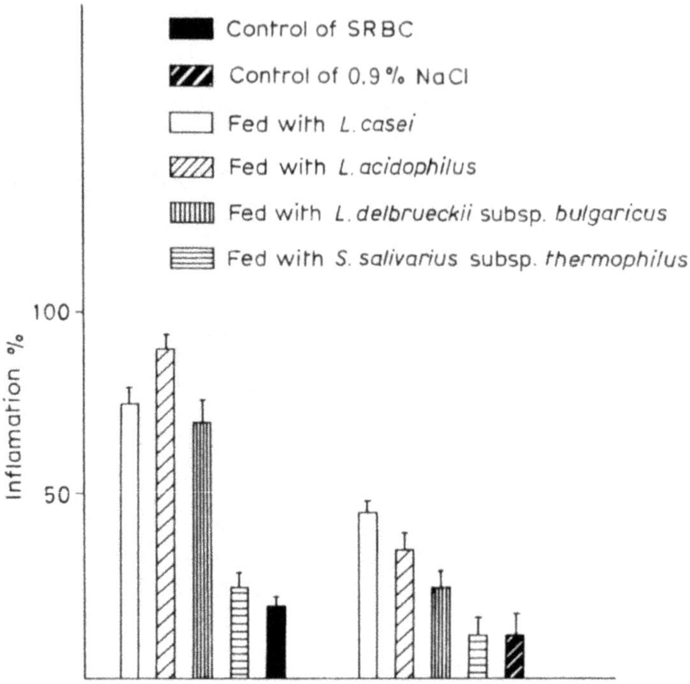

Figure 7.6 Delayed hypersensitivity response (DTH); (a) to sheep red blood cells (SRBC) and (b) to lactic acid bacteria. Mice were fed for 7 consecutive days with the microorganisms. At the end of the feeding period, animals were injected intraperitoneally with SRBC and on the 4th day post-inoculation DTH assay for SRBC was made. DTH assay against the different lactic acid bacteria was analysed at the end of feeding period. Results were expressed as $(A - S)/s \times 100$, where A = thickness of the footpad injected with SRBC or lactic acid bacteria and S = thickness of the footpad injected with 0.9%. NaCl.

humoral response. The results obtained open up a promising perspective in the use of adjuvants for the human being. Of all the previously proposed substances, MDP has the least harmful side-effects. However, if such substances are to be effective they must be administered with others that act as vehicles (Tsujimoto *et al.*, 1986).

7.2.3 Effect of Long-term Administration on the Induction of Side-effects

A successful immunomodulator must not only enhance the immune response, but must also be free from adverse side-effects. Usually the side-effects act on organs such as the liver and spleen. In the present cases lactic acid bacteria are administered orally, and it is important to find out whether or not they effect an increase in the inflammatory response in the gut. This effect will be analysed in the

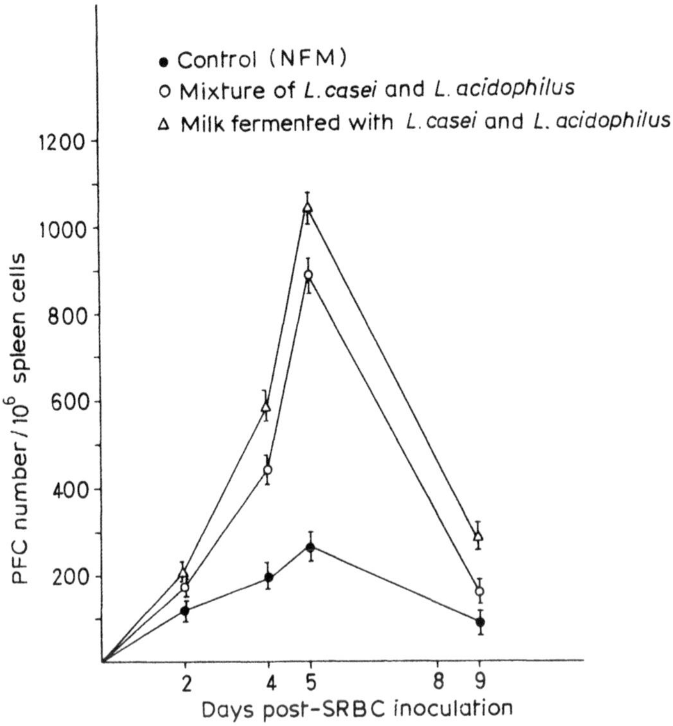

Figure 7.7 Production of IgM plaque-forming cells (PFC) in spleen against sheep red blood cells (SRBC) in mice fed with a mixture of *L. casei* and *L. acidophilus* and milk fermented with these microorganisms, at a dose of 2.4×10^9 cells day^{-1} $mouse^{-1}$ during 7 consecutive days. SRBC were inoculated on the 8th day and the mice killed on the 2nd, 4th, 5th and 9th day following SRBC inoculation. Points and bars represent mean of the 5 mice ± standard deviation. The control group was fed with non-fat milk (NFM).

section concerning the action of lactic acid bacteria on immunity in the intestinal mucosae.

To investigate whether feeding with lactic acid bacteria could produce side-effects such as hepatomegaly and splenomegaly, six groups of 15 mice, all having the same weight (30 g), were studied. Four groups were fed with the four species of lactic acid bacteria under study, another group was fed with milk at 10% concentration, and another control group with water. All groups were fed *ad libitum*.

After treatment for 7 consecutive days, the animals were sacrificed and body, spleen and liver weights were measured, results being expressed as a percentage of the body weight. Results showed (Figure 7.8) no significant differences in any of the animals fed with water, milk or lactic acid bacteria. These results would indicate that the oral administration of *L. casei*, *L. acidophilus*, *L. delbrueckii* subsp. *bulgaricus* and *S. salivarius* subsp. *thermophilus* do not produce either

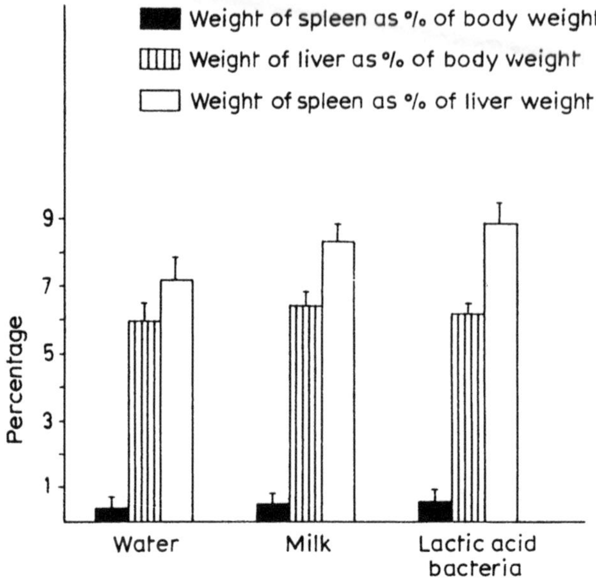

Figure 7.8 Relationships between body and spleen or liver weights in mice fed with milk or different lactic acid bacteria for 7 consecutive days. All groups were fed *ad libitum*. The liver and spleen weights were expressed as a percentage of the body weight. The ratio of spleen weight to liver weight is also expressed as a percentage value.

hepatomegaly or splenomegaly. There were differences only between body weight in the mice fed with milk and with the different lactic acid bacteria (weight: 3.31 ± 1.02 g) and the controls to which water had been given (weight: 0.71 ± 0.25 g).

The variations between body, spleen and liver weights observed were identical in the four species assayed. We believe that the increase in body weight that was observed may be due to the nutritional properties of the proteins in milk.

7.3 EFFECT OF ORAL ADMINISTRATION ON THE SECRETORY IMMUNE SYSTEM

The mucosal surfaces of mammals are in direct contact with the environment, so that they are exposed to antigens. The internal secretions in these surfaces are involved in the defence of the host and it has been demonstrated that the resistance to infection is directly related to the mucosal antibody secretions, rather than serum antibodies.

Secretory IgA (S-IgA) is the main type of immunoglobulin found in secretions, while serum shows a predominance of IgG. S-IgA is extremely important because it constitutes the first line of defence

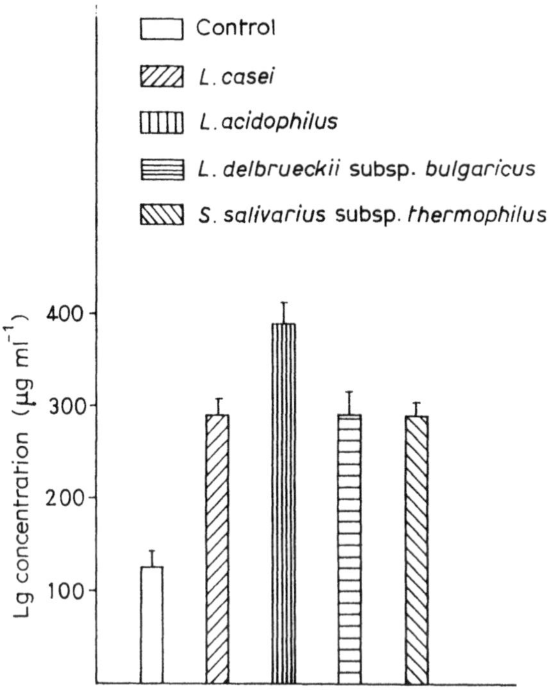

Figure 7.9 Effect of feeding with different lactic acid bacteria at a dose of 1.2×10^9 cells day^{-1} $mouse^{-1}$ for 7 consecutive days, on the immunoglobulin concentration in intestinal fluids measured by radial immunodiffusion (RID) using goat anti-mouse whole immunoglobulins. Diameters of precipitation rings were measured after 24 h.

against the penetration of viruses and pathogenic microorganisms, as well as from that of macromolecules and foreign antigens (Tomasi and Bienenstock, 1968).

The antigens that penetrate by the oral route use the migration of cells of the gut-associated lymphoreticular tissue. All the cellular elements involved in the immune response such as macrophages, dendritic cells and regulatory cells (T_s- and T_h-lymphocytes) and B-lymphocytes (precursors of IgA-producing plasma cells) are present in the gut.

When the antigen reaches the gut, the cells that interact with the antigen migrate through the mesenteric lymph nodes into the thoracic duct and the blood. This cellular migration constitutes the cellular cycle of the IgA (Phillips *et al.*, 1983). It is hormonally influenced and the number of IgA-producing cells is greater during the breast-feeding period. The above cycle has revealed the existence of an immunological system common for all mucosae. The clonal expansion of the precursors of plasma cells of the IgA type in Peyer's patches would be caused by the antigenic stimulation that reaches the gut; these plasma cells might come from the normal or pathological intestinal flora, or from foods (Husband and Gowans, 1978; Cahill *et al.*, 1976; Ahlstedt *et al.*, 1977). Therefore, oral ingestion of the antigen is extremely important for maintaining the mucosal immune system response. The ingestion of the antigen may produce synthesis of IgA locally and/or at different secretory sites (Husband and Gowans, 1978), or it may produce systemic tolerance (Challacombe and Tomasi, 1980; Kagnoff, 1982). The lipopolysaccharides (LPS) of the cell wall of the Gram-negative bacteria of the normal flora play an important role in the tolerance mechanisms, because they might be involved in the maturation of the T-cell precursors which act as suppressors in the gastrointestinal tract (Kiyono *et al.*, 1982). This fact led to the hypothesis that germ-free or conventional mice that responded to LPS could not induce suppressor T-cells and, contrary to the normal behaviour, oral stimulation would cause the production of helper T-cells (Michalek *et al.*, 1983). The oral tolerance mechanism would be a protective mechanism that would prevent the synthesis of a large variety of antibodies against the numerous antigens that reach the gut.

In order to study the action of lactic acid bacteria on the intestinal mucosal immunity, animals were fed for 2, 5 and 7 consecutive days with each of the species under study, while the control was fed with sterile milk at 10% concentration. At the end of each feeding period, mice were sacrificed and the intestinal fluid was extracted by the method of Lim *et al.* (1981). Total concentration of immunoglobulins in the fluid was determined by the radial immunodiffusion (RID) method and the presence of lysosomal enzymes (β-glucuronidase and β-galactosidase) was also determined in order to find out whether the lactic acid bacteria induced side-effects in the gut by increasing the inflammatory response. The above enzymes are relevant factors in inflammatory processes (Schorlemmer *et al.*, 1977), together with other substances such as prostaglandins (PG) and fundamentally important in such processes. They are released by the immune system cells involved in the inflammatory response. We also examined histological sections of the small intestine in animals treated with *L. casei* in order to observe the possible modifications in the epithelium and in the *lamina propria* of the gut, as a consequence of continuous feeding with this bacterium. *L. casei* was chosen because, besides being a strain capable of survival in the gut, it showed the highest degree of efficiency in the enhancement of the systemic immune response. *L. acidophilus* activated the non-specific defence mechanisms of the host moderately well.

The results obtained in the RID assay (Perdigón *et al.*, 1990 a) showed that feeding for 7 days with the different lactic acid bacteria induced an increase in the total immunoglobulin in the intestinal fluid. Highest

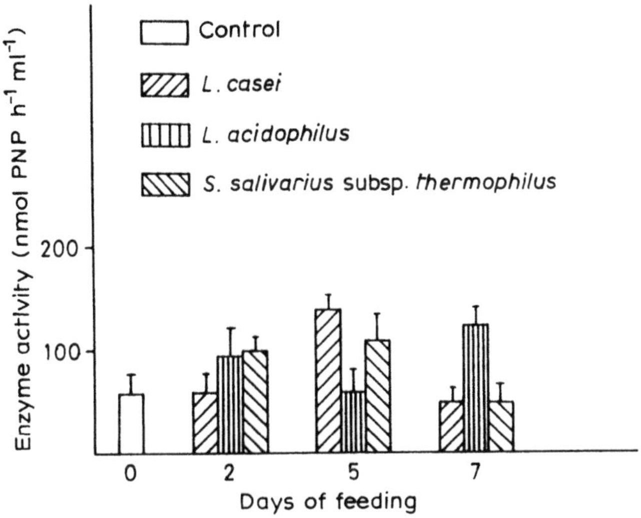

Figure 7.10 ß-Glucuronidase levels in intestinal fluid. Mice were fed with *Lactobacillus casei*, *L. acidophilus* and *Streptococcus salivarius* subsp. *thermophilus* at a dose of 1.2×10^9 cells day^{-1} $mouse^{-1}$ for 2, 5 and 7 consecutive days. Enzymatic activity was expressed as nmol PNP h^{-1} ml^{-1}. Control group were mice without lactobacilli or streptococcus administration.

values, which were three times the normal ones, were obtained by feeding with *L. acidophilus* (Figure 7.9). It is important to notice that the increase observed, as will be seen later, did not wholly agree with an increase in the Secretory-IgA, and that certain amounts of IgG were detected, probably due to an increase in the inflammatory response, which enhances the serum IgG uptake by the gut.

The levels of β-galactosidase were higher than those of β-glucuronidase. This might be due to the fact that, in the intestinal contents, not only the levels of β-galactosidase produced by inflammatory cells were determined, but also the presence of the enzyme synthetized by the intestinal epithelium.

β-Glucuronidase showed significant increase with respect to the control after 5 days feeding with the lactic acid bacteria (Figure 7.10). An increase in the levels of β-galactosidase was also observed for the same feeding period (Figure 7.11). These results seem to indicate that there would be a certain dose (the one reached after 5 days, i.e. 6×10^9 cells) that would induce an inflammatory response. *L. casei* would not increase with a longer feeding period (7 days), but would revert to practically normal values. In order to understand these results, it would be necessary to find out whether these enzymatic values correlate with the number of lymphoid cells present in the *lamina propria*. The histological sections made from the gut of mice fed with *L. casei* for 2,

5 and 7 days showed that the number of lymphoid cells that infiltrate the *lamina propria* increase with the feeding period, reaching a peak after 5 days, and then decreasing. Changes in the intestinal epithelium were also observed.

During an inflammatory process, there occurs in the tissues an accumulation of cells, predominantly neutrophils and monocytes or macrophages, with T lymphocytes also being involved (Campbell, 1986). We believe that on the fifth feeding day the cells present in the *lamina propria* are TL suppressors that have increased as a defence mechanism of the host in order to diminish the inflammatory response by immunosuppression, and thus prevent a worsening in the condition of the intestinal mucosae. It would be of interest to identify the cells present in the *lamina propria* in order to prove our hypothesis.

These studies have demonstrated the importance of the dose of lactic acid bacteria in relation to undesirable effects in the gut, especially when these microorganisms are to be used as adjuvants of the secretory immune system. This is important when using them as probiotics for the prevention of intestinal infections, and even though the effect observed might be transitory, the period of administration should be taken into

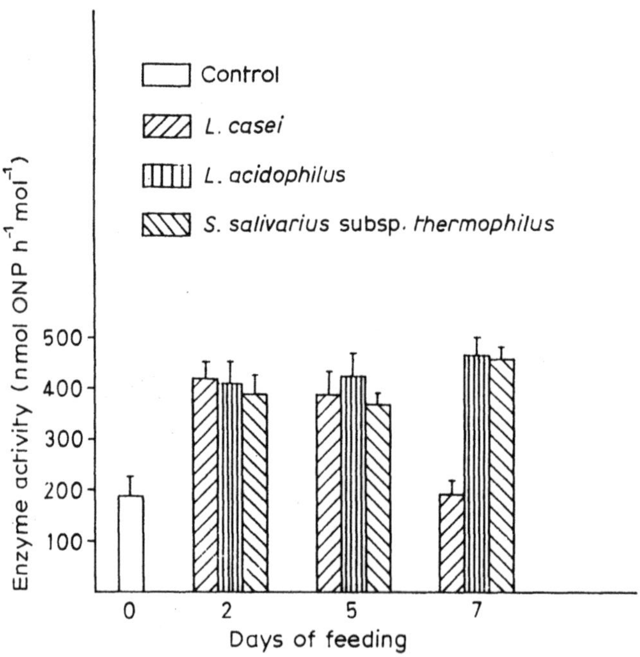

Figure 7.11 ß-Galactosidase level in intestinal fluid. Mice were fed with *Lactobacillus casei, L. acidophilus* and *Streptococcus salivarius* subsp *thermophilus* for 2, 5 and 7 consecutive days at a dose of 1.2×10^9 cells day^{-1} mouse^{-1}. Enzymic activity was expressed as nmol ONP h^{-1} mil^{-1}. Control group consisted of mice without lactobacillus or streptococcus administration.

account in order to prevent side-effects, or a possible effect on the immune system as a consequence of a long-term administration of the probiotic.

7.4 EFFECT ON THE PROTECTION AGAINST ENTERIC INFECTIONS

Enteric infections of bacterial aetiology, which produce symptoms of diarrhoea, are a worldwide medical problem. They are one of the major causes of infant mortality in developing countries and constitute a permanent risk for visitors from developed countries. Protection against some enteropathogens can be obtained by vaccination, but, at present, oral vaccines using bacteria have proved ineffective.

There are numerous mechanisms employed by the host for the defence against bacterial aggression of the mucosae, some of which are: the physical barrier of epithelial cells and their mucous coat, the physical removal of material by cilia and the peristaltic action in the intestinal tract, the pH of the mucosal environment, and protective factors in the mucosal environment, such as bile salts and metabolites of the normal indigenous flora. The disruption of any of these normal host defence mechanisms obviously has the potential for contributing to bacterial invasion of mucosal surfaces.

In recent years, with the increased knowledge concerning mucosal immunity, renewed emphasis has been laid on oral vaccination, since the antibodies secreted at the intestinal level (S-IgA) play an important role in the defence against pathogenic agents. They may act by agglutinating and immobilizing pathogens, preventing their attachment to the mucosa and/or by neutralizing the toxins produced by such pathogens (Cantey, 1978).

Enteric infections frequently occur in newborn animals during their first months and also in post-weaning stages; these infections are a serious economic problem.

Different strategies have been tried using non-viable and non-virulent bacteria, for the purpose of developing a protective immunity in the gastrointestinal mucosae. However, one of the practical problems encountered in vaccine development has been that oral administration of antigens to animals or humans often evokes little or no detectable immunological response, presumably because the mucosal immune system has conflicting requirements for responses to enteric antigens. It should be remembered that while a positive immune response must be elicited in response to pathogens, it would be undesirable for the host to respond with a vigorous immune response to antigens present in the normal indigenous flora or in food antigens. Metchnikoff (1907) was the first to report the beneficial effect of lactic acid bacteria for the prevention or treatment of intestinal disorders. There is now a renewed interest in using these bacteria as food additives to prevent diarrhoea.

The most often used bacteria in the prevention of diarrhoea in newborn piglets are, *Enterococcus faecium* and L.*acidophilus*, usually for the control of infection with *Escherichia coli* (Underdahl *et al.*, 1982). Other workers (Hitchins *et al.*, 1985) have demonstrated that feeding with yoghurt may increase the survival rate in mice infected with *Sal. typhimurium*.

It is important to find a treatment that will increase immunity in the mucosa of newborn animals, including the human baby. The great susceptibility of newborn children and animals to diarrhoea is due to the large number of mannose-receptor sites in the intestinal epithelium, to which the enteropathogenic microorganisms become attached (Israel and Walker, 1987). The number of such sites decreases in the adult intestine, thus diminishing the possibility of attachment of the bacteria and, consequently, the risk of infections, which occur only when there is an alteration of the surface of the microvilli due to medication or to lack of adequate nutrients.

In our experimental model we analysed the effect of the strains under study on the protective capacity against an infection with *Sal. typhimurium*. We also studied the behaviour of a product fermented with *L. casei* and *L. acidophilus* on the protective capacity against the same pathogen.

The effect of lactic acid bacteria administered prior to or simultaneously with the pathogen was analysed. The relationship between the protective capacity and the presence of antimicrobial substances, produced and secreted into the intestinal lumen as a consequence of feeding with lactic acid bacteria, was also investigated.

The protective capacity was determined by using spleen and liver colonization assays, and measuring the levels of antienteropathogenic S-IgA present in the intestinal fluid, by the ELISA test. The survival rate was not used because, during the course of an infection, a normal animal can produce the immune response required to counteract it, and even though the animal becomes ill it can recover. We believe that, in order to suggest the use of a certain substance or microorganism as a probiotic agent, this fact has to be taken into account. It is the extremely good intestinal response to antibodies, which prevents the colonization of the pathogen and its spreading to other organs and producing disease. In other words, the invasive capacity to the pathogen can be suppressed at the intestinal level by impaired attachment of the pathogen to the epithelial surfaces. Naturally, the above factors are not valid in the case of mice or a host in which the reactive capacity of the immune system is lowered (immunosuppressed); in such cases the survival degree must be taken into consideration when analysing a probiotic substance since, besides an increase in the local response, there must also be an increase in the systemic immune response.

The preventive assay was carried out with *L. acidophilus*, *L. casei* and *S. salivarius* subsp. *thermophilus* orally administered for 2, 5 and 7 consecutive days using the same dose as that for the previous studies (1.2×10^9 cells $day^{-1}mouse^{-1}$). In the case of *L. delbrueckii* subsp. *bulgaricus*, feeding was for 7 days. At the end of each feeding period, mice were orally infected with 20 LD_{50} *Sal. typhimurium*. This dose allowed good colonization in liver and spleen and made the animals ill for 15 days, at the end of which period some died but, on the whole, most of them survived with no *Salmonella* being detectable in liver and spleen. The colonization assay was carried out on the 2nd, 4th and 7th days post-infection because the kinetics of *Salmonella* invasion in the control group indicated that no effect was likely beyond this time. After 7 days the response to the pathogen is independent of the previous treatment, for the reason explained before.

Results showed (Perdigón *et al.*, 1989, 1990a, b) that previous feeding with *L. delbrueckii* subsp. *bulgaricus* for 7 days was not effective for protection, and that the levels of anti-*Salmonella* S-IgA were similar to those of the controls. Feeding for 2, 5 and 7 days with *L. acidophilus* was not effective for protection. The colonization values obtained for feeding for 2 days were significantly lower than those of the control (Figure 7.12). The concentration of specific IgA for the pathogen in these microorganisms was lower than that of the control at 5 and 7 days of feeding; S-IgA levels for 2 days were higher, thus agreeing with the lower degree of colonization for this administration period (Figure 7.13).

We believe that the lack of protection against infection, even with the previous administration of *L. acidophilus*, is due to an increase in the inflammatory response because, as indicated in the previous section, high levels of β-glucuronidase and β-galactosidase were observed. This higher inflammatory response would increase the pathogen–host interaction, favouring invasion by the pathogen. Although it seems likely that S-IgA was produced, it was not detected, possibly due to increased proteolytic activity. S-IgA may have suffered alterations that prevented it from adequately performing its biological function of antigen neutralization. Another possibility in the case under consideration is that, although the selected *L. acidophilus* and *L. delbrueckii* subsp. *bulgaricus* strains may behave in the manner described above, other strains might have exerted a beneficial effect. In the light of the evidence presented, we cannot ignore the importance of the inflammatory effect

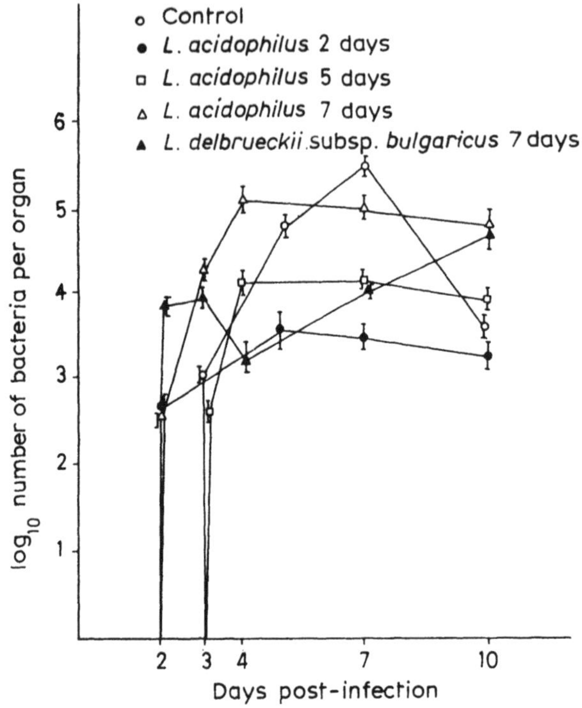

Figure 7.12 Effect of feeding with *Lactobacillus acidophilus* for 2, 5 and 7 consecutive days and *L. delbrueckii* subsp. *bulgaricus* for 7 days, on the protection against *Salmonella typhimurium* infection. Mice were challenged at the end of each period of feeding and on the 2nd, 4th, 7th and 10th day post-challenge the number of bacteria in livers and spleens were determined. Each point represents the mean number of bacteria ± standard deviation. Number of salmonellae in liver was the same as that in the spleen.

that these lactobacilli provoke in the host, which make them ineffective for protection against pathogens and unable to control infection at the intestinal level.

Previous feeding with S. *salivarius* subsp. *thermophilus* for 2 or 5 days was not protective, but feeding for 7 days was (Figure 7.14). Even in the mice fed for 7 days, colonization in liver and spleen was observed on the 2nd day post-infection. When analysing the levels of pathogen-specific S-IgA in the intestinal fluid of treated mice, it was found that they were very low compared with those of the control group (Figure 7.15). Although titres of anti-*Salmonella* IgA were low, S. *salivarius* subsp. *thermophilus* was the only treatment in which anti-S. *salivarius* subsp. *thermophilus* antibodies were detected in the intestinal fluid. Although we do not as yet know the reason why this microorganism can protect against infections with *Sal. typhimurium*, the effect may be due to several factors:

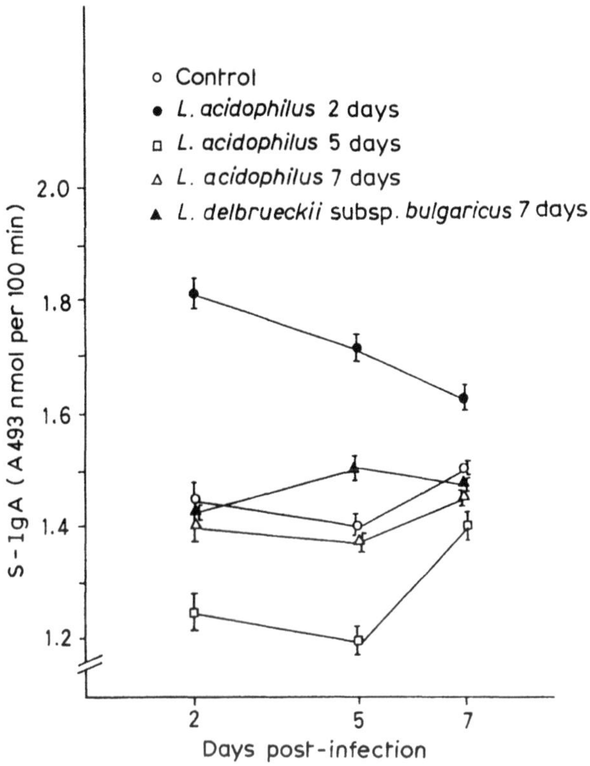

Figure 7.13 Intestinal secretion S–IgA antisalmonella response, measured in mice treated with *Lactobacillus acidophilus* and *L. delbrueckii* subsp. *bulgaricus* for 2, 5, 7 days and 7 days, respectively. Then the animals were challenged with the pathogen, and the ELISA test was performed on the 2nd, 5th and 7th day post-infection. Points and bars represent the mean ± standard deviation of each group of mice.

1. There is an activation of immunocompetent cells in the *lamina propria*, without secretion of antibodies. Ernst *et al.* (1985) have shown that the infectious agent is eliminated only by cellular mechanisms, provided by specialized intraepithelial lymphocytes, macrophages or mast cells.
2. The anti-S. *salivarius* subsp. *thermophilus* antibodies may play an important role in the protection, possibly because the hydrocarbonated portion of the anti-S. *salivarius* subsp. *thermophilus* S-IgA non-specifically binds the pathogen, preventing its attachment, and although this is not such an efficient mechanism as specific S-IgA production, it may help or protect against infection. A similar mechanism has been described in the case of infection with *E. coli* and *S. pneumoniae*, in which the pathogen is not only bound by the Fab fragment of the immunoglobulin, but also

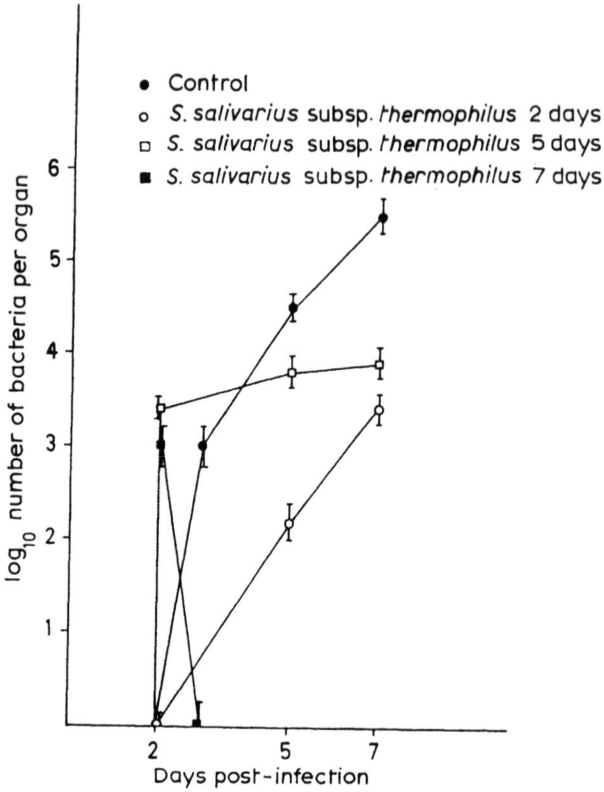

Figure 7.14 Number of salmonellae in livers and spleens of mice fed with *Streptococcus salivarius* subsp. *thermophilus* for 2, 5 and 7 consecutive days and then challengeed with *Salmonella typhimurium*. Points and bars represent the mean of 5 mice in each group ± standard deviation.

by the hydrocarbonated portion of the antibodies (Svanborg-Eden *et al.*, 1987).

3. There is a non-specific elimination either by mucins that coat the mucosal wall thus providing a protective effect (Lake *et al.*, 1979), or by antibiotic substances produced by *S. salivarius* subsp. *thermophilus* (McNabb and Tomasi, 1981).

Previous feeding with *L. casei* was effective only when it was carried out for 2 and 7 days, a marked colonization, similar to that of the controls, being observed on the 5th day. After 2 days there was no colonization. After 7 days, although there was slight colonization on the 2nd day post-challenge, it was controlled because *L. casei* induced a good systemic response so that the Kupffer liver cells and lymphoid spleen cells were activated and rapidly eliminated the pathogen. The effectiveness of the protection by the administration of *L. casei* was also demonstrated for another enteropathogen, E. coli. In this case colonization was observed in mice fed for 5 days (Table 7.5). The above effect is significant if we bear in mind the fact that the accumulated dose of *L. casei* obtained after feeding for 5 days was the one that induced the good activation in the systemic immune response.

Table 7.5 Protective capacity of *L.actobacillus casei* against *Salmonella typhimurium* (20 LD_{50}) and *Escherichia. coli.* (10^7 cells) infections.

		Number of bacteria per organ			
			Administration of *L. casei*		
Infection with	Days post-infection	Control	2 days	5 days	7 days
	2	11 300 ± 500	0	12 589 ± 1 300	12 685 ± 1 200
Salmonella	5	31 623 ± 1 900	0	22 387 ± 1 800	0
typhimurium	7	316 228 ± 3 200	0	25 119 ± 2 100	0
	2	14 780 ± 1 300	0	51 000 ± 2 000	0
Escherichia	5	9 480 ± 1 200	0	48 000 ± 1 900	0
coli	7	0	0	0	0

Number of *Salmonellae* and *E. coli* in livers and spleens of controls and treated mice with *L. casei* during 2, 5 and 7 consecutive days. They were challenged with the pathogens at the end of each feeding period. Differences between number of pathogens in liver and spleen were not observed. Values represent mean of 5 mice ± standard deviation.

From the results, it would seem that when there is a large activation of immunocompetent cells the inflammatory response increases, as revealed by the detection of the enzymes involved in the inflammatory processes. This would make bacterial penetration easier and would support the hypothesis that the low colonization after 7 days was due to a lower inflammatory response, as shown by the decrease in the level of]-glucuronidase and]-galactosidase.

The ELISA test supported the above results, because the levels of anti-*Salmonella* S-IgA were significantly higher than those of the controls after 2 and 7 days of feeding, and lower than the latter after 5 days previous feeding – results which are in close agreement with those obtained for the colonization assay. The levels of anti-*E. coli* S-IgA also showed a pattern similar to that obtained for *Salmonella*, though the absorbance values in this case were lower (Figure 7.6). If we relate these results to the ones obtained with the histological sections, we might conclude that the large lymphoid infiltration observed in the *lamina propria* in those mice fed for 5 days with *L. casei*, and which produce a higher inflammatory response, would be of T-suppressor cells. This hypothesis would be supported by the result obtained with the ELISA test in which, for this feeding period, no antibodies to enteropathogens were detected. Moreover, there is work which demonstrates that the induction of inflammation and the development of clinical impairment can be prevented by immunosuppression, specifically by the inhibition of T-cell responsiveness with dominance of the CD^{+}_{8} TL suppressor subset (Doherty *et al.*, 1990). The improvement in the IgA levels obtained after 7 days feeding would be due to the normalization in the number of CD_4 TL helper cells. This would mean that, even though the administration of *L. casei* exerts beneficial effects in the protection against pathogens, there would be a negative stage as a consequence of the entrance into the gut of a large number of bacterial antigens, which would provoke an exaggerated inflammatory response. This effect would be counteracted by the appearance of TL suppressors, which would prevent an undesirable effect in the gut, but that would provoke a decrease in the synthesis of antibodies. The later normalization, even when increasing the dose of lactic acid bacteria, would be due to the decrease in the host of inflammatory response with a lower degree of activation of the immunocompetent cells as a result of the activation of the immunoregulatory mechanisms.

From the above, we might conclude that the dose administered plays an important role in the protection against infection. We think that the effect of the dose would vary according to whether the host is a normal adult, as in our experimental model, or a newborn host animal which is immunologically immature. The dose may also be different in animals who are immunosuppressed by long administration of medicines, or by malnutrition. In these cases a long-term administration of lactic acid bacteria might either exert a positive effect or worsen the condition of the host. Consequently, it would be extremely important to determine the optimum dose in each case. Although our experiences with *L. acidophilus* proved to be negative as regards protection, we know that they are very useful in the prevention of infections with *E. coli* in newborn piglets. However, this is a different case from that of mice, since pigs and calves are born without immunoglobulins because they are not transferred through the placenta. In the case of piglets and calves, the administration of a bacteria used as a probiotic would be of great usefulness, probably because its antigenic stimulation would favour the maturation of the secretory immune system thus preventing infection.

Feeding for 7 days with milk fermented with *L. casei* and *L.*

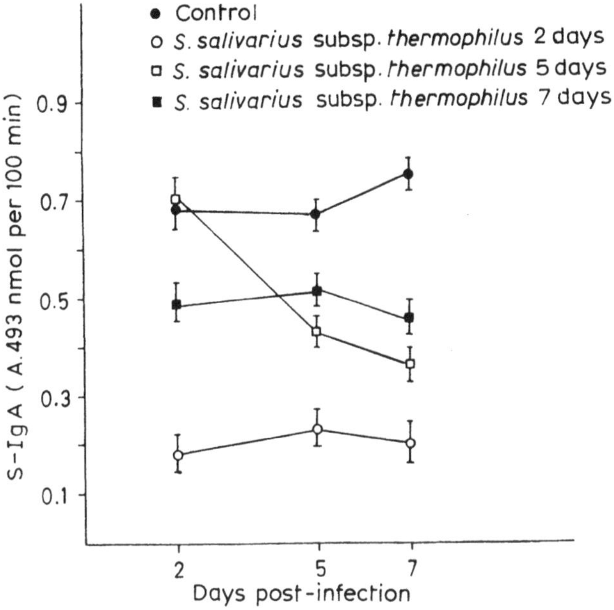

Figure 7.15 S–IgA anti-salmonellae levels in intestinal secretion, measured by ELISA test. Mice were fed with *Streptococcus salivarius* subsp. *thermophilus* in different periods and then challenged with *Salmonella typhimurium*. On the 2nd, 5th and 7th day post-challenge the ELISA test was performed. Points and bars represent the mean ± standard deviation of each group of mice.

acidophilus was just as effective as when using only *L. casei*; the same colonization in liver and spleen and level of the S-IgA was observed for the same feeding period. The fermentation process and the addition of *L. acidophilus*, while improving the nutritional properties, did not increase the capacity of *L. casei* to induce antienteropathogen secretory IgA.

Taking into account the fact that, among the lactic acid bacteria tested, *L. casei* administered prior to challenge was the most effective in the protection against infection, we analysed its protective capacity against *Sal. typhimurium* when *L. casei* was administered together with the pathogen (unpublished data).

Mice were challenged with 20 LD_{50} *Sal. typhimurium* and feeding with *L. casei* under the described conditions was started simultaneously for 2, 5 and 7 days. At the end of each feeding period, mice were sacrificed and colonization and antibody detection assays in intestinal fluid were performed. The results obtained showed a colonization similar to that of the controls for all the periods assayed (Table 7.6). The ELISA test showed slightly higher IgA levels, though not significantly different from those of the controls. These results would indicate that, although *L. casei* stimulated the synthesis of antienteropathogen S-IgA protecting

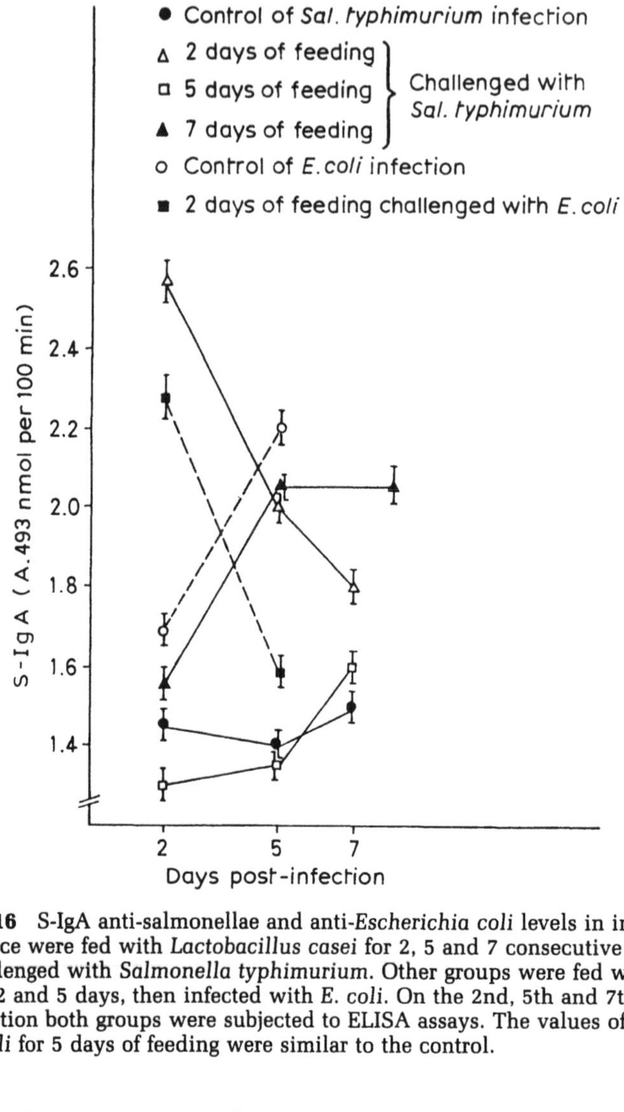

Figure 7.16 S-IgA anti-salmonellae and anti-*Escherichia coli* levels in intestinal fluids. Mice were fed with *Lactobacillus casei* for 2, 5 and 7 consecutive days, then challenged with *Salmonella typhimurium*. Other groups were fed with *L. casei* for 2 and 5 days, then infected with *E. coli*. On the 2nd, 5th and 7th day post-infection both groups were subjected to ELISA assays. The values of S-IgA anti-*E. coli* for 5 days of feeding were similar to the control.

against infection, it was only effective when administered prior to the infection.

One explanation might be that the early administration activated the immune cells before the antigenic impact of the pathogen, behaving as a true adjuvant. Such cells immediately responded by antibody synthesis, which inhibited attachment of the pathogen. An activated cell produces a larger release of lymphokines, favouring clonal expansion and differentiation of BL to antibody-producing plasma cells. Simultaneous administration did not inhibit pathogen attachment, and though it later induced a slight increase in S-IgA, no effectiveness was observed. At present, the role of S-IgA, when the pathogen is already attached, is not known; it probably prevents a renewal of infection.

We also studied (unpublished data) the production of antimicrobial substances, secreted into the intestinal lumen, induced by feeding with *L. casei* and *L. acidophilus*. These lactobacilli were used as they proved to be capable of resisting the action of the gastric and intestinal fluids. The purpose of these experiments was to try to establish a relationship between the protective capacity of lactobacilli against infection, the state of activation of the secretory immune system, and also with the *in vivo* production of antimicrobial substances.

The strain of *L. acidophilus* used showed good *in vitro* antimicrobial activity; this was not so with *L. casei* but, as demonstrated above, this microorganism was able to afford protection against infection.

Animals were fed for 2, 5 and 7 days, at the end of which periods the intestinal contents were collected and used for the inhibition test

Table 7.6 Protective capacity of *Lactobacillus casei* against *Salmonella typhimuriun* infection administered simultaneously.

Days post-infection with *Sal. typhimurium*	Number of bacteria per organ: Control	Number of bacteria per organ: Fed with *L. casei* simultaneously to the infection
2	11 300 ± 500	9 582 ± 1 000
5	31 623 ± 1 900	14 755 ± 1 300
7	316 228 ± 3 200	45 782 ± 2 000

Number of salmonellae in livers and spleens of mice challenged with 20 LD_{50} of *Sal. typhimurium* and simultaneously fed with *L. casei* for 2, 5 and 7 consecutive days. The same number of salmonellae in liver and spleen were observed. Values represent means of 5 mice ± standard deviation.

against *Salmonella* and *E. coli*, using the same method as for the *in vitro* determination (Reddy *et al.*, 1984).

We were unable to detect antimicrobial substances in the intestinal contents probably because, during the period assayed, the number of lactic acid bacteria in the gut required to produce antimicrobial substances capable of detection was not reached. If they are synthesized, as most of these substances are peptides, they could have been destroyed by enzyme action. As already mentioned, the level of proteolytic enzymes was increased as a consequence of feeding with lactic acid bacteria.

We were not able to determine whether the *in vitro* antimicrobial activity also occurred in the intestine. However, we were able to prove that, for the administration period assayed, the immune system is undoubtedly involved in the protection against enteropathogenic agents.

Bearing in mind the above considerations, lactic acid bacteria, especially *L. casei*, could therefore be used for preventive purposes in intestinal infections and also as immunomodulators, thus opening a very promising possibility not only as regards their use as oral adjuvants in human and animal enteric vaccines, but also as protection against other diseases in which the immune system is involved.

ACKNOWLEDGEMENTS

The author wish to thank Marta G. Medici for technical assistance. This work was supported by grants, PID 3–19300/85 and 3–127100/88 from Consejo Nacional de Investigaciones Científicas y Técnicas (CONICET), Argentina.

REFERENCES

Ahlstedt, S., Carlsson, B., Fallstrom, S.P., *et al.* (1977) Antibodies in human serum and milk induced by enterobacteria and food proteins, in *Immunology of the Gut*, Ciba Foundation Symposium **46**, p. 115.

Ayebo, A., Shahani, K., Dam, R. and Friend, B. (1982) Ion exchange separation of the antitumor component(s) of yogurt dialyzate. *J. Dairy Sci.*, **65**, 2388–90.

Bauer, H., Paronetto, F., Burnsaud, W. and Einheber, A. (1966) The enhancing effect of the microbial flora on macrophage function and the immune response. *J. Exp. Med.*, **123**, 1013–26.

Bloksma, N., de Heer, E., van Dijk, M. and Willers, M. (1979) Adjuvanticity of lactobacilli. I. Differential effects of viable and killed bacteria. *Clin. Exp. Immunol.*, **37**, 367–75.

Bogdanov, I., Davlev, P., Gurevich, A. *et al.* (1975) Antitumor glycopeptides from *Lactobacillus bulgaricus* cell wall. *FEBS Lett.*, **57**, 259–63.

Cahill, R.N.P., Frost, H. and Trnka, Z. (1976) The effects of antigen on the migration of recirculating lymphocytes through single lymph nodes. *J. Exp. Med.*, **143**, 870.

Campbell, P. (1986) Are inflammatory phagocytes responsible for resistance to facultative intracellular bacteria? *Immunol. Today*, **7** (3), 70–2.

Cantey, J. (1978) Prevention of bacterial infections of mucosal surfaces by immune secretory IgA. *Adv. Exp. Med. Biol.*, **107**, 461–70.

Conchie, J., Findlay, O. and Levy, G. (1959) Mammalian glucosidases. Distribution in the body. *Biochem. J.*, **71**, 318–25.

Challacombe, S.J. and Tomasi, T.B. Jr (1980) Systemic tolerance and secretory immunity after oral immunization. *J. Exp. Med.*, **152**, 1459.

De Simone, C., Bianchi-Salvadori, B., Negri, R. *et al.* (1986) The adjuvant effect of yogurt on production of gamma-interferon by ConA-stimulated human peripheral blood lymphocytes. *Nutr. Repts Int.*, **33**, 419–43.

Doherty, P., Allan, J., Lynch, F. and Ceredig, R. (1990) Dissection of an inflammatory process induced by CD8+ T-cells. *Immunol. Today*, **11** (2), 55–9.

Ernst, P., Befus, A. and Bienenstock, J. (1985) Leucocytes in the intestinal epithelium: An unusual immunobiological compartment. *Immunol. Today*, **6**, 50–3.

Enders, G., Gottawald, T. and Brendel, W. (1986) Induction of oral tolerance in rats without Peyer's patches. *Immunology*, **58**, 311–14.

Gasser, F. and Gasser, C. (1971) Immunological relations among lactic acid dehydrogenases in the genera *Lactobacillus* and *Leuconostoc*. *J. Bacteriol.*, **106**, 113–25.

Gemsa, D., Kubelka, C., Debatin, K. and Krammers, P. (1984) Activation of macrophages by lymphokines from T-cell clones, evidence for different macrophage-activating factors. *Mol. Immunol.*, **21**, 1267–76.

Goldin, B. and Gorbach, M. (1980) Effect of *Lactobacillus acidophilus* dietary supplements on 1, 2-dimethylhydrazine dihydrochloride induced intestinal cancer in rats. *J. Natl Cancer Inst.*, **64**, 263–5.

Hashimoto, S., Seyama, Y., Yokokura, T. and Mutai, M. (1985) Cytotoxic factor production by Kupffer cells elicited with *Lactobacillus casei* and *Corynebacterium parvum*. *Cancer Immunol. Immunother.*, **20**, 117–21.

Hitchins, A., Well, P., McDonough, F. and Wong, N. (1985) Amelioration of the adverse effect of a gastrointestinal challenge with *Salmonella entiritidis* on weanling rats by yogurt diet. *Amer. J. Clin. Nutr.*, **41**, 92–100.

Husband, A.J. and Gowans, J.L. (1978) The origin and antigen dependent distribution of IgA containing cells in the intestine. *J. Exp. Med.*, **148**, 1140.

Ianello, D., Bonina, L., Delfino, D. *et al.* (1984) Effect of oral administration of a variety of bacteria on depressed macrophage functions in tumor-bearing rats. *Ann. Immunol. (Paris)*, **135C**, 345–52.

Israel, E. and Walker, W.A. (1987) Development of intestinal mucosal barrier function to antigens and bacterial toxins. *Adv. Exp. Med. Biol.*, **216B**, 673–83.

Iwasaki, I., Yumoto, N., Iwase, H. and Ide, G. (1983). Potentiation of large intestinal tumorigencity of cycasin derivative by high fat diet and *Lactobacillus* in germ free mice. *Acta Pathol. Japan*, **33**, 1197–204.

Kagnoff, M.F. (1982) Immunological unresponsiveness after enteric antigen administration, in *Recent Advances in Mucosal Immunity* (ed. W. Strober, *et al.*), Raven Press, New York, pp 95–111.

Kato, I., Kohayashi, S., Yokokura, T. and Mutai, M. (1981) Antitumor activity of *Lactobacillus casei* in mice. *Gann*, **72**, 517–23.

Kato, I., Yokokura, T. and Mutai, M. (1983) Macrophage activation by *Lactobacillus casei* in mice. *Microbiol. Immunol.*, **27** (7), 611–18.

Kato, I., Yokokura, T. and Mutai, M. (1984) Augmentation of mouse natural killer cell activity by *Lactobacillus casei* and its surface antigens. *Microbiol Immunol.*, **28**, 209–17.

Kato, I., Yokokura, T. and Mutai, M. (1985) Induction of tumoricidal peritoneal exudate cells by administration of *Lactobacillus casei*. *Int. J. Immunopharmacol.*, **7**, 103–9.

Kishimoto, T., Hirai, Y., Nakanishi, K. *et al.* (1979) Regulation of antibody response in different immunoglobulin classes. I. Selective supression of IgE response by administration of antigen conjugated muramyl-peptides. *J. Immunol.*, **123(B)**, 2709–15.

Kiyono, H., McGhee, J.R., Wannemuehler, M.J. and Michalek, S.M. (1982) Lack of oral tolerance in C_3H/H_eJ mice. *J. Exp. Med.* **155**, 605.

Lake, A., Bloch, M., Neutra, M. and Walker, W. (1979) Intestinal goblet cell mucus release. II. *In vivo* stimulation by antigen in the immunized rat. *J. Immunol.*, **122**, 834–7.

Lim, T.S., Messiha, N. and Watson, R.R. (1981) Immune components of the intestinal mucosae of ageing and protein deficient mice. *Immunology*, **43**, 401–7.

Lowy, I., Theze, J. and Chedid, L. (1980) Stimulation of the *in vivo* dinitrophenyl antibody response to the DNP conjugate of L-glutamic acid-L-alanine-L-tyrosine (GAT) polymer by a synthetic adjuvant activity and isotypic pattern of MDP stimulated response. *J. Immunol.*, **124** (1), 100–4.

Matsuzaki, T., Yokokura, T. and Azuma, I. (1985) Antitumor activity of *Lactobacillus casei* on Lewis Lung carcinoma and line-10 hepatoma in syngeneic mice and guinea pig. *Cancer Immunol. Immunother.*, **20**, 18–22.

McNabb, P. and Tomasi, T. (1981) Host defense mechanisms at mucosal surfaces. *Ann. Rev. Microbiol.*, **35**, 477–96.

Metchnikoff, E. (1907) *The Prolongation of Life. Optimistic Studies*, Putnam, New York.

Michalek, S.M., McGhee, J.R., Kiyono, H. *et al.* (1983) The IgA response: inductive aspects, regulatory cells, and effector functions. *Ann. N.Y. Acad. Sci.*, **409**, 48.

Miller, S. and Hanson, D. (1979) Inhibition of specific immune responses by feeding protein antigens. *Infect. Immun.*, **123**, 2344–50.

Moreau, M.C., Corthier, G., Muller, M.C. *et al.* (1986) Relationships between Rotavirus diarrhea and intestinal microflora establishment in conventional and gnotobiotic mice. *J. Clin. Microbiol.*, **23** (5), 863–8.

Namba, Y., Hidaka, J., Taki, K. and Morimoto, T. (1981) Effect of oral administration of lysozyme or digested bacterial cell wall on immunostimulation in guinea pigs. *Infect. Immun.*, **31**, 580–3.

Perdigón, G., Alvarez, S., Nader de Macías, M.E, *et al.* (1986a) Lactobacilli administered orally induce release of enzymes from peritoneal macrophages in mice. *Milchwiss.*, **41** (6), 344–8.

Perdigón, G., Nader de Macías, M.E., Alvarez, S. *et al.* (1986b) Actividad imunopotenciadora de bacterias lácticas admintradas por vía oral. Su efecto beneficioso en diarreas infantiles. *Medicina (Buenos Aires)*, **46**, 751–4.

Perdigón, G., Nader de Macías, M.E., Alvarez, S., *et al.* (1986c) Effect of a mixture of *Lactobacillus casei* and *Lactolbacillus acidophilus* administered orally on the immune system in mice. *J. Food Protect.*, **49** (12), 986–9.

Perdigón, G., Nader de Macías, M.E., Alvarez, S. *et al.* (1986) Effect of perorally administered lactobacilli on macrophage activation in mice. *Infect. Immuni.*, **53** (2), 404–10.

Perdigón, G., Nader de Macías, M.E., Alvarez, S. *et al.* (1987) Enhancement of immune response in mice fed with *Streptococcus thermophilus* and *Lactobacillus acidophilus*. *J. Dairy Sci.*, **70**, 919–26.

Perdigón, G., Alvarez, S., Nader de Macías, M.E., P. de Ruiz Holgado, A.A. (1988a) Activadad adyuvante de bacterias lácticas: perspectivas para su uso en vacunas orales. *Rev. Argentina Microbiol.*, **20**, 141–6.

Perdigón, G., Nader de Macías, M.E., Alvarez, S. *et al.* (1988b) Systemic augmentation of the immune response in mice by feeding fermented milks with *Lactobacillus casei* and *Lactobacillus acidophilus*. *Immunology*, **63**, 17–23.

Perdigón, G., Alvarez, S., Nader de Macías, M.E. and Medici, M. (1989) Effect of lactic acid bacteria orally administered and of yoghurt on the immune system. *Les Laits Fermentés. Actualité de la recherche*, John Libbey, London pp. 77–84.

Perdigón, G., Alvarez, S., Nader de Macías, M.E. *et al.* (1990a) The oral administration of lactic acid bacteria increases the mucosal intestinal immunity in response to enteropathogens. *J. Food Protect.* **53** (5), 404–10.

Perdigón, G., Nader de Macías, M.E., Alvarez, S. *et al.* (1990b) Prevention of gastrointestinal infection using immunobiological methods with milk fermented with *Lactobacillus casei* and *Lactobacillus acidophilus*. *J. Dairy Res.*, **57**, 255–64.

Phillips., Quagliata, J.M., Roux, M.E. *et al.* (1983) Migration and regulation of B-cells in the mucosal immune system. *Sec. Immune System*, **409**, 194–203.

Reddy, G. V., Friend, B.A. Shahani, K.M. and Farmer, R. (1983) Antitumor activity of yogurt components. *J. Food Protect.*, **46**, 8–11.

Reddy, G.V., Shahani, K.M., Friend, B.A. and Chandan, R.C (1984) Natural antibiotic activity of *Lactobacillus acidophilus* and *Lactobacillus bulgaricus*. *Cultured Dairy Products J.* 7–11.

Saito, I., Sato, K., Horikawa, Y. *et al.* (1983) Enhanced humoral antibody production and delayed type hypersensitivity response in mice by *Lactobacillus casei*. *Hiroshima J. Med. Sci.*, **32**, 223–5.

Schorlemmer, H., Davies, P., Hylton, W. *et al.* (1977) The selective release of lysosomal acid hydrolases from mouse peritoneal macrophages by stimuli of chronic inflammation. *J. Exp. Pathol.*, **58**, 315–25.

Sekine, K., Toida, T., Saito, M. *et al.* (1985) A new morphologically characterized cell wall preparation (whole peptidoglycan) from *Bifidobacterium infantis* with a higher efficacy on the regression of an established tumor in mice. *Cancer Res.*, **45**, 1300–7.

Shahani, K., Friend, B. and Bailey, P. (1983) Antitumor activity of fermented colostrum and milk. *J. Food Protect.*, **46**, 385–6.

Sharpe, M., Hill, R. and Lapage, S. (1973) Pathogenic lactobacilli. *J. Med. Microbiol.*, **6**, 281–6.

Slobodianik, N., Langini, S., Rio, M.E. *et al.* (1984) Effect of casein on the recovery of lymphoid organs surface and intracellular markers in growing rats. *Comun. Biológicas*, **3**, 191.

Stossel, T.P. (1980) Phagocytosis, in *Manual of Clinical Immunology*, 2nd edn. (eds H. Rose and H. Friedman), Amer. Soc. Microbiol, Washington, D C, p. 313.

Svanborg-Eden, C., Andersson, B., Aniansson, G. *et al.* (1987) Glucoconjugate receptors for bacteria attaching to mucosa sites: examples for *Escherichia coli* and *Streptococcus pneumoniae*. *Recent Adv. Mucosal Immun.*, **216(A)**.

Tanaka, A., Nagao, S., Imai, K. and Mori, R. (1980) Macrophage activation by muramyl dipeptide as measured by macrophage spreading and attachment. *Microbiol Immunol.*, **24**, 547–57.

Tolone, G., Del Guercio, P., Binaghi, R. and Biozzi, G. (1970) Kinetics of phagocytosis of bacteria opsonized with natural or immune antibodies. *Int. Arch. Allergy*, **37**, 344–50.

Tomasi, T.B. and Bienenstock, J. (1968) Secretory immunoglobulins. *Adv. Immunol.* **9**, 1.

Tsujimoto, M., Kotani, S., Kinoshita, F. *et al.* (1986) Adjuvant activity of 6-*o*-acyl-muramyldipeptides to enhance primary cellular and humoral immune responses in guinea pig: Adaptability to various vehicles and pyrogenicity. *Infect. Immun.*, **53**, 511–16.

Underdahl, H., Torres Medina, A. and Doster, A. (1982) Effect of *Streptococcus faecium* C-68 in control of *Escherichia coli* induced diarrhea in gnotobiotic pigs. *Amer. J. Vet. Res.*, **43**, 2227–32.

Yasutake, N., Kato, I., Ohwaki, M. *et al.* (1984a) Host mediated antitumor activity of *Lactobacillus casei* in mice. *Gann*, **75**, 72–80.

Yasutake, N., Ohwaki, M., Yokokura, T. and Mutai, M. (1984b) Comparison of antitumor activity of *Lactobacillus casei* with other bacterial immunopotentiators. *Med. Microbiol. Immunol.*, **173**, 113–25.

Chapter Eight

Genetic manipulation of gut microorganisms

GERALD W. TANNOCK

8.1 INTRODUCTION

The gastrointestinal tract ecosystem of vertebrate animals contains a variety of habitats for microbes. These habitats are diverse and include the liquid fraction of the gut contents, the surfaces of particulate material in the digesta, the mucus associated with the surface of secretory epithelia lining the tract, and the surface of epithelial cells themselves. The organic molecules available as a source of nutrients for gastrointestinal microbes is also diverse since they comprise molecules present in the host animal's diet, as well as molecules derived from the host's own secretions and cells. This diversity of habitats and complex chemical environment is reflected in the range of different metabolic and morphological types of microbes that can be detected in the gastrointestinal tract of animals. For example, at least 400 types of bacteria have been detected in the faeces of humans (Table 8.1). Thirty to forty species constitute the major part of the microbial population (reviewed by Drasar and Barrow, 1985; and by Savage, 1977). The collection of microbes colonizing the healthy gastrointestinal tract, referred to as the normal microflora, thus represents a rich field of endeavour for microbiologists. The metabolically diverse character of the gastrointestinal microflora also provides a source of potentially exploitable microbes for the biotechnologist. One area of commercial exploitation may be the utilization of gastrointestinal microbes or their products as agents for the promotion of animal health.

The production of preparations containing living microbial cells that, when ingested, are said to improve the health of animals is by no means a novel concept (reviewed by Tannock, 1984). Such 'probiotics' are widely available and have been reviewed in other chapters of this book. Yet the production of scientifically valid probiotic preparations must surely be regarded at present as a difficult task because a number

Table 8.1 The 25 most prevalent (numerically dominant) species of bacteria detected in the faeces of Americans (after Finegold *et al.*, 1974)

Rank	Bacterial species
1	*Bacteroides vulgatus*
2	*Bacteroides* spp.
3	*Bacteroides fragilis*
4	*Bacteroides thetaiotaomicron*
5	*Peptostreptococcus micros*
6	*Bacillus* spp.
7	*Bifidobacterium adolescentis* D
8	*Eubacterium aerofaciens*
9	*Bifidobacterium infantis*
10	*Ruminococcus albus*
11	*Bacteroides distasonis*
12	*Peptostreptococcus intermedius*
13	*Peptostreptococcus* sp.
14	*Peptostreptococcus productus*
15	*Eubacterium lentum*
16	*Streptococcus* sp.
17	*Fusobacterium russii*
18	*Bifidobacterium adolescentis* A
19	*Bifidobacterium adolescentis* C
20	*Bacteroides clostridiiformis* subsp. *clostridiiformis*
21	*Peptococcus prevotii*
22	*Bifidobacterium infantis* subsp. *liberorum*
23	*Clostridium indolis*
24	*Streptococcus faecium*
25	*Bifidobacterium longum* subsp. *longum*

of microbial characteristics must be present in the ideal probiotic culture:

1. The microbial strain should persist in the gastrointestinal tract for some time. The ability of the bacteria to adhere or associate with mucosal surfaces would be important in this respect. Bacterial growth rate and efficient utilization of nutrients are also thought to be important microbial colonization characteristics (Tannock *et al.*, 1990).
2. The microbial strain should produce a substance inhibitory to gastrointestinal pathogens or stimulate host immunity so as to increase the resistance of the animal to intestinal infection.
3. The microbe should contribute to the nutrition of the host animal by synthesizing essential nutrients that thus become more readily

available to the host, and/or by digesting dietary substances that the host is physiologically ill-equipped to utilize.
4. The microbial strain must be suitable for cultivation under large-scale industrial conditions, and must be amenable to preservation for storage prior to retailing and use in the field.
5. The bacterial strain must be avirulent and devoid of metabolic characteristics that could compromise animal health. This aspect of probiotic development is seldom considered, yet some members of the genus *Lactobacillus*, bacteria that are generally regarded as safe for consumption by animals, produce potentially harmful substances: bile salt hydrolase and 3-methylindole (Honeyfield and Carlson, 1990; Tannock *et al.*, 1989; Yokoyama and Carlson, 1981).
6. All the above properties must be stable characteristics of the bacterial cells.

The detection of such a prototype strain in nature would require a large and expensive screening programme and, indeed, would be unlikely to succeed. The alternative is to derive a microbial strain by genetic manipulation that meets the above-listed criteria. For example, a microbial strain known to colonize the gastrointestinal tract efficiently could be genetically modified so that genes encoding undesirable attributes were inactivated, and genes encoding characteristics desirable in a probiotic culture could be introduced using recombinant DNA technology.

8.2 MICROBES OF POTENTIAL INTEREST

The type of microbe chosen to be developed as a probiotic strain would depend upon the required characteristics of the commercial product. Considerations in this respect include (1) the animal species for which a probiotic is required and (2) the desired nature of the probiotic.

8.2.1. The animal species for which a probiotic is required

The distribution of the normal microflora within the gastrointestinal tract differs between animal species. Thus a large part of the human gastrointestinal tract is free from permanent colonization by microbes and only the terminal ileum and large bowel harbour microbial communities of appreciable size. Factors that prevent the colonization of the stomach, duodenum and jejunum include gastric acidity, a mucus secreting mucosa lining the entire gastrointestinal tract, and the relatively rapid, peristaltic propulsion of the digesta through the small bowel (reviewed by Savage, 1977). Humans, therefore, have evolved mechanisms for reducing competition between themselves and their normal microflora in regions of the digestive tract where digestion of food and absorption of nutrients occurs. Microbial strains for use in products designated for human use should, therefore, be drawn from genera known to be numerically dominant members of the microflora of the ileum and large bowel.

In contrast to humans, pigs and fowl harbour large populations of lactobacilli in the proximal regions of their digestive tracts. These large populations can be achieved in the digestive tract of pigs and fowl because the lactobacilli can associate with the stratified, squamous epithelial surface lining part of the proximal digestive tract of these animal species (reviewed by Tannock, 1990). The crop of fowl and the *pars oesophagea* of pigs have this type of epithelium. Being non-secretory, the squamous epithelium is not covered by a layer of mucus, and lactobacilli can adhere to the animal cells. Replication of adhering bacterial cells results in a relatively thick layer of lactobacilli on the epithelial surface. Lactobacillus cells shed from this layer continually inoculate the digesta passing through the proximal region of the tract and a large lactobacillus population can be detected in the stomach and small bowel contents. Lactobacilli are also present in the large bowel contents of these animals, but in this site they are outnumbered by obligately anaerobic bacteria (Tables 8.2 and 8.3). A case can, therefore, be made that a probiotic product for these animal species should contain a *Lactobacillus* strain capable of colonizing the proximal digestive tract. Such a strain would be shed from the colonized epithelium into the digesta and thus permeate all levels of the gastrointestinal tract.

8.2.2 The desired nature of the probiotic

This is determined by whether the probiotic strain is to provide long-term colonization of the animal and to continually exert its influences on the host, or whether the existence of the microbes in the gastrointestinal tract is to be transitory and the cells to act as delivery packages of chemicals to an appropriate level of the tract. If colonization is required, consideration of the composition of the normal microflora of the host would be appropriate and a numerically dominant member of the microflora chosen. *Bacteroides* species would thus be an appropriate choice in the case of humans (Table 8.1). As indicated above, a *Lactobacillus* species would be an appropriate choice for use with pigs and fowl (Tables 8.2 and 8.3). The choice of microbes to act as non-colonizing delivery systems is of course wider, and any microbe demonstrated to be harmless upon ingestion, and amenable to industrial cultivation and genetic manipulation, might be appropriate. Lactobacilli, however, might be attractive choices here also, since certain species have been used in the production of dairy products for many decades and the

Table 8.2 Distribution of members of the normal microflora in the digestive tract of fowl.

Organ	Bacteria	Population level*
Crop	Lactobacilli	10^9
	Streptococci	10^4
	E. coli	10^2
Small bowel	Lactobacilli	10^8
	Steptococci	10^4
	E. coli	10^2
Large bowel	Lactobacilli	10^9
	Streptocci	10^7
	E. coli	10^6
	Yeasts	10^2
	Obligate anaerobes†	10^{10}

*CFU g^{-1} of organ contents (wet weight).
†Anaerobic cocci, *Eubacterium*, *Clostridium*, *Gemmiger*, *Fusobacterium* and *Bacteroides* species.
(After Barnes *et al.*, 1980; Salanitro *et al.*, 1974, 1978; Smith, 1965.)

Table 8.3 The distribution of members of the normal microflora in the digestive tract of pigs.

Organ	Bacteria	Population level*
Stomach	Lactobacilli	10^9
	Streptococci	10^6
	E. coli	10^2
	Yeasts	10^4
Small bowel	Lactobacilli	10^7
	Streptococci	10^4
	E. coli	10^4
	Yeasts	10^4
Large bowel	Lactobacilli	10^9
	Streptococci	10^7
	E. coli	10^7
	Yeasts	10^4
	Obligate anaerobes†	10^{10}

*CFU g^{-1} of organ contents (wet weight).
†*Eubacterium*, *Clostridium*, *Propionibacterium*, *Peptostreptococcus*, *Peptococcus*, *Megasphaera*, *Bacteroides* species.
(After Russell, 1979; Salanitro *et al.*, 1977; Smith, 1965.)

industrial technology for their large-scale cultivation is already available. Probiotic products containing lactobacilli could thus be devised for use with all animal species, including humans, since the bacteria would not be expected to colonize the gastrointestinal tract, but merely to lyse upon reaching an appropriate region of the tract.

Members of the genera *Lactobacillus* and *Bacteroides* appear, therefore, to be microbes of potential interest in the derivation of genetically modified strains for use in probiotic preparations. Both of these bacterial genera have been studied at the genetic level in recent years.

8.3 MOLECULAR GENETICAL STUDIES

Molecular genetical studies can be divided into several broad areas: plasmids, cloning (recombinant DNA), and mutagenesis. The impetus for molecular genetical studies with lactobacilli has been mainly derived from dairy microbiology, since some *Lactobacillus* species are used in the production of dairy products. *Bacteroides* species of intestinal origin have been studied because of their potential importance as sources of antibiotic resistance determinants and of enzymes involved in the catabolism of plant structural materials, but also because determination of bacterial colonization attributes requires the derivation of isogenic strains.

8.3.1 Lactobacillus plasmids

Plasmids are commonly detected in lactobacilli, including strains isolated from the gastrointestinal tract. The majority of plasmids detected in lactobacilli are cryptic (do not have a known, associated phenotype), but in certain strains *N*-acetyl-D-glucosamine fermentation, proteolysis, lactose metabolism, maltose utilization, cysteine uptake, bacteriocin production, or antibiotic resistance are encoded by plasmid-borne genes (Axelsson, *et al.*, 1988; Liu *et al.*, 1988; Muriana and Klaenhammer, 1987; Shay *et al.*, 1988; reviewed by Tannock, 1990). Only a few examples of conjugative plasmids in lactobacilli have been reported (reviewed by Chassy, 1987; Muriana and Klaenhammer, 1987). Several lactobacillus plasmids have been mapped by restriction endonuclease analysis and, in some cases, nucleotide sequencing has also been accomplished. Shuttle plasmids capable of replication in both lactobacilli and in *Escherichia coli* have been constructed (Lee-Wickner and Chassy, 1985; Damiani *et al.*, 1987; Morelli *et al.*, 1987; Shrago and Dobrogoscz, 1988; Bates and Gilbert, 1989; Bouia *et al.*, 1989; Bringel *et al.*, 1989; Jossen *et al.*, 1989; Skaugen, 1989; Takiguchi *et al.*, 1989).

The collection of plasmid types harboured by a *Lactobacillus* strain (its plasmid profile) has been shown to be a useful characteristic that can be used as an ecological tool to monitor the presence or absence of the particular strain in gastrointestinal or silage samples (Hill and Hill, 1986; Tannock *et al.*, 1990). Purified plasmid DNA, labelled with biotin, has been used in hybridization experiments to detect the presence of specific *Lactobacillus* strains in gastric specimens from mice (Tannock, 1989).

8.3.2 Bacteroides plasmids

Small cryptic plasmids less than 7 kb pr occur commonly in *Bacteroides* species (Beul *et al.*, 1985; Callihan *et al.*, 1983). Several conjugative plasmids encoding antibiotic resistance have also been detected. The best-studied of these conjugative plasmids is plasmid pBF4, a 41 kb plasmid detected in *Bacteroides fragilis*. This plasmid contains transposon Tn*4351* that encodes two antibiotic-resistance genes. One gene encodes resistance to macrolide/lincosamide antibiotics that are expressed in *Bacteroides* species but not in *E. coli*. The other gene encodes tetracycline resistance which is expressed in aerobically cultivated *E. coli*, but not in *E. coli* or *Bacteroides* under anaerobic conditions. Genetic elements capable of transferring antibiotic resistance in the genus *Bacteroides*, but that do not appear to be typical plasmids, have also been described. Some of these elements, thought to be located on the bacterial chromosome, transfer tetracycline resistance only, while others transfer both tetracycline and erythromycin resistance. The elements may be similar to the conjugative transposons encountered in Gram-positive cocci (reviewed by Odelson *et al.*, 1987 and Salyers *et al.*, 1987).

Transformation with plasmid DNA, mediated by polyethylene glycol, can be achieved with *Bact. fragilis*. Several plasmids suitable for recombinant DNA work have been derived for use with this transformation system. The plasmid vectors are based on pBI143, a 2.7 kb cryptic plasmid that belongs to the class I homology group (reviewed by Odelson *et al.*, 1987). Salyers *et al.*, (1987), however, have pointed out the advantages of using conjugation, rather than transformation, in the genetical modification of obligate anaerobes such as bacteroides: conjugal transfer of genes can be achieved in several *Bacteroides* species that are non-transformable; restriction barriers in the recipient cell can be circumvented by conjugation because the DNA enters in single-stranded form (double-stranded in chemically induced transformation); conjugation is easily performed under anaerobic conditions, whereas it is more difficult to maintain anaerobiosis with transformation techniques.

A number of shuttle plasmids that can be mobilized from E *coli* to *Bacteroides* spp. have been constructed. In general, they contain two independent origins of replication, each of which can be recognized by the appropriate bacterial host, at least one selectable marker suitable for each host, and a transfer origin that is recognized by the mobilizing element. The IncP plasmids pRK231 and R751 have been used to mobilize the shuttle vectors, but the non-plasmid, transmissible tetracycline-resistance elements mentioned above will also perform this function. One such element, Tc^{r}ERL, not only transfers itself among *Bacteroides* species, but also transfers *Bacteroides–E. coli* shuttle vectors from *Bacteroides* donors to *E. coli* or *Bacteroides* recipients. Transfer frequency is enhanced by growth of the donor in the presence of tetracycline (Shoemaker *et al.*, 1986b). In the case of *Bact. uniformis* 0061, grown in the presence of tetracycline, it appears that Tc^{r}ERL mediates the excision of two plasmid-like elements from the *Bacteroides* chromosome (Shoemaker and Salyers, 1988). A further element, Tn*4399*, does not encode a known antibiotic resistance but can transpose from chromosome to chromosome and to plasmids, and mobilizes a non-conjugal plasmid from *Bact. fragilis* to *E. coli* (Hecht and Malamy, 1989).

8.3.3 Cloned *Lactobacillus* DNA

Cloning of several *Lactobacillus* genes has been accomplished, the experimental details of which can be summarized as follows.

(a) Genes involved in lactose metabolism by Lactobacillus casei

Lactose metabolism in *L. casei* commonly occurs *via* phosphoenolpyruvate-dependent phosphotransferase uptake of lactose and subsequent cleavage of lactose-6-phosphate by β-D-phosphogalactoside galactohydrolase (phosphogalactosidase). The phosphogalactosidase gene has been cloned from plasmids in two different *L. casei* strains. The gene was located on a 2.4 to 2.9 kb fragment of plasmid DNA. A phosphogalactosidase gene has been sequenced, revealing an open reading frame of 1422 base pairs capable of encoding a protein of 474 amino acids. The protein has a high degree of homology to the proteins whose sequences have been deduced from the nucleotide sequence of the phosphogalactosidase genes of *Staphylococcus aureus* and *Lactococcus lactis*, suggesting a common origin for the phosphogalactosidases of these three organisms. A cryptic phosphogalactosidase gene located on the chromosome of *L. casei* MSK248 has also been cloned in *E. coli* (Chassy and Alpert, 1989; Lee *et al.*, 1982; Porter and Chassy, 1988; Shimizu-Kadota, 1988).

The lactose-specific factor III, a cytoplasmic protein capable of binding to the cytoplasmic membrane and involved in the phosphoenolpyruvate-dependent phosphotransferase system, is encoded by an open reading frame of 336 base pairs on the *L. casei* plasmid pLZ64. The gene has been cloned in *E. coli*. The factor III protein of *L. casei* has high homology with that of *Staph. aureus* (Alpert and Chassy, 1988).

(b) β-Galactosidase of **L. delbrueckii** *subsp.* **bulgaricus**

The *L. delbrueckii* subsp. *bulgaricus* β-galactosidase gene has been cloned in *E. coli*, and nucleotide sequenced. From amino acid sequence alignments, the lactobacillus β-galactosidase has a 30 to 34% similarity to the *E. coli lacZ* gene, *E. coli eqbA* and *Klebsiella pneumoniae lacZ* β-galactosidase enzymes. Active-site amino acids are conserved among the enzymes, and regions of high homology between the β-galactosidases suggest that they evolved from a common ancestral gene (Schmidt *et al.*, 1989).

(c) Malolactic fermentation **by L. delbrueckii**

The gene responsible for the malolactic fermentation of wine has been cloned in *E. coli* and in the yeast *Saccharomyces cerevisiae*. The gene encodes the malolactic enzyme that catalyses the conversion of L-malate to L-lactate. The gene was cloned in *E. coli* using plasmid pBR322 and two clones able to convert L-malate to L-lactate were selected for further studies. Both clones contained a 5-kilobase segment of *L. delbrueckii* chromosome. The segment was subcloned to an *E.coli*–yeast shuttle vector, and gene expression was assessed for both types of host. *E. coli* cells containing the malolactic gene converted about 10% of L-malate in the medium to L-lactate, *Sac. cerevisiae* converted about 1% of L-malate, while *L. delbrueckii* converted 25% (Williams *et al.*, 1984).

(d) Dihydrofolate reductase in **L. casei**

The nucleotide sequence of the dihydrofolate gene of a methotrexate-resistant strain of *L. casei* has been determined. Comparison of the amino acid sequence of the enzyme produced by *L. casei*, *E. coli* and *Enterococcus faecium* suggests a strong conservation of amino acid sequences (Andrews *et al.*, 1985).

(e) Histidine decarboxylase of **Lactobacillus** *30A*

The structural gene *hdcA* that encodes the enzyme histidine decarboxylase of *Lactobacillus* 30A has been cloned and sequenced. A second gene, *hcdB*, that forms an operon with *hcdA* has also been cloned and sequenced (Copeland, *et al.*, 1989; Vanderslice *et al.*, 1986).

(f) D-2-Hydroxyisocaproic acid dehydrogenase from L. casei

An oligonucleotide probe corresponding to the *N*-terminal 23 amino acids of the enzyme was synthesized and used to locate a region of the chromosome containing the dehydrogenase gene. The gene was cloned using pBR322, in *E. coli*. The gene encodes a protein subunit of 38 kDa (Lerch *et al*., 1989).

(g) Tryptophan genes of L. casei

Five *trp* genes were cloned in auxotrophic mutants of *E. coli*. The genes appear to constitute an operon and are located in the order *trpD*, *trpC*, *trpF*, *trpB* and *trpA* in a segment of DNA of 6468 base pairs. The entire nucleotide sequence of this segment was determined. Five contiguous open-reading frames in this segment encode proteins containing 341, 260, 199, 406 and 266 amino acids respectively. The amino acid sequences of these proteins exhibit 25 to 50% homology with the sequences of the corresponding *trp* enzymes of *E. coli* (Natori *et al*., 1990).

(h) Serine t-RNA gene of L. delbrueckii *subsp.* bulgaricus

A gene encoding a serine t-RNA with the anticodon CGA was isolated from a clone bank in *E. coli*, a host in which the gene is expressed and specifically complements the *E. coli leuB6* mutation. In other studies, a cloned fragment of chromosomal DNA from *L. delbrueckii* subsp. *bulgaricus* N123 was observed to complement the *leu291* lesion of *E. coli* GE891. The cloned DNA, in addition, served as a specific probe to recognize *L. delbrueckii* subspp. *bulgaricus*, *delbrueckii* and *lactis* (Delley *et al*., 1990; Hottinger *et al*., 1987).

A major difficulty in studying the molecular genetics of lactobacilli has been their lack of a natural competency for transformation with exogenous DNA. The recent development of electrotransformation (electroporation) techniques for use with lactobacilli has overcome this problem. In electrotransformation, a high-voltage electric discharge is passed through a suspension of bacterial cells. Transient ‘pores’ are formed in the cell membrane at an electric field above a certain threshold level, permitting the entry of DNA molecules. Bacterial strains differ in the voltage and other factors that must be used to obtain transformation. Optimal electrotransformation conditions must therefore be determined for each bacterial strain. Transduction of lactobacilli with plasmid DNA has also been accomplished so that, together with conjugation, the three classical methods of introducing DNA into bacterial cells are available for use in studies of lactobacilli (reviewed by Chassy, 1987; Luchansky *et al*., 1988; Raya *et al*., 1989).

8.3.4 Cloned Bacteroides DNA

Some *Bacteroides* genes have been cloned, and randomly cloned chromosomal fragments have been used as DNA probes for the detection of *Bacteroides* in digestive tract samples.

(a) The glutamine synthetase gene of Bact. fragilis

A glutamine synthetase gene, *gln*A, from *Bact. fragilis* was cloned in *E. coli*. The cloned gene enabled *gln*A deletion mutants to utilize ammonium sulphate as a sole source of nitrogen (Southern *et al.*, 1986).

(b) The pilin gene of Bact. nodosus

In an attempt to derive a simplified vaccine to immunize sheep against foot rot, the gene encoding the synthesis of pilin, the monomeric unit of which the *Bact. nodosus* pili are composed, was cloned in *E. coli*. Pilin was synthesized by the *E. coli* host, but surface structural pili were not produced (Elleman *et al.*, 1986).

(c) The chondroitin l yase II gene of Bact. thetaiotaomicron

One of the numerically predominant *Bacteroides* species inhabiting the human colon, *Bact. thetaiotaomicron*, is capable of fermenting a variety of carbohydrates, including chondroitin sulphate. Mucopolysaccharides such as chondroitin sulphate are likely to be substrates for *Bact. thetaiotaomicron* inhabiting the intestinal tract because they are constituents of the host's mucosa. Bacterial metabolism of chondroitin sulphate requires a number of inducible proteins, including two unrelated lyases (I and II) that break the polysaccharide into disaccharides. The gene (3.3 kb) for chondroitin lyase II has been cloned in *E. coli* and active enzyme was synthesized by the *E. coli* host. The protein produced by the *E. coli* host differed, however, in some properties compared to the enzyme synthesized by *Bact. thetaiotaomicron*. The reasons for these differences are not known (Guthrie *et al.*, 1985; Kotarski *et al.*, 1985; Linn *et al.*, 1983).

(d) Endoglucanases and xylanases of Bact. (Fibrobacter) succinogenes

A *Bact. succinogenes* DNA library was prepared in *E. coli* and clones were screened for β-glucanase activity. Six clones giving activity on β-glucans containing β-(1–3) (1–4) linkages were obtained and were found to contain 5.2 kb inserts of *Bact. succinogenes* DNA (Irvin and Teather, 1988). In another study, an endoglucanase gene from *Bact. succinogenes* was cloned in *E. coli* that then exhibited high enzymic activity on carboxymethylcellulose, *p*-nitrophenylcellobioside, and lichenan (Taylor *et al.*, 1987). A gene encoding a xylanase has also been cloned from *Bact. succinogenes* into *E. coli*. The gene expressed in *E. coli* is fully functional and encodes an enzyme with similar properties to that of the bacteroides (Sipat *et al.*, 1987).

(e) DNA probes

DNA probes specific for the detection of *Bact. uniformis, Bact. distasonis, Bacteroides* group 3452-A, *Bact. ovatus* and *Bact. thetaiotaomicron* have been prepared by cloning random fragments of purified chromosomal DNA from each of these species in pBR322. The probes were used to enumerate the species in human faecal samples, an important development since the population sizes of these species had not previously been determined due to technical difficulties in differentiating *Bacteroides* isolates by conventional biochemical tests (Kuritza and Salyers, 1985; Kuritza *et al.*, 1986). A cloned fragment of chromosomal DNA has also been used as a probe specific for the detection of *Bact. ruminicola* in the rumen (Attwood *et al.*, 1988).

8.3.5 Mutagenesis

Chemical substances that alter bases already incorporated into DNA and thus change their hydrogen bonding cause mutations. N-methyl-N-nitro-N-nitrosoguanidine (NNG) is well known in this respect since it is possible, using this alkylating agent, to produce a bacterial population in which 1% is mutant for any particular gene. A disadvantage of chemical mutagenesis, however, is that a bacterial cell is likely to contain mutations in more than one gene. The desired mutation must therefore be separated from unwanted mutations by further genetic manipulations. More precise methods of mutagenesis are preferable: those in which the nucleotide base sequence of a specific region of DNA can be altered by the insertion of exogenous DNA. In this respect, transposon-mediated mutagenesis and targeted, insertional mutagenesis are two methods of potential importance for the genetic modification of gastrointestinal bacteria.

Transposons are segments of DNA that are endowed with the ability to transpose from site to site in DNA of chromosomal, plasmid, or bacteriophage origin. Composite transposons are composed of two identical insertion sequence elements (IS elements) flanking additional DNA that often encodes antibiotic resistance. Transposons can cause mutations when they insert into genes, and bacterial strains that have acquired a transposon can be easily recognized and selected if the transposon confers an antibiotic resistance phenotype on the bacterial cells. Labelled-DNA probes prepared from the transposon can then be used to pinpoint mutations, and consequently genes encoding specific phenotypes, in cloned fragments of bacterial DNA. The introduction of transposons into bacterial cells can be achieved by using plasmids as vectors, but some transposons from Gram-positive bacteria are capable of self-transmission i.e are conjugative (Clewell and Gawron-Burke, 1985).

Targeted, insertional mutagenesis utilizes constructed plasmid vectors that are unable to replicate in the host cell into which they are introduced for the purposes of mutagenesis. The vector cannot replicate because, by design, it lacks an origin of replication that will function in the new host cell. Such plasmids are often referred to as 'suicide' plasmids. Although lacking an appropriate means of initiating replication, the plasmid can insert, by recombination, into the bacterial chromosome because it contains cloned DNA homologous with a region of the recipient cell's chromosome. In addition, the vector encodes a selectable phenotype so that cells that have been transformed with the plasmid, and in which insertion of the plasmid into the bacterial chromosome has occurred, can be recognized. Insertion of the vector into the chromosome disrupts the normal nucleotide base sequence so that a mutation results. The insertion event can be targeted to a particular region of the chromosome by virtue of the choice of fragment of DNA that is cloned in the vector.

8.3.6 Lactobacilli and transposons

The conjugative transposon Tn*919* has been shown to transfer from *Ent. faecalis* GF590 to a strain of *L. plantarum* and, from the same donor strain, to *L. curvatus* Lc 2-c (Knauf *et al.*, 1989; Hill *et al.*, 1985). The transposition of Tn*917* into indigenous plasmid DNA of *L. plantarum* strains following electrotransformation with plasmid pTV1 has been reported on two occasions (Aukrust and Nes, 1988; Cosby *et al.*, 1989). Transposon-mediated mutagenesis has not yet been reported, however, for lactobacilli.

8.3.7 Bacteroides and transposons

Transposon-mediated mutagenesis has been used successfully in *Bacteroides* species. Transposon Tn*4351* was introduced into *Bact. uniformis* 0061 on suicide plasmid vectors. The plasmids were transferred from *E. coli* to bacteroides-recipient cells by conjugation, and erythromycin-resistant transconjugants were selected. In the resistant transconjugants, Tn*4351* was found to have transposed from the suicide vector to the chromosome and, in some isolates, the transposon had mediated the integration of the entire plasmid vector into the chromosome (Shoemaker *et al.*, 1986a). Mutants of *Bact. thetaiotaomicron*, unable to utilize chondroitin sulphate, have been derived by the transposition of Tn*4351* to the *Bacteroides* chromosome. The mutants were subsequently used in experiments utilizing gnotobiotic animals to determine the significance of chondroitin sulphate utilization as a colonization factor (Salyers *et al.*, 1988).

8.3.8 Bacteroides and targeted, insertional mutagenesis

As described above, *Bact. thetaiotaomicron* produces two chondroitin lyase enzymes (I and II). Chondroitin lyase II accounts for about two-thirds of the chondroitin lyase activity in the cell. To answer the question as to whether both lyases are required to degrade chondroitin sulphate, a mutant strain unable to synthesize lyase II was derived by targeted, insertional mutagenesis. This was accomplished by cloning a segment of the chondroitin lyase II gene into a constructed suicide plasmid vector pE-3. Specifically, plasmid pE-3 contained a 4.4 kb cryptic *Bact. eggerthii* plasmid (pB8-51), the *E. coli* cloning vector pBR328, and the *Eco*RI D fragment from the conjugative *Bact. fragilis* plasmid pBF4. A 0.8 kb fragment of DNA from the centre of the *Bact. thetaiotaomicron* chondroitin lyase II gene was inserted into pE-3 to create plasmid pEG817. Although pEG817 was stably maintained in *E. coli* and could be mobilized into *Bact. thetaiotaomicron* by the IncP plasmid R751, pEG817 was not maintained as a plasmid in *Bacteroides* species. Instead, plasmid pEG817, under conditions of selective antibiotic pressure, inserted into the chondroitin lyase II gene by homologous recombination. The mutant bacteroides so obtained did not synthesize chondroitin lyase II, yet were still able to grow with chondroitin sulphate as the sole carbon source. Thus chondroitin lyase I appears to be sufficient for growth of the bacteria on chondroitin sulphate, at least under *in vitro* conditions (Guthrie and Salyers, 1986).

8.4 STABILITY OF GENETIC DETERMINANTS

Although much remains to be done in developing cloning, DNA transfer methodologies and mutagenesis for members of the gastrointestinal microflora, the basic technologies are more or less in place and only time and financial commitment are required for genetically modified members of the normal microflora to be derived. Emphasis in the future needs to be placed on the development of methods by which bacterial characteristics, resulting from the introduction of novel genes, can be stably inherited by the genetically modified strains. In other words, the newly introduced genes must be maintained stably by the bacterial host and transmitted to its progeny under both industrial conditions of culture and in the gastrointestinal environment. Recombinant plasmids capable of autonomous replication in the bacterial cytoplasm will probably not prove a satisfactory means of deriving genetically modified strains for use in probiotic products. This is because recombinant plasmids are often not stably maintained in bacterial cells, even under laboratory conditions, and are therefore likely to be quickly lost under the harsher conditions prevailing in natural habitats (Davies and Gasson, 1981). Retention of recombinant plasmids would be encouraged, however, if a genetic determinant essential to the survival of the bacterium was incorporated into the recombinant molecule along with the DNA encoding novel characteristics. Recent investigations of the transmission of antibiotic resistance determinants between bacterial genera, however, have demonstrated that intergeneric transfer of DNA is common (Trieu-Cuot *et al.*, 1987; Roberts and Hillier, 1990). Thus recombinant plasmids might be transferred widely within the normal microflora of the gastrointestinal tract; a consideration unlikely to please governmental or scientific agencies concerned with the regulation of the release of genetically modified bacteria.

A more satisfactory solution to the problem of stability would be to insert the novel genetic determinants into the chromosome of the bacterial host (Leenhouts *et al.*, 1990). The newly acquired DNA would then be replicated as part of the bacterial chromosome and hence would be partitioned between, and maintained by, daughter cells. Insertable (integrable) plasmid vectors can be used for cloning and stablizing the inheritance of heterologous genes. These vectors are used to clone exogenous genes and to then transform bacterial cells. Plasmid integration vectors are not maintained in transformed recipient cells as autonomous replicons because they lack an origin of replication that can function in the bacterial recipient. In other words, they are suicide vectors. The plasmids become established as an inheritable factor, however, following recombination between DNA sequences, also cloned in the plasmid, that are homologous with a region of the recipent's chromosome. An additional benefit of chromosomal insertion is that multiple copies of the inserted DNA may eventuate (gene amplification). A specific example is provided by the integration and expression of α-amylase and endoglucanase genes in the chromosome of a *L. plantarum* strain. In this work, a strain of *L. plantarum* was electrotransformed with plasmids containing an α-amylase gene from *Bacillus stearothermophilus* and an endoglucanase gene from *Clostridium thermocellum*. Both genes were expressed in the lactobacillus and the active enzymes were excreted by the bacterial cells into the culture medium. The plasmids, however, were of low stability in the lactobacillus host. Plasmids were, therefore, constructed that contained fragments (2–5 kb pr) of lactobacillus chromosome, an origin of replication recognized only by Gram-negative bacteria, and the α-amylase and endoglucanase genes. Following electrotransformation of the *L. plantarum* strain with this type of plasmid, the plasmid DNA constructs inserted into the chromosome by means of single homologous recombinational events (Campbell-type recombination). The majority of the transformants had a stable phenotype, and endoglucanase activity was detected in the culture supernatants. α-Amylase expression, however, was poor. Obvious secondary mutations were not generated. Two transformants contained up to ten copies of the DNA construct in their chromosome (Scheirlinck *et al.*, 1989). Further plasmid-based integration vectors for lactobacilli could doubtless be constructed using undefined chromosomal fragments, or DNA sequences from *Lactobacillus* genes that have already been cloned. Insertion sequence elements (a DNA sequence flanked by short inverted nucleotide sequence repeats) such as ISL1 of *L. casei*, and transposons, additionally offer potential for plasmid integration into any region of the chromosome containing an insertion site for the transposable DNA (Shimizu-Kadota *et al.*, 1988). Integration of a plasmid vector can also be facilitated using DNA from a lysogenic bacteriophage to provide homologous sequences for plasmid–chromosome recombination as described for *Lactococcus lactis* (Chopin *et al.*, 1989). Lysogeny is widespread in the genus *Lactobacillus* (Sozzi *et al.*, 1981; Stetter, 1977; Yokokura *et al.*, 1974) and it is likely that bacteriophage DNA could be used to direct site-specific integration of plasmid vectors into the lactobacillus chromosome.

8.5 POSSIBLE DEVELOPMENTS

The types of probiotic that could be developed through the use of molecular genetics can be grouped as follows.

8.5.1 Immunizing strains

A suitable member of the normal microflora could be genetically modified so that its cells synthesized an immunogen characteristic of a particular intestinal pathogen. Colonization of the gastrointestinal tract with such a strain could result in a continuous exposure of the intestinal mucosa to the immunogen so that secretory IgA antibodies would be synthesized by the host animal. This could result in immunity of the host to the specific intestinal pathogen. The genetically modified member of the microflora could be transmitted from mother to infant, thus perpetuating the cycle of immunization without further attention from a medical or veterinary practitioner. The concept of genetically modifying bacteria or viruses so that they synthesize a variety of immunogens has already been applied to avirulent *Salmonella* strains and to vaccinia virus, but the use of members of the normal microflora may be aesthetically more pleasing to the general public (Clements *et al.*, 1986; Formal *et al.*, 1981; Tartaglia *et al.*, 1990).

8.5.2 Delivery strains

Biotechnological research has led to the derivation of genetically modified bacterial and yeast strains that can be cultivated in large volume under industrial conditions. The modified organisms synthesize large amounts of substances of medical or industrial significance and these substances are subsequently harvested and purified from the cultures. It may be possible to derive strains of gastrointestinal or other microbes that similarly synthesize novel substances following genetic modification, but that can be used to deliver molecules with biological activity to particular regions of the intestinal tract without the need to industrially purify the beneficial substance in question. For example, a microbial strain whose cells lysed under specific conditions of surfactant concentration and/or pH might be derived for incorporation into foods so that the microbial cells would deliver a product to a specific level of the intestinal tract (Klaenhammer and Kleeman, 1981).

8.5.3 Nutrition

The contribution of the rumen microflora to the nutritional well-being of ruminants is well known. Consortia of microbes inhabiting the rumen degrade plant materials ingested by the animal host and ferment the degradation products to produce short-chain fatty acids that constitute the major energy source for ruminants. Microbial cells washed into the abomasum and intestinal tract are digested by host processes, liberating the contents of bacterial cells so that they become available for host nutrition. Thus ruminants, because of the biochemical versatility of their gastrointestinal microbes, can thrive on forage of relatively low quality (Hobson and Wallace, 1982).

The microbial digestion of plant materials also occurs in the large bowel of animals, including humans. The fermentation products produced in the large bowel are similar to those detected in the rumen. The nutritional value of the large bowel fermentation is questionable in the case of non-coprophagic animals, however, since the microbial products are produced distal to the small bowel where most absorption of nutrients occurs. The caecum of some animal species makes a major contribution to total body weight, however, suggesting that retention of a large, sacculated bowel harbouring numerous microbes and food residues is of advantage to the animal host (McBee, 1977).

Probiotics to aid host nutrition might best be developed for use with ruminants since the rumen ecosystem has been investigated intensively from both biochemical and microbiological aspects, and the importance of microbial activities in digesting components of the host's food is clear. Microbial strains that would permit the ruminant to more fully utilize the energy contained within its diet could be derived and introduced into the rumen. The derivation of microbes that produce increased amounts of cellulolytic enzymes, for example, would be of considerable interest since only about 60% of cellulose entering the rumen is degraded. Other possibilities include the derivation of microbial strains that synthesize large quantities of amino acids that are growth-limiting for the ruminant, or strains that have enhanced ability to convert lactic acid to propionic acid thus avoiding lactic acidosis in the host under conditions of radical change in diet, e.g. high fibre to high grain diets (Teather, 1985).

Genetically modified microbial strains could also be useful in the inactivation of toxic molecules liberated from the diet during the digestion of food in the gastrointestinal tract. *Leucaena leucocephala* is a tropical leguminous shrub that produces high yields of protein-rich forage in tropical and subtropical regions. Diets rich in *Leucaena* have been associated with ruminant health problems in Northern Australia because of a breakdown product of the amino acid mimosine present in the plant material. Autolytic or microbial degradation of mimosine produces a potent goitrogen, 3-hydroxy-4 (1H) pyridone (DHP). Ruminants ingesting a *Leucaena* diet in Hawaii or Indonesia, however, do not exhibit the symptoms of hypothyroidism seen in Australian ruminants and excrete very little DHP in their urine. It is likely that Hawaiian and Indonesian ruminants harbour rumen microbes capable of detoxifying DHP since inoculation of Australian ruminants with rumen fluid from Indonesian animals permits the ingestion of a *Leucaena* diet without toxicity problems (Allison *et al.*, 1990; Jones and Lowry, 1984). These observations raise the interesting possibility of manipulating the rumen microflora of an animal so that it is protected from the toxic substances present in potentially useful pasture plants introduced from another country. The genetic modification of rumen bacteria so that they can inactivate molecules toxic to the host is likely to be part of this process.

8.6 RELEASE OF GENETICALLY MODIFIED MICROBES

The use of genetically modified bacteria as probiotics will involve the deliberate release of the microbes into the natural environment.

Although microbes in nature have all of the molecular equipment necessary to do their own genetic engineering, and although substantial gene rearrangements are known to occur under natural conditions, fears have been expressed that microbes genetically engineered in the laboratory will be potentially hazardous to mankind (Jain and Sayler, 1987). Scientists themselves first raised concerns that recombinant DNA technology might have undesirable environmental and public health consequences. These concerns were summarized and discussed, in relation to agriculture and environmental usage, by the Committee for Scientific and Technological Policy of the Organization for Economic Cooperation and Development (Anon., 1986) as follows.

> Concerns about micro-organisms derived by rDNA techniques released into the environment include the possibility that the genetic modification might affect their host range, affect their capacity to utilize substrates such as nitrogen or lignin, convert them into pathogens, and/or alter the balance between them and ecologically interrelated populations in the ecosystem.

Since pathogenic organisms have the most obvious impact on human and agricultural systems, a major concern is the possibility that genetic modifications might convert non-pathogenic micro-organisms to pathogens or alter host range or virulence of pathogens used to control plant, insect, or other pests. Studies with pathogens have demonstrated that many genes must interact appropriately for a microbe to cause disease. The pathogen must possess and appropriately express characteristics such as recognition factors, adhesion ability, toxigenicity, and resistance to the host defence system. Single gene modifications of organisms with no pathogenic potential or history, or introduction of several genes not contributing to pathogenicity, do not appear likely to result in unanticipated pathogenicity. Moreover, wide experience and extensive data on pathogenicity exist which can be used to define the parameters of concern when considering the effects of rDNA modifications.
The committee concluded that, while recombinant DNA technology may be used to derive organisms with combinations of genes not observed in nature, the genetic changes will have greater predictability compared with traditional techniques of deriving strains with altered phenotype (hybrids). This is because of the greater precision that recombinant DNA technology affords to specific modifications. Finally, the committee concluded that the risks associated with the release of genetically modified organisms into the environment could be assessed in generally the same way as those associated with the release of non-modified organisms.

Many governments have established regulations or guidelines, and regulatory bodies to assess the risks involved in releasing genetically modified organisms. In New Zealand, for example, risk assessment is carried out by the Interim Assessment Group for the Field Testing and Release of Genetically Modified Organisms. The Assessment Group considers both the characteristics of the organism and its ecological and environmental characteristics. In the case of field trials involving genetically modified plants and animals, emphasis is placed on monitoring and on measures to minimize the risk of heritable material such as pollen from leaving the trial site. Genetically modified bacteria are assessed in a two-stage process: first by the Advisory Committee on Novel Genetic Techniques that categorizes the potential risk at the laboratory trial stage, and then by the Interim Assessment Group. The Brenner categorization scheme is used whereby three aspects of risk are assessed.

(a) Access factor: the probability of the organism escaping and entering the human body, surviving and entering susceptible tissues.
(b) Expression factor: an estimate of how efficiently the new gene will be expressed.
(c) Damage factor: the probability that the product of the foreign gene will cause physiological damage in the body of the individual to which it gains access.

In the period August 1988 to June 1990, the Interim Assessment Group has assessed 21 applications for trials involving the release of genetically modified organisms, an indication of the growing interest in the use of recombinant technology in agriculture and biotechnology. Most of these proposals concerned modified plant species; three concerned genetically modified strains of *E. coli* (Roberts, 1990). An important concern of the Interim Assessment Group is that the public of New Zealand are notified of proposed trials and time is allowed for submissions to be made. Judging from recent large surveys of the attitudes of New Zealanders and Americans to genetic engineering, a more positive attitude towards the utilization of genetically modified organisms in specific environmental or therapeutic applications is emerging (Couchman and Fink-Jensen, 1990). The acceptance of the use of recombinant microbes in the preparation of foods (chymosin from *E. coli*; the mycoprotein retailed as 'Quorn'; genetically modified yeast for breadmaking) provides a source of optimism for the use of genetically modified microbes in the development of probiotics. It is apparent, however, that responsibility for increased public awareness and approval of recombinant DNA technology rests with the scientific community which must develop more effective means of dialogue with the general population.

8.7 CONCLUSIONS

1. The characteristics of 'ideal' probiotic strains can be predicted.
2. The potential for the derivation of ideal probiotic strains using recombinant DNA technology exists.
3. The technology by which genetically modified strains of gastrointestinal bacteria can be derived, and by which novel genes can be integrated into the bacterial host's genome, are available.
4. Specific applications of recombinant technology in developing probiotic strains can be proposed.
5. Public fears concerning the release of genetically modified organisms into the environment must be reduced by education of the public concerning the regulatory procedures that must be followed before trials can be undertaken, the specific experiments to be undertaken, and the specific benefits that will accrue as a result of successful trials.

REFERENCES

Allison, M.J., Hammond, A.C. and Jones, R.J. (1990) Detection of ruminal bacteria that degrade toxic dihydroxypyridine compounds produced from mimosine. *Appl. Environ. Microbiol.*, **56**, 590–4.

Alpert, C.–A. and Chassy, B.M. (1988) Molecular cloning and nucleotide sequence of the factor III^{lac} gene of *Lactobacillus casei*. *Gene*, **62**, 277–88.

Andrews, J., Clore, G.M., Davies, R.W. *et al.* (1985) Nucleotide sequence of the dihydrofolate reductase gene of methotrexate-resistant *Lactobacillus casei*. *Gene*, **35**, 217–22.

Anon. (1986) *Recombinant DNA Safety Considerations*, OECD, Paris.

Attwood, G.T., Lockington, R.A., Xue, G.–P. and Brooker, J.D. (1988) Use of a unique gene sequence as a probe to enumerate a strain of *Bacteroides ruminicola* introduced into the rumen. *Appl. Environ. Microbiol.*, **54**, 534–9.

Aukrust, T. and Nes, I.F. (1988) Transformation of *Lactobacillus plantarum* with the plasmid pTV1 by electroporation. *FEMS Microbiol. Lett.*, **52**, 127–32.

Axelsson, L.T., Ahrne, S.E.I., Andersson, M.C. and Stahl, S.R. (1988) Identification and cloning of a plasmid-encoded erythromycin resistance determinant from *Lactobacillus reuteri*. *Plasmid*, **20**, 171–4.

Barnes, E.M., Impey, C.S. and Cooper, D.M. (1980) Manipulation of the crop and intestinal flora of the newly hatched chick. *Amer. J. Clin. Nutr.*, **33**, 2426–33.

Bates, E.E.M. and Gilbert, H.J. (1989) Characterization of a cryptic plasmid from *Lactobacillus plantarum*. *Gene*, **85**, 253–8.

Beul, H.A, Eichel-Streiber, C., Schreiner, M. *et al.* (1985) Characterization of cryptic plasmids in clinical isolates of *Bacteroides fragilis*. *J. Med. Microbiol.*, **20**, 39–48.

Bouia, A., Bringel, F., Frey, L. *et al.* (1989) Structural organization of pLP1, a cryptic plasmid from *Lactobacillus plantarum* CCM 1904. *Plasmid*, **22**, 185–92.

Bringel, F., Frey, L. and Hubert, J.–C. (1989) Characterization, cloning, curing, and distribution in lactic acid bacteria of pLP1, a plasmid from *Lactobacillus plantarum* CCM 1904 and its use in shuttle vector construction. *Plasmid*, **22**, 193–202.

Callihan, D.R., Young, F.E. and Clark, V.L. (1983) Identification of three homology classes of small, cryptic plasmids in intestinal *Bacteroides* species. *Plasmid*, **9**, 17–30.

Chassy, B.M. (1987) Prospects for the genetic manipulation of lactobacilli. *FEMS Microbiol. Rev.*, **46**, 297–312.

Chassy, B.M. and Alpert, C.–A. (1989) Molecular characterization of the plasmid-encoded lactose-PTS of *Lactobacillus casei*. *FEMS Microbiol. Rev.*, **63**, 157–66.

Chopin, M.–C, Chopin, A., Rouault, A. and Galleron, N. (1989) Insertion and amplification of foreign genes in the *Lactococcus lactis* subsp. *lactis* chromosome. *Appl. Environ. Microbiol.*, **55**, 1769–74.

Clements, J.D., Lyon, F.L, Lowe, K.L., *et al.* (1986) Oral immunization of mice with attenuated *Salmonella enteritidis* containing a recombinant plasmid which codes for production of the B subunit of heat labile enterotoxin of *Escherichia coli*. *Infect. Immun.*, **53**, 685–92.

Clewell, D.B. and Gawron-Burke, C. (1986) Conjugative transposons and the dissemination of antibiotic resistance in streptococci. *Ann. Rev. Microbiol.*, **40**, 635–59.

Copeland, W.C., Domena, J.D. and Robertus, J.D. (1989) The molecular cloning, sequence and expression of the *hcdB* gene from *Lactobacillus* 30A. *Gene*, **85**, 259–65.

Cosby, W.M., Axelsson, L.T. and Dobrogosz, W.J. (1989) Tn917 transposition in *Lactobacillus plantarum* using the highly temperature-sensitive plasmid pTV1Ts as a vector. *Plasmid*, **22**, 236–43.

Couchman, P.K. and Fink-Jensen, K. (1990) *Public Attitudes to Genetic Engineering*, DSIR Crop Research, Christchurch.

Damiani, G., Romagnoli, S., Ferretti, L. *et al.* (1987) Sequence and functional analysis of a divergent promoter from a cryptic plasmid of *Lactobacillus acidophilus* 168 S. *Plasmid*, **17**, 69–72.

Davies, F.L. and Gasson, M.J. (1981) Reviews of the progress of dairy science: genetics of lactic acid bacteria. *J. Dairy Res.*, **48**, 363–76.

Delley, M., Mollet, B. and Hottinger, H. (1990) DNA probe for *Lactobacillus delbrueckii*. *Appl. Environ. Microbiol.*, **56**, 1967–70.

Drasar, B.S. and Barrow, P.A. (1985) *Intestinal Microbiology*, Van Nostrand Reinhold, Wokingham.

Elleman, T.C., Hoyne, P.A., Emery, D.L. *et al.* (1986) Expression of the pilin gene from *Bacteroides nodosus* in *Escherichia coli*. *Infect. Immun.*, **51**, 187–92.

Finegold, S.M., Attebery, H.R. and Sutter, V.L. (1974) Effect of diet on human fecal flora: comparison of Japanese and American diets. *Amer. J. Clin. Nutr.*, **27**, 1456–69.

Formal, S.B., Baron, L.S., Kopecko, D.J. *et al.* (1981) Construction of a potential bivalent vaccine strain: introduction of *Shigella sonnei* Form I antigen genes into the galE *Salmonella typhimurium* Ty21a typhoid vaccine strain. *Infect. Immun.*, **34**, 746–50.

Guthrie, E.P., Shoemaker, N.B. and Salyers, A.A. (1985) Cloning and expression in *Escherichia coli* of a gene coding for a chondroitin lyase from *Bacteroides thetaiotaomicron*. *J. Bacteriol.*, **164**, 510–515

Guthrie, E.P. and Salyers, A.A. (1986) Use of targeted insertional mutagenesis to determine whether chondroitin lyase II is essential for chondroitin sulfate utilization by *Bacteroides thetaiotaomicron*. *J. Bacteriol.*, **166**, 966–71.

Hecht, D.W. and Malamy, M.H. (1989) Tn*4399*, a conjugal mobilizing transposon of *Bacteroides fragilis*. *J. Bacteriol.*, **171**, 3603–8

Hill, C., Daly, C. and Fitzgerald, G.F. (1985) Conjugative transfer of the transposon Tn*919* to lactic acid bacteria. *FEMS Microbiol Lett.*, **30**, 115–19.

Hill, H.A. and Hill, J.E. (1986) The value of plasmid profiling in monitoring *Lactobacillus plantarum* in silage fermentations. *Curr. Microbiol.*, **13**, 91–4.

Hobson, P.N. and Wallace, R.J. (1982) Microbial ecology and activities in the rumen, Parts I and II. *CRC Crit. Rev. Microbiol.*, **9**, 165–225 and 253–320.

Honeyfield, D.C. and Carlson, J.R. (1990) Effect of indoleacetic acid and related indoles on *Lactobacillus* sp. strain 11201 growth, indoleacetic acid catabolism, and 3-methylindole formation. *Appl. Environ. Microbiol.*, **56**, 1373–7.

Hottinger, H., Ohgi, T., Zwahlen, M.–C. *et al.* (1987) Allele-specific complementation of an *Escherichia coli leuB* mutation by a *Lactobacillus bulgaricus* tRNA gene. *Gene*, **60**, 75–83.

Irvin, J.E. and Teather, R.M. (1988) Cloning and expression of a *Bacteroides succinogenes* mixed-linkage β-glucanase (1, 3–1, 4-β-D-glucan 4-glucanohydrolase) gene in *Escherichia coli*. *Appl. Environ. Microbiol.*, **54**, 2672–6.

Jain, R.K. and Sayler, G.S. (1987) Problems and potential for *in situ* treatment of environmental pollutants by engineered microorganisms. *Microbiol. Sci.*, **4**, 59–63.

Jones, R.J. and Lowry, J.B. (1984) Australian goats detoxify the goitrogen 3-hydroxy-4(1H) pyridone (DHP) after rumen infusion from an Indonesian goat. *Experientia*, **40**, 1435–6.

Josson, K., Scheirlinck, T., Michiels, F. *et al.* (1989) Characterization of a gram-positive broad-host-range plasmid isolated from *Lactobacillus hilgardii*. *Plasmid*, **21**, 9–20.

Klaenhammer, T.R. and Kleeman, E.G. (1981) Growth characteristics, bile sensitivity, and freeze damage in colonial variants of *Lactobacillus acidophilus*. *Appl. Environ. Microbiol.*, **41**, 1461–7.

Knauf, H.J., Vogel, R.F. and Hammes, W.P. (1989) Introduction of the transposon Tn*919* into *Lactobacillus curvatus* Lc2-c. *FEMS Microbiol. Lett.*, **65**, 101–4.

Kotarski, S.F., Linz, J., Braun, D.M. and Salyers, A.A. (1985) Analysis of outer membrane proteins which are associated with growth of *Bacteriodes thetaiotaomicron* on chondroitin sulfate. *J. Bacteriol.*, **163**, 1080–6

Kuritza, A.P. and Salyers, A.A. (1985) Use of a species-specific DNA hybridization probe for enumerating *Bacteroides vulgatus* in human feces. *Appl. Environ. Microbiol.*, **50**, 958–64.

Kuritza, A.P., Shaughnessy, P. and Salyers, A.A. (1986) Enumeration of polysaccharide-degrading *Bacteroides* species in human feces by using species-specific DNA probes. *Appl. Environ. Microbiol.*, **51**, 385–90.

Lee, L.–J, Hansen, J.B., Jagusztyn-Krynicka, E.K. and Chassy, B.M. (1982) Cloning and expression of the βD-phosphogalactoside galactohydrolase gene of *Lactobacillus casei* in *Escherichia coli* K-12. *J. Bacteriol.*, **152**, 1138–46.

Lee-Wickner, L.–J. and Chassy, B.M. (1985) Characterization and molecular cloning of cryptic plasmids isolated from *Lactobacillus casei*. *Appl. Environ. Microbiol.*, **49**, 1154–61.

Leenhouts, K.J., Kok, J. and Venema, G. (1990) Stability of integrated plasmids in the chromosome of *Lactococcus lactis*. *Appl. Environ. Microbiol.*, **56**, 2726-35.

Lerch, H.–P., Blocker, H., Kallwass, H. *et al.* (1989) Cloning, sequencing and expression in *Escherichia coli* of the D-2-hydroxyisocaproate dehydrogenase gene of *Lactobacillus casei*. *Gene*, **78**, 47–57.

Linn, S., Chan, T., Lipeski, L. and Salyers, A.A. (1983) Isolation and characterization of two chondroitin lyases from *Bactoroides thetaoiotaomicron*. *J. Bacteriol.*, **156**, 859–66.

Liu, M.–L., Kondo, J.K., Barnes, M.B. and Bartholomew, D.T. (1988) Plasmid-linked maltose utilisation of *Lactobacillus* sp. *Biochimie*, **70**, 351–5.

Luchansky, J.B., Muriana, P.M. and Klaenhammer, T.R. (1988) Application of electroporation for transfer of plasmid DNA to *Lactobacillus, Lactococcus, Listeria, Pediococcus, Bacillus, Staphylococcus, Enterococcus*, and *Propionibacterium*. *Mol. Microbiol.*, **2**, 637–46.

McBee, R.H. (1977) Fermentation in the hindgut, in *Microbial Ecology of the Gut*, Academic Press, London, pp. 185–222.

Morelli, L., Cocconcelli, P.S., Bottazzi, V. *et al.* (1987) *Lactobacillus* protoplast transformation. *Plasmid*, **17**, 73–5.

Muriana, P.M. and Klaenhammer, T.R. (1987) Conjugal transfer of plasmid-encoded determinants for bacteriocin production and immunity in *Lactobacillus acidophilus* 88. *Appl. Environ. Microbiol.*, **53**, 553–60.

Natori, Y., Kano, Y. and Imamoto, F. (1990) Nucleotide sequences and genomic constitution of five tryptophan genes of *Lactobacillus casei*. *J. Biochem.*, **107**, 248–55.

Odelson, D.A., Rasmussen, J.L., Smith, C.J. and Macrina, F.L. (1987) Extrachromosomal systems and gene transmission in anaerobic bacteria. *Plasmid*, **17**, 87–109.

Porter, E.V. and Chassy, B.M. (1988) Nucleotide sequence of the β-D-phosphogalactoside galactohydrolase gene of *Lactobacillus casei*: comparison to analogous *pbg* genes of other gram-positive organisms. *Gene*, **62**, 263–76.

Raya, R.R., Kleeman, E.G. Luchansky, J.B. and Klaenhammer, T.R. (1989) Characterization of the temperate bacteriophage øadh and plasmid transduction in *Lactobacillus acidophilus* ADH. *Appl. Environ. Microbiol.*, **55**, 2206–13.

Roberts, L. (1990) *Interim Assessment Group for the Field Testing and Release*

of Genetically Modified Organisms, Report, Ministry for the Environment, Wellington.
Roberts, M.C. and Hillier, S.L. (1990) Genetic basis of tetracycline resistance in urogenital bacteria. *Antimicrob. Agents Chemother.*, **34**, 261–4.
Russell, E.G. (1979) Types and distribution of anaerobic bacteria in the large intestine of pigs. *Appl. Environ. Microbiol.*, **37**, 187–93.
Salanitro, J.P., Blake, I.G. and Muirhead, P.A. (1974) Studies on the cecal microflora of commercial broiler chickens. *Appl. Microbiol.*, **28**, 439–47.
Salanitro, J.P., Blake, I.G. and Muirhead, P.A. (1977) Isolation and identification of fecal bacteria from adult swine. *Appl. Environ. Microbiol.*, **33**, 79–84.
Salanitro, J.P., Blake, I.G., Muirhead, P.A. *et al.* (1978) Bacteria isolated from the duodenum, ileum, and cecum of young chicks. *Appl. Environ. Microbiol.*, **35**, 782–90.
Salyers, A.A., Shoemaker, N.B. and Guthrie, E.P. (1987) Recent advances in *Bacteroides* genetics. *CRC Crit. Rev. Microbiol.*, **14**, 49–71.
Salyers, A.A., Pajeau, M. and McCarthy, R.E. (1988) Importance of mucopolysaccharides as substrates for *Bacteroides thetaiotaomicron* growing in intestinal tracts of exgermfree mice. *Appl. Environ. Microbiol.*, **54**, 1970–6.
Savage, D.C. (1977) Microbial ecology of the gastrointestinal tract. *Ann. Rev. Microbiol.*, **31**, 107–33.
Scheirlinck, T., Mahillon, J., Joos, H. *et al.* (1989) Integration and expression of α-amylase and endoglucanase genes in the *Lactobacillus plantarum* chromosome. *Appl. Environ. Microbiol.*, **55**, 2130–7.
Schmidt, B.F., Adams, R.M., Requadt, C. *et al.* (1989) Expression and nucleotide sequence of the *Lactobacillus bulgaricus* β-galactosidase gene cloned in *Escherichia coli*. *J. Bacteriol.*, **171**, 625–35.
Shay, B.J., Egan, A.F., Wright, M. and Rogers, P.J. (1988) Cysteine metabolism in an isolate of *Lactobacillus sake*: plasmid composition and cysteine transport. *FEMS Microbiol. Lett.*, **56**, 183–8.
Shimizu-Kadota, M. (1988) Cloning and expression of the phospho-β-galactosidase genes on the lactose plasmid and the chromosome of *Lactobacillus casei* C257 in *Escherichia coli*. *Biochimie*, **70**, 523–9.
Shimizu-Kadota, M., Flickinger, J.L. and Chassy, B.M. (1988) Evidence that *Lactobacillus casei* insertion element ISL1 has a narrow host range. *J. Bacteriol.*, **170**, 4976–8.
Shoemaker, N.B., Getty, C., Gardner, J.F. and Salyers, A.A. (1986a) Tn*4351* transposes in *Bacteroides* spp. and mediates the integration of plasmid R751 into the *Bacteroides* chromosome. *J. Bacteriol.*, **165**, 929–36.
Shoemaker, N.B., Getty, C., Guthrie, E.P. and Salyers, A.A. (1986b) Regions in *Bacteroides* plasmids pBFTM10 and pB8-51 that allow *Escherichia coli–Bacteroides* shuttle vectors to be mobilized by IncP plasmids and by a conjugative *Bacteroides* tetracycline resistance element. *J. Bacteriol.*, **166**, 959–65.
Shoemaker, N.B. and Salyers, A.A. (1988) Tetracycline-dependent appearance of plasmidlike forms in *Bacteroides uniformis* 0061 mediated by conjugal *Bacteroides* tetracycline resistance elements. *J. Bacteriol.*, **170**, 1651–7.

Shrago, A.W. and Dobrogosz, W.J. (1988) Conjugal transfer of Group B streptococcal plasmids and comobilization of *Escherichia coli–Streptococcus* shuttle plasmids to *Lactobacillus plantarum*. *Appl. Environ. Microbiol.*, **54**, 824–6.

Sipat, A., Taylor, K.A., Lo, R.Y.C. *et al.* (1987) Molecular cloning of xylanase gene from *Bacteroides succinogenes* and its expression in *Escherichia coli*. *Appl. Environ. Microbiol.*, **53**, 477–81.

Skaugen, M. (1989) The complete nucleotide sequence of a small cryptic plasmid from *Lactobacillus plantarum*. *Plasmid*, **22**, 175–9.

Smith, H.W. (1965) Observations on the flora of the alimentary tract of animals and factors affecting its composition. *J. Path. Bact.*, **89**, 95–122.

Southern, J.A., Parker, J.R. and Woods, D.R. (1986) Expression and purification of glutamine synthetase cloned from *Bacteroides fragilis*. *J. Gen. Microbiol.*, **132**, 2827–35

Sozzi, T., Watanabe, K., Stetter, K. and Smiley, M. (1981) Bacteriophages of the genus *Lactobacillus*. *Intervirology*, **16**, 129–35.

Stetter, K.O. (1977) Evidence for frequent lysogeny in lactobacilli: temperate bacteriophages within the subgenus *Streptobacterium*. *J. Virol.*, **24**, 685–9.

Takiguchi, R., Hashiba, H., Aoyama, K. and Ishii, S. (1989) Complete nucleotide sequence and characterization of a cryptic plasmid from *Lactobacillus helveticus* subsp. *jugurti*. *Appl. Environ. Microbiol.*, **55**, 1653–5.

Tannock, G W. (1984) Control of gastrointestinal pathogens by normal flora, in *Current Perspectives in Microbial Ecology*, American Society for Microbiology, Washington, DC, pp. 374–82.

Tannock, G.W. (1989) Biotin-labeled plasmid DNA probes for detection of epithelium-associated strains of lactobacilli. *Appl. Environ. Microbiol.*, **55**, 461–4.

Tannock, G W. (1990) The microecology of lactobacilli inhabiting the gastrointestinal tract. *Adv. Micro. Ecol.*, **11**, 147–71.

Tannock, G.W., Dashkevicz, M.P. and Feighner, S.D. (1989) Lactobacilli and bile salt hydrolase in the murine intestinal tract. *Appl. Environ. Microbiol.*, **55**, 1848–51.

Tannock, G.W., Fuller, R. and Pedersen, K. (1990) Lactobacillus succession in the piglet digestive tract demonstrated by plasmid profiling. *Appl. Environ. Microbiol.*, **56**, 1310–16.

Taylor, K.A., Brosby, B., McGavin, M. *et al.* (1987) Characteristics of the endoglucanase encoded by a *cell* gene from *Bacteroides succinogenes* expressed in *Escherichia coli*. *Appl. Environ. Microbiol.*, **53**, 41–6.

Tartaglia, J., Pincus, S. and Paoletti, E. (1990) Poxvirus-based vectors as vaccine candidates. *Crit. Rev. Immunol.*, **10**, 13–30.

Teather, R M. (1985) Application of gene manipulation to rumen microflora. *Can. J. Anim. Sci.*, **65**, 563–74.

Trieu-Cuot, P., Arthur, M. and Courvalin, P. (1987) Origin, evolution and dissemination of antibiotic resistance genes. *Microbiol. Sci.*, **4**, 263–6.

Vanderslice, P., Copeland, W.C. and Robertus, J.D. (1986) Cloning and nucleotide sequence of wild type and a mutant histidine decarboxylase from *Lactobacillus* 30a. *J. Biol. Chem.*, **261**, 15186–91.

Williams, S.A., Hodges, R.A., Strike, T.L. *et al.* (1984) Cloning the gene for the malolactic fermentation of wine from *Lactobacillus delbrueckii* in *Escherichia coli* and yeasts. *Appl. Environ. Microbiol.*, **47**, 288–93.

Yokokura, T., Kodaira, S., Ishiwa, H. and Sakurai, T. (1974) Lysogeny in lactobacilli. *J. Gen. Microbiol.*, **84**, 277–84.

Yokoyama, M.T. and Carlson, J.R. (1981) Production of skatole and *para*-cresol by a rumen *Lactobacillus* sp. *Appl. Environ. Microbiol.*, **41**, 71–6.

Chapter Nine

Selection of strains for probiotic use

ROBERT HAVENAAR, BART TEN BRINK AND JOS H.J. HUIS IN 'T VELD

9.1 INTRODUCTION

A probiotic can be defined as 'a viable mono- or mixed culture of microorganisms which, applied to animal or man, beneficially affects the host by improving the properties of the indigenous microflora' (Havenaar and Huis in 't Veld, 1992). This implies that the term 'probiotic' is restricted to products which (a) contain live microorganisms, e.g. as freeze-dried cells or in a fermented product; (b) improve the health status of man or animals and exert their effects in the mouth or gastrointestinal tract (e.g. included in food or administered as capsules), in the upper respiratory tract (aerosol) or in the urogenital tract (by local application).

In the other chapters it has been shown that the indigenous microflora is host- and location-specific, very complex in composition and generally possesses properties that are beneficial to the host. However, we must realize that we do not precisely know which species or strains of microorganisms play a crucial role in these beneficial properties. Besides the 'natural' benefits of a sound indigenous microflora other health claims are proposed for several microorganisms. For farm animals the most important claims are:

- growth promotion
- improved feed conversion
- health control such as the prevention of intestinal disturbances, especially in young animals
- predigestion of antinutritional factors, e.g. trypsin inhibitors, phytic acid, glucosinolates.

For man it is suggested that specific microbial strains could play an important role in:

- colonization resistance in the intestinal, respiratory and urogenital tracts

- cholesterol metabolism
- inhibiting the carcinogenesis, directly or indirectly, by stimulation of the immune system
- the metabolism of lactose, the absorption of calcium and the synthesis of vitamins.

The crucial question now is whether specific microbial strains (possible probiotics) manifest one or more of these specific beneficial properties and, if so, how can one identify and select the most beneficial strains? It is clear that the selection of strains for probiotic use must be based on criteria which are coherent with the specific health claim the probiotic is used for. One can call these 'specific selection criteria'. For example, when the purpose of a probiotic is 'inhibition of cholesterol metabolism' the probiotic strain should be selected on properties which relate to this activity. These **specific** claims are dealt with in other chapters in this book. It must also be remembered that probiotic strains can only be used, and are only active, on or in the body of the host if they fulfil a large number of **general** criteria. These general criteria are concerned with:

- bio-safety aspects
- production and processing aspects
- the method of administering the probiotic
- the location on/in the body where the microorganisms of the probiotic product must be active.

9.2 AIM OF THIS CHAPTER

In this chapter we try to evaluate a list of general criteria that can be used for the selection of microbial strains (bacteria, fungi and yeasts) for probiotic use. This list of criteria is summarized in a decision-tree, as shown in Figure 9.1. Next to these criteria, suggestions are made for *in vitro* experiments to investigate the strains on these specific criteria. These experiments are evaluated for their extrapolative value to practical circumstances and the *in vivo* situation.

The succession of criteria and experiments are chosen in such a way that the product or the microorganisms can be submitted to tests which grow gradually more and more complex. It starts with the origin (section 9.3) and the determination and viability (section 9.4) of the microbial strain(s). In section 9.5 selection criteria are discussed which are of importance for the processing and storage of a probiotic product. The next sections evaluate criteria concerning resistance to environmental conditions of the probiotic strains on or in the body of the host (section 9.6), adherence and colonization (section 9.7) and antimicrobial activity (section 9.8).

Based on these predetermined criteria it should be possible to select the best commercially available probiotics. But, more important, these

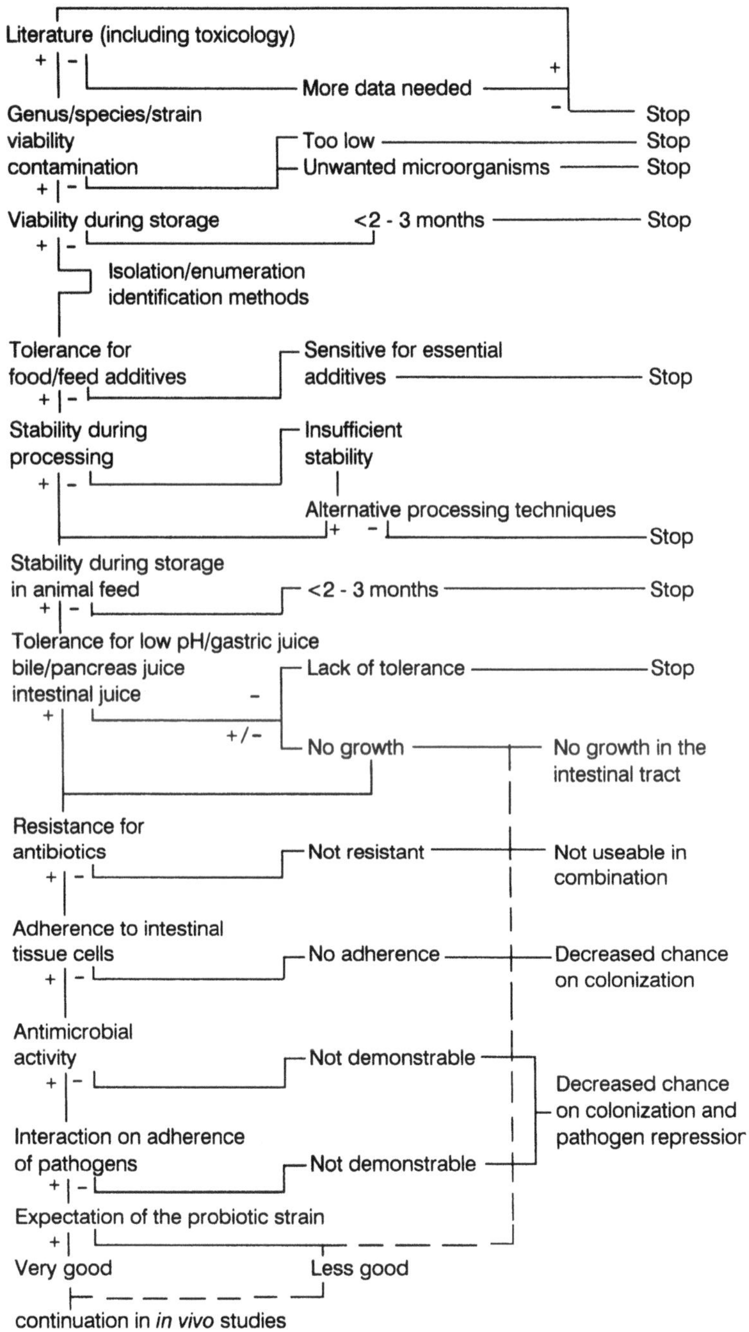

Figure 9.1 Flow diagram for the *in vitro* selection of strains for probiotic use.

criteria and validated experiments could be very helpful in tracing the course of action in a research project for the development of a new probiotic.

9.3 FIRST STEPS IN THE CHOICE OF MICROBIAL STRAINS

There is no doubt about the first step in the selection of microbial strains for probiotic use: it must be representative of microorganisms that are Generally Recognized As Safe (GRAS microorganisms), such as *Lactobacillus* species or some *Bifidobacterium* and *Streptococcus* (*Enterococcus*) species. The use of other microorganisms is very problematic; if there is any doubt about the safety of a microbial strain, short- and long-term toxicological studies could be necessary before acceptance of the probiotic. These types of study are very expensive and time consuming.

The next important basic step in the selection procedure is the choice of the origin of the strain. Do we select a strain from a culture collection, e.g. an industrial fermentation process, or a strain that is freshly isolated from man or animal. This choice will mainly be determined by the specific purpose of the probiotic. For example, if it is not required that the probiotic strain will adhere to and/or colonize the inner or outer body surface of the host (see section 9.7), the origin of the strain is not very important. However, if colonization is essential to gain the ultimate goal of the application of the probiotic, species and location specificity play an important role. In this case it could be of paramount importance that the origin of the strain is selected from the host species. Probably it is even important that it is isolated from the same location as where the activity of the strain has to exhibit in the host, e.g. a specific site in the intestinal tract.

9.4 SPECIES AND VIABILITY OF PROBIOTIC MICROORGANISMS

By definition a probiotic contains viable microorganisms (section 9.1). However, this does not exclude a 'non-viable' product from also possessing health-promoting properties. Hale and Newton (1979) claimed a significant scouring reduction on pigs by a non-viable lactobacillus fermentation product. The mechanism behind live or dead microorganisms will be discussed in the other chapters.

On the packing or in the information leaflet of commercially available probiotics, the genus/species or even the strain(s) and the number of culturable microorganisms have to be mentioned. However, it appears to be important to check the identification and numbers of the microorganisms. Gilliland (Gilliland and Speck, 1977; Gilliland, 1981) examined 30 lactobacillus-containing probiotics, including dairy and pharmaceutical products, products bought in health food stores and animal feed supplements. Although the labels on all products indicated that these products contained live *Lactobacillus acidophilus*, less than half contained viable lactobacilli. Only the dairy and pharmaceutical products contained *L. acidophilus* in high quantities. Of all products obtained from the health food stores and as feed supplements, only one product contained *L. acidophilus*, but in low quantities ($1 \times 10^4 g^{-1}$). It is very likely that during storage, especially under unfavourable conditions, the viability of the microorganisms declines rapidly. It is therefore of 'vital' importance that storage conditions and the ultimate consumption date are stated on the packing.

Contamination of probiotic products with undesirable microorganisms is also possible, especially in uncontrolled fermentation procedures. Until now this has not been reported in the literature, but it may be worth while to examine probiotic products for contaminants, to minimize health risks.

It is important to determine the criteria and procedures for the quality control of probiotic products. To determine the types and numbers of microorganisms in the product, serial dilutions of the product should be plated onto selective agar media for the probiotic species as well as for possible contaminants. Examination of the viability of the probiotic strains during storage under different conditions – e.g. temperature, relative humidity – is essential to determine the optimal storage conditions and the ultimate application date.

9.5 PROCESSING OF VIABLE MICROORGANISMS TO END-PRODUCTS

In the development of probiotics it is important to investigate the viability and resistance of the microorganisms during processing. The purpose for which the probiotic will be used determines the method of administration to man or animal. Oral administration can be effected by tablets, capsules, fermented milk products or powders mixed with the animal feed. Respiratory administration can take place via aerosols – for instance in poultry flocks – which provides a rapid method when administering to large quantities of animals. Local application is possible by unguents or suppositories on the skin or in the urogenital tract, respectively. Each method of administration carries specific demands with regard to the processing of the products and thus on the properties of the probiotic microorganisms. During the selection procedures it is advisable to examine the microbial strains with regard to these properties.

One of the first steps in producing a probiotic is large-scale culturing, washing and drying of the microorganisms. Most lactic acid bacteria can be cultured in large-scale fermentors and are rather resistant to centrifugation. In general lactobacilli are insensitive to freezing and frozen storage at −20 °C or lower (Klaenhammer and Kleeman, 1981), but are less resistant to freeze-drying and especially to spray-drying. The composition of the growth and suspension medium after centrifugation have a profound effect on cell stability during these procedures. For example, calcium carbonate (0.1%) in the growth medium increased the survival of *L. acidophilus* during freezing (−20 °C), but not during freeze-dying, while 5% glycerol in the suspension medium increased survival during lyophilization (Bozoglu and Gurakan, 1989). Staab and Ely (1987) obtained better freeze-drying results of anaerobes (e.g. *Bifidobacterium* and *Peptostreptococcus*) suspended in chopped meat broth with 12% sucrose as compared to double strength skim milk. In cases of storage over long periods and/or under unfavourable conditions, encapsulation of the microorganisms could be considered. Dziezak (1988) published a review of microencapsulation and drying processes commonly used in the food industry.

In the animal feed industry it is also very common to compress the mealy feed into pellet form. During this process the temperature in the feed rises to about 60 – 80°C over a limited period of time. The viability of the microorganisms can decrease dramatically during pelleting, thus preservation or protective measures are absolutely necessary for most microorganisms, with the possible exception of bacterial spores.

The resistance of the microorganisms against chemical compounds in the end-product should be investigated. In the case of probiotics for farm animals, special attention must be paid to growth-promoting feed additives, for example antibiotics. These antibiotics are used in low concentrations, but probiotic strains can be very sensitive. In dry feed, with low a_w values, the activity of the antibiotics will not be very high. But, when the humidity of the feed increases, such as under poor storage conditions or when the feed is eaten by the animals, the probiotic strain can be killed. Antibiotic sensitivity can be examined in agar diffusion tests.

9.6 RESISTANCE TO *IN VIVO* CONDITIONS

After administration of the probiotic the microorganisms should not be killed by the defence mechanisms of the host. Depending on the administration site of the microorganisms, they should be resistant to the specific conditions occurring on or in that location of the body. This means, for example, that the microorganisms in a probiotic for oral use should be resistant to the enzymes in the oral cavity (e.g. amylase, lysozyme), to the enzymes (pepsin, lipase) and the low pH value (high HCl concentration) in the stomach and the concentration of bile, pancreatic juice and mucus in the small intestine. Thus, for orally applied probiotics the conditions in the oro-gastrointestinal tract are the major selection criteria for microbial strains.

Gilliland (1979) showed that mainly Gram-positive bacteria are sensitive to lysozyme, but that *Lactobacillus* and *Streptococcus* are more resistant than other Gram-positive bacteria.

Lactic acid bacteria are considered to be acid tolerant. Conway *et al.* (1987) demonstrated that *L. acidophilus* has a higher tolerance to gastric juice than *L.delbrueckii* subsp. *bulgaricus*, which in turn was more resistant than *S.salivarius* subsp. *thermophilus*. *In vitro* resistance to low pH values is dependent on the type of buffer (Hood and Zottola, 1988) and on the presence of food/feed such as milk (Conway *et al.* 1987). This is in agreement with the *in vivo* studies described by Robins-Browne *et al.* (1981). In their studies the survival of *L. bulgaricus* and *L. acidophilus* in jejunal fluid after passage through the stomach was much lower in fasting subjects than in non-fasting subjects. Conway *et al.* (1987) showed that the duodenal and jejunal persistence of *L. acidophilus* was higher than that of *L.delbrueckii* subsp. *bulgaricus*.

It is important to note from these findings the high correlation between the results of the *in vitro* and the *in vivo* studies. This means that part of the selection of probiotic strains can be based on *in vitro* experiments. The survival of the strains during passage of the stomach (2–3 hours) can be studied *in vitro* in gastric juice (e.g. from slaughtered animals) at different pH values in combination with different types and quantities of food/feed.

If metabolic activity, multiplication or colonization in the small intestine is required for optimal activity of the probiotic strains, then a tolerance for bile is an essential criterion in the selection of microbial strains. According to Gilliland and Speck (1977) non-intestinal bacteria such as *L. bulgaricus* and *L. lactis* are very sensitive to bile: concentrations lower than 0.05% are inhibitory. However, even within the bacterial species large differences in tolerance can exist, as shown by Gilliland *et al.* (1985) for *L. acidophilus*. Studies by Klaenhammer and Kleeman (1981) showed that lactobacilli growing as smooth colonies possess a higher bile resistance than rough colonies. In the presence of bile the colonial morphology changed to rhizoid and the cell morphology showed protrusions of the cytoplasmic membrane induced by gaps in the cell wall. This could mean that colonial and cellular morphology are valuable parameters for the selection of *Lactobacillus* strains.

Selection of bile-resistant (lactic acid) bacteria can be made by culturing on selective agar medium with various levels of bile (Gilliland and Speck, 1977). Quantitative selection can be made by growth curves in broth with different concentrations of bile. According to Gilliland *et al.* (1984) there is a good correlation between the selection of bile-resistant *L. acidophilus* strains based on *in vitro* experiments and the numbers of lactobacilli in the jejunum of calves (but not with the numbers in the ileum).

For these *in vitro* experiments it is important to allow for variations in the composition of bile. Not only do differences occur between various animal species, but also between cannulated bile and bile sampled from the gall bladder (e.g. in pigs, 10% higher concentrations of bile salts). It is hard to say to which concentration of bile the selected strains should be resistant. The mean daily bile flow in pigs is very high (around 2 litres for a 40 kg pig) but varies during the day depending on feed consumption and the composition of the feed (Kulig *et al.*, 1989; Payne *et al.*, 1989).

In addition to studying bile tolerance, it might be of interest in the selection of strains to check the tolerance for pancreatic juice and/or total intestinal juice. In a basic medium, supplemented with various concentrations of (centrifuged and sterilized) intestinal juice, growth curves can be made to investigate the potential of the strains to multiply under '*in vitro* intestinal conditions'. This type of experiment can be made even more complex by combining intestinal variables with components of the feed, such as growth promoters.

9.7 ADHERENCE AND COLONIZATION

For prolonged survival of probiotic strains on or in the body of the host – for instance, in the gastrointestinal tract – the microorganisms should have a short generation time and/or the ability to colonize the internal body surfaces, otherwise the strains will be removed by the contractions of the gut.

In general, adherence and colonization is considered an important property of probiotic strains, comparable with viability and metabolic activity. However, it is not quite clear whether (daily application or consumption of) non-colonizing, viable microorganisms are without any health-promoting value. It is conceivable that these microorganisms might also possess immune-stimulating properties.

Adherence can be considered as the first step of colonization. However, Hood and Zottola (1988) showed that *L. acidophilus* cells, killed by low pH values without affecting the acidic polysaccharide layer, were still able to adhere to cultivated monolayers of intestinal tissue cells as effectively as viable *L. acidophilus* cells. This indicates that active metabolism is no prerequisite for adherence, although it is essential for growth and colonization. Selection of strains with the capacity to adhere to gastrointestinal cells can be based on *in vitro* tests (e.g. the methods described by Ellen and Gibbons, 1974 and Fuller *et al.*, 1978; Conway and Kjelleberg, 1989), but it is hard to extrapolate the results to the *in vivo* situation. The main factor controlling adherence and colonization *in vivo* is the animal species specificity of microorganisms. Although the same *Lactobacillus* and *Bifidobacterium* species were isolated from humans and animals, within these bacterial species various biotypes were characterized depending on the host from which they were isolated (Mitsuoka, 1969a; 1969b). This indicates that a bacterial strain isolated from the indigenous microflora of one animal species will not necessarily colonize the same site in an other animal species (Fuller, 1973; Suegara *et al.*, 1975; Tannock *et al.*, 1982). Beside the animal species specificity for adherence, one can also recognize a strain specificity within the bacterial species. Barrow *et al.*, (1980) demonstrated that the degree of adhesion to squamous epithelial cells of the stomach of pigs was different for various strains within the same species of lactobacilli. This may be caused by the fact that some *Lactobacillus* species do not consist of a genetically homologous group. For example, according to DNA-hybridization techniques, *L. acidophilus* forms a heterogeneous group of bacteria (Johnson *et al.*, 1987). Host-species specificity of the intestinal microflora has also become clear from characterization studies (differences in biotypes, serotypes and plasmid patterns). In a study by Conway *et al.* (1987), the tested lactic acid bacteria showed comparable adhesion patterns for human and pig ileal cells. On the other hand, it is not surprising that Mäyra-Mäkinen *et al.* (1983) showed that lactobacilli from plant material and cultured milk and cheese did not adhere *in vitro* to epithelial cells of pigs and calves. Another factor affecting adherence is the relation with the consumption of food/feed. In chickens, *Lactobacillus* strains did not adhere to the crop wall after fasting for 12 hours in contrast to non-fasting animals (Fuller and Turvey, 1971).

The degree of *in vitro* adherence can be influenced by laboratory conditions such as pH and the presence of enzymes (Barrow *et al.*, 1980) and by the composition of the growth medium (1% skim milk did increase the adherence) and suspension buffer (Conway *et al.*, 1987). Although the *in vitro* test conditions can be chosen resembling the circumstances on or in the host as much as possible, the results must be interpreted carefully.

Adherence *in vitro* is no guarantee for adherence *in vivo* and subsequent colonization. Colonization of a probiotic strain in an already existing microbial ecosystem requires more than adherence alone. After settlement on the mucosa of the host the strain must be able to multiply itself, in spite of the presence of defence mechanisms of the host and interactions with the surrounding microflora. Microorganisms with the ability to compete with other microorganisms, for example by the production of antimicrobial substances, have the best chance to colonize the ecosystem.

9.8. ANTIMICROBIAL ACTIVITY

When a probiotic is used to repress potential pathogenic or other disadvantageous microorganisms in the host, it would be of benefit if the strain possessed antagonistic properties. Lactic acid bacteria, which are frequently used as probiotics, possess a number of antagonistic properties which operate by:

- decreasing the pH by the production of lactic acid
- consumption of available nutrients
- decreasing the redox potential
- production of hydrogen peroxide (under aerobic conditions)
- production of specific inhibitory components, such as bacteriocins.

By definition, bacteriocins are of a proteinaceous nature and have a bactericidal mode of action with a narrow inhibitory spectrum: they are only active against organisms that are closely related to the producer (Koninsky, 1982). However, bacteriocins of Gram-positive organisms usually possess a somewhat broader inhibitory spectrum and are also active against other Gram-positive species (Tagg *et al.*, 1976). The best-known bacteriocin from lactic acid bacteria is nisin, which is produced by strains of *Lactococcus lactis* (Hurst, 1983). In the past decades a large number of reports have been published describing *Lactobacillus* strains that are claimed to produce special inhibitory components. A selection is listed in Table 9.1.

Some of the antimicrobial agents (Table 9.1) are true bacteriocins: lactocin 27, lactacin B, plantaricin A and helveticin J are only active against *Lactobacillus* species that are closely related to the producer strain. Sakacin A, produced by *L. sake*, also inhibits *Listeria monocytogenes* (Schillinger and Lucke, 1989). None of these bacteriocins is inhibitory against Gram-negative bacteria. This means that these microorganisms cannot compete with Enterobacteriaceae involved in gut infections.

The potential to produce bacteriocins that are only active against other lactic acid bacteria could still be an important property for probiotic strains, since it could help these strains to reach and maintain a sufficiently high number in the gastrointestinal tract. The antimicrobials

Table 9.1 Antimicrobial activity of various *Lactobacillus* species

Species	Product	References
L. acidophilus	acidolin	Hamdan and Mikolajcik (1974)
	acidophilin	Shahani *et al.* (1976, 1977)
	lactacin B	Barefoot and Klaenhammer (1983, 1984)
L. bulgaricus	bulgaricin	Shahani *et al.* (1976); Reddy *et al.* (1983)
L. helveticus	lactocin 27	Upreti and Hinsdill (1973)
	helveticin J	Joerger and Klaenhammer (1986)
L. plantarum	plantacin B	West and Warner (1988)
	plantaricin A	Daeschel *et al.*(1990)
	plantaricin SIK 83*	Andersson (1986); Andersson *et al.* (1988)
L. reuteri	reuterin	Talarico and Dobrogosz (1989); Talarico; *et al.* (1988); Axelsson *et al.* (1989); Chung *et al.* (1989)
L.sake	sakacin A	Schillinger and Lucke (1989)
	lactosin S	Mortvedt and Nes (1990)

**Note:* The organism used was probably *Lactococcus lactis* producing a compound very similar to nisin (Andersson *et al.*, 1988).

acidolin (Hamdan and Mikolajcik, 1974), acidophilin (Shahani *et al.*, 1976; 1977) and bulgarican (Shahani *et al.*, 1976; Reddy *et al.*, 1983) have been reported to possess a much broader inhibitory spectrum; many Gram-positive and Gram-negative bacteria (including pathogens such as staphylococci, salmonella, shigella, pseudomonads) were said to be inhibited by the purified products. However, it is quite conceivable that these products – low molecular molecules of a highly acidic nature which were isolated from methanol-acetone extracted fermented milks – are not responsible for the inhibition that was observed during the initial screening procedure.

Recently, Dobrogosz and co-workers described a new broad spectrum antimicrobial substance produced by *L. reuteri* (Talarico *et al.*, 1988; 1989; Axelsson *et al.*, 1989; Chung *et al.*, 1989). This product, termed reuterin, is not only inhibitory towards all of the tested bacteria (including enteropathogens) but also towards yeasts, fungi and even the protozoan *Trypanosoma cruzi*. Reuterin is produced during glycerol metabolism and consists of a mixture of β-hydroxy-propionaldehyde in monomeric and cyclic dimeric form. Reuterin might play an important role in controlling the gastrointestinal ecosystem.

The validity of some published data is questionable since in some cases methods are used which allow false positive results, mostly because sufficient control experiments are absent. When the producer strain and the indicator strain are grown simultaneously in liquid media or on solid media, an observed inhibitory effect can be caused by any one of the above-mentioned antagonistic properties of lactobacilli. In most papers the method of deferred antagonism is used, where the inhibitor is accumulated prior to testing. When solid media are used the producer organism is grown first and then the indicator strain may be cross-streaked or added in the form of an overlaying layer of seeded agar after which the plate is incubated further. When the producer organism is grown in liquid media a sample of the fermented broth is transferred onto an agar plate freshly seeded with the indicator organism. After a period of prediffusion the plate is incubated to allow growth of the indicator organism. Also with this method there are some possible sources of error: the lactic acid still present or produced by remaining viable lactobacilli as well as competition for nutrients during the growth of the remaining lactobacilli and the indicator organism may be responsible for the inhibitory effect.

To select probiotic strains for antimicrobial activity *in vitro*, a good screening procedure should therefore exclude pH effects (use buffered media or neutralize before testing) and active lactobacillus cells (centrifugation, filtration, killing by heat or exposure to chloroform vapour). Inhibitory effects caused by hydrogen peroxide can be neutralized by the addition of catalase. When inhibitory effects are observed in spite of these precautions, a biochemical and functional characterization of the antimicrobial compound involved should be performed. Antimicrobial spectrum, molecular size, pH and temperature resistance, sensitivity towards various enzymes and mode of action can be determined in order to evaluate the potential probiotic use of an antimicrobial producing strain of *Lactobacillus*.

Although lactic acid bacteria selected for the production of antagonistic substances could be used as probiotic strains, some considerations have to be made: under *in vivo* conditions the production of the antimicrobial agent might be lower and/or inactivation could occur; part of the naturally occurring target organisms may become resistant to the antimicrobial agent.

9.9 GENE TECHNOLOGY

In the previous sections of this chapter a large number of general selection criteria are discussed. It is possible that a potential probiotic strain might fulfil a number of these criteria, but lacks one important property. By (induced) mutation, genetically determined properties can be changed, but the addition of one specific property, without changing others, is very difficult. In this scope the use of gene technology for the construction and improvement of microorganisms probably provides great promise for future development of probiotic strains. Conditions for gene technology are: the possibility of isolating the genes that are responsible for a specific factor, such as adherence, and to transfer these genes to a probiotic strain without altering other beneficial properties. McCarthy *et al.* (1988) demonstrated genetic transformation for *L. acidophilus*: a pig strain was transformed to colonize in mice.

9.10 CONCLUSION

For the selection of microbial strains to use as probiotics, criteria can be determined in relation to bio-safety, production, administration, survival and/or colonization in the host. A number of *in vitro* experiments are available to investigate whether the microbial strains fulfil these criteria. Based on these general selection criteria validated in *in vitro* experiments, it should be possible to screen microorganisms on their potential value as probiotic strains. *In vivo* experiments, such as the use of laboratory animals, should only be performed with those microorganisms that meet the minimal selection criteria.

REFERENCES

Andersson, R. (1986) Inhibition of *Staphylococcus aureus* and spheroplasts of gram-negative bacteria by an antagonistic strain of *Lactobacillus plantarum*. *Int. J. Food Microbiol.*, **3**, 149–60.

Andersson, R.E., Daeschel, M.A. and Hassan, H.M. (1988) Antibacterial activity of plantaricin SIK-83, a bacteriocin produced by *Lactobacillus plantarum*. *Biochimie*, **70**, 381–90.

Axelsson, L.T., Chung, T.C. Dobrogosz W.J. and Lindgren S.E. (1989) Production of a broad spectrum antimicrobial substance by *Lactobacillus reuteri*. *Microb. Ecol. Health Dis.*, **2**, 131–6.

Barefoot, S.F. and Klaenhammer T.R. (1983) Detection and activity of lactacin B, a bacteriocin produced by *Lactobacillus acidophilus*. *Appl. Environ. Microbiol.*, **45**, 1808–15.

Barefoot, S.F. and Klaenhammer T.R. (1984) Purification and characterization of the *Lactobacillus acidophilus* bacteriocin lactacin B. *Antimicrob. Agents Chemother.*, **26**, 328–34.

Barrow, P.A., Brooker, B.E. Fuller R. and Newport M.J. (1980) The attachment of bacteria to the gastric epithelium of the pig and its importance in the microecology of the intestine. *J.Appl.Bacteriol.*, **48**, 147–54.

Bozoglu, T.F. and Gurakan G.C. (1989) Freeze-drying injury of *Lactobacillus acidophilus*. *J.Food Protect.*, **52**, 259–60.

Chung, T.C., Axelsson, L. Lindgren S.E. and Dobrogosz W.J. (1989) In vitro studies on reuterin synthesis by *Lactobacillus reuteri*. *Microb Ecol. Health Dis.*, **2**, 137–44.

Conway, P.L., Gorbach S.L. and Goldin B.R. (1987) Survival of lactic acid bacteria in the human stomach and adhesion to intestinal cells. *J.Diary Sci.*, **70**, 1–12.

Conway, P.L. and Kjelleberg S. (1989) Protein-mediated adhesion of *Lactobacillus fermentum* strain 737 to mouse stomach epithelium. *J.Gen.Microbiol.*, **135**, 1175–86.

Daeschel, M.A., Kennedy M.C. and McDonald L.C. (1990) Bacteriocidal activity of *Lactobacillus plantarum* C-11. *Food Microbiol.*, **7**, 91–8.

Dziezak, J.D. (1988) Microencapsulation and encapsulated ingredients. *Food Technol.*, April, 136–57.

Ellen, R.P. and Gibbons R.J. (1974) Parameters affecting the adherence and tissue tropisms of *Streptococcus pyogenes*. *Infect. Immun.*, **9** 85.

Fuller, R. (1973) Ecological studies on the lactobacillus flora associated with the crop epithelium of the fowl. *J. Appl. Bacteriol.*, **36**, 131–9.

Fuller R. and Turvey A. (1971) Bacteria associated with the intestinal wall of the fowl (*Gallus domesticus*). *J. Appl. Bacteriol.*, **34**, 617–22.

Fuller, R., Barrow P.A. and Brooker B.E. (1978) Bacteria associated with the gastric epithelium of neonatal pigs. *Appl. Environ. Microbiol.*, **35**, 582–91.

Gilliland, S.E. (1979) Beneficial interrelationships between certain microorganisms and humans: candidate microorganisms for use as dietary adjuncts. *J. Food Protect.*, **42**, 164.

Gilliland, S.E. (1981) Enumeration and identification of lactobacilli in feed supplements marketed as sources of *Lactobacillus acidophilus*. *Oklahoma Agric Exp. Sta. Misc. Publ..*, **108**, 161

Gilliland, S.E. and Speck M.L. (1977) Enumeration and identification of lactobacilli in dietary products. *J. Food Protect.*, **40**, 760.

Gilliland, S.E., Staey, T.E. and Bush, L.J. (1984) Importance of bile tolerance of *Lactobacillus acidophilus* used as dietary adjunct. *J. Diary Sci.*, **67**, 3045.

Gilliland, S.E., Nelson, C.R. and Maxwell, C. (1985) Assimilation of cholesterol by *Lactobacillus acidophilus*. *Appl. Environ. Microbiol.*, **49**, 377–81.

Hale, O.M. and Newton G.L. (1979) Effects of a nonviable *Lactobacillus* species fermentation product on performance of pigs. *J.Anim.Sci.*, **48**, 770–5.

Hamdan, I.Y. and Mikolajcik E.M. (1974) Acidolin: an antibiotic produced by *Lactobacillus acidophilus*. *J. Antibiot* **27**, 631–6.

Havenaar, R. and Huis in't Veld, J.H.J. (1992) Probiotics: a general view, in *Lactic Acid Bacteria in Health and Disease*, vol. 1 (ed. J.B.J. Wood), Elsevier Applied Science Publishers (in press).

Hood, S.K. and Zottola E.A. (1988) Effect of low pH on the ability of *Lactobacillus acidophilus* to survive and adhere to human intestinal cells. *J.Food Sci.*, **53**, 1514–16.

Joerger, M.C. and Klaenhammer T.R. (1986) Characterization and purification of helveticin J and evidence for a chromosomally determined bacteriocin produced by *Lactobacillus helveticus* 481. *J. Bacteriol.*, **167**; 439–46.

Johnson, M.C., Ray B. and Bhowmik T. (1987) Selection of *Lactobacillus acidophilus* strains for use in 'acidophilus products'. *Ant. Leeuwenhoek* **53**; 215–31.

Klaenhammer, T.R. and Kleeman E.G. (1981) Growth characteristics, bile sensitivity, and freeze damage in colonial variants of *Lactobacillus acidophilus. Appl.Environ.Microbiol.*, **41**; 1461–7.

Koninsky, J. (1982) Colicins and other bacteriocins with an established mode of action. *Ann. Rev. Microbiol.*, **36**; 125–44.

Kulig, W., Kirchgessner, M. Erhardt, W. *et al.* (1989) Zur Messung von Galleflussrate und Gallelipiden an gallengangskanuelierten Schweinen mittels einer vollautomatischen Rueckfuerapparatur. *J.Anim. Physiol. Anim. Nutr.*, **62**; 93–100.

McCarthy, D.M., Lin, J.H.C., Rincker, L.A. and Savage, D.C. (1988) Genetic transformation in *Lactobacillus* sp. strain 100–33 of the capacity to colonize the nonsecreting gastric epithelium in mice. *Appl. Environ.Microbiol.*, **54**; 416–22.

Mitsuoka, T. (1969a) Vergleichende untersuchungen über die Laktobazillen aus den Faeces von Mensch, Schweines und Hühner. *Zentralbl. Bacteriol. Microbiol. Hyg. [A]*, **210**, 32–51.

Mitsuoka, T. (1969b) Vergleichende untersuchungen über die Bifidobakterien aus dem Verdauungstrakt von Mensch und Tier. *Zentralbl. Bacteriol. Microbiol. Hyg. [A]* **210**; 52–64.

Mortvedt, C.I. and Nes, I.F. (1990) Plasmid-associated bacteriocin production by a *Lactobacillus sake* strain. *J. Gen. Microbiol.*, **136**; 1601–7.

Mäyra-Mäkinen, A., Manninen, M. and Gyllenberg H. (1983). The adherence of lactic acid bacteria to the columnar epithelial cells of pigs and calves. *J.Appl.Bacteriol.*, **55**, 241–5.

Payne, D., Juste, C. Corring T. and Fevrier C. (1989) Effects of wheat bran on bile secretion in the pig. *Nutr. Rep. Int.*, **40**, 761–71.

Reddy, G.V., Shahani, K.M. Friend B.A. and Chandan R.C. (1983) Natural antibiotic activity of *Lactobacillus acidophilus* and *bulgaricus*: III. Production and partial purification of bulgarican from *Lactobacillus bulgaricus. Cult. Dairy Prod. J.* **18**, 15–19.

Robins-Browne, R.M., Path F.F. & Levine M.M. (1981) The fate of ingested lactobacilli in the proximal small intestine. *Amer J.Clin.Nutr.*, **34**; 514–19.

Schillinger, U. and Lucke F.K. (1989) Antibacterial activity of *Lactobacillus sake* isolated from meat. *Appl. Environ. Microbiol.*, **55**, 1901–6.

Shahani, K.M., Vakil J.R. and Kilara A. (1976) Natural antibiotic activity of *Lactobacillus acidophilus* and *bulgaricus*: I. Cultural conditions for the production of antibiosis. *Cult. Dairy Prod. J.*, **11**, 14–17.

Shahani, K.M., Vakil J.R. and Kilara A. (1977) Natural antibiotic activity of *Lactobacillus acidophilus* and *bulgaricus*: II. Isolation of acidophilin from *L. acidophilus. Cult. Dairy Prod. J.*, **12**, 8–11.

Staab, J.A. and Ely J.K. (1987) Viability of lyophilized anaerobes in two media. *Microbiology*, **24**, 174–8.

Suegara, N., Morotomi, M. Watanabe, T. *et al.* (1975) Behavior of microflora in the rat stomach: adhesion of lactobacilli to the keratinized epithelial cells of the rat stomach *in vitro. Infect. Immun.*, **12**, 173–9.

Tagg, J.R., Dajani A.S. and Wannaker L.W. (1976) Bacteriocins of gram-positive species. *Bacteriol. Rev.*, **40**, 722–56.

Talarico, T.R., Casas, I.A., Chung T.C. and Dobrogosz W.J. (1988) Production and isolation of reuterin, a growth inhibitor produced by *Lactobacillus reuteri*. *Antimicrob. Agents Chemother.*, **32**, 1854–8.

Talarico, T.L. and Dobrogosz W.J. (1989) Chemical characterization of an antimicrobial substance produced by *Lactobacillus reuteri*. *Antimicrob. Agents Chemother.*, **33**, 674–9.

Tannock, G.W., Szylit, O. and Raibaud P. (1982). Colonization of tissue surfaces in the gastrointestinal tract of gnotobiotic animals by *Lactobacillus* strains. *Can. J. Microbiol.*, **28**, 1196–8.

Upreti, G.C. and Hinsdill R.D. (1973) Isolation and characterization of a bacteriocin from a homofermentative lactobacillus. *Antimicrob. Agents Chemother.*, **4**, 487–94.

West, C.A. and Warner, P.J. (1988) Plantacin B, a bacteriocin produced by *Lactobacillus plantarum* NCDO 1193. *FEMS Microbiol. Lett.*, **49**, 163–5.

Chapter Ten

Probiotics for chickens

PAUL A. BARROW

10.1 INTRODUCTION

As with other animals, the development of probiotics for poultry has developed out of our increasing understanding of the microflora of the alimentary tract. Although Pasteur's observations suggested that the host and its intestinal microflora were interdependent, the intellectual origins of probiosis started with Metchnikoff (1907). He thought that, on balance, the intestinal microflora was detrimental to the host because of absorption of toxic bacterial metabolites. By correlating the longevity of some of the rural inhabitants of the Caucasus with their ingestion of fermented milk throughout their life, he suggested that health status and longevity could be improved by consumption of milks fermented by lactobacilli.

This description of the intestinal microflora in adversarial terms was perpetuated by Dubos *et al.* (1965) who divided the indigenous microflora into the autochthonous organisms, such as lactobacilli and *Bacteroides*, which had developed an evolutionary symbiotic relationship with the host and allochthonous organisms such as *Escherichia* and *Clostridium* which were potential pathogens. These, together with non-enteric organisms acquired from the environment, comprised the normal intestinal flora.

These descriptions are far too simplistic and must be seen as early models attempting to describe several highly complex ecosystems. For example, microbial opportunism and true commensalism are largely ignored. Regarding the flora as a climax community in which every niche is occupied is also patently inaccurate. Our inadequate understanding of microbial taxa at that time presumably led to regarding *Escherichia coli* as a potential pathogen although many strains may be beneficial to the host and can certainly be used in that way (Linton *et al.*, 1978; Duval-Iflah *et al.*, 1983). However, these hypotheses provided an important stimulus to studying the microecology of the alimentary tract. These early models had a profound effect on the development of probiotics. Many preparations currently used for poultry and other animals are based on the assumption that the early hypotheses are correct with the result that the approach to probiosis is often too simplistic.

10.2 THE NORMAL INTESTINAL FLORA OF POULTRY

The gut microflora of poultry is complex and the interactions between different types of organisms must be very complicated. Those interactions that are best understood are probably the most simple. It is clear that the basis upon which many probiotic preparations have been selected and used is often also too simplistic. Thus, in reviewing work on the use of probiotics for poultry the scientific basis for their selection and use must be analysed. This cannot be done without first reviewing our current knowledge of gut microorganisms and what is known of their interactions both with the host and with each other.

Our knowledge of the components of the gut flora of chickens is continually increasing. Information for other poultry species is much less detailed. The flora and its activities must be studied in relation to the host anatomy and physiology and to the conditions existing in individual sections of the alimentary tract.

Early studies on the normal flora were hampered by the use of highly selective media with aerobic incubation producing a very inaccurate picture of the organisms present. Anaerobic jars have now been replaced by methods which prevent any oxygen from coming into contact with samples or cultures. These techniques, together with complex non-selective media, allow a more detailed examination of the highly oxygen-sensitive organisms present in the caeca. However, even now some organisms, such as the budding bacteria, cannot be cultured and others such as methanogens have not been found even though there is chemical and metabolic evidence for their existence. This must presumably indicate nutritional inadequacies in the techniques and media used.

Despite the fact that the flora can be subdivided most conveniently according to the area of the alimentary tract involved it must be remembered that the flora is almost continuous throughout the length of the gut. Microorganisms from the crop which survive the low pH of the gizzard generally multiply in the small intestine. Organisms from this organ may be taken into the caeca. The microbial content of the cloaca and faeces depends on whether they contain material from the small intestine or from the caeca. Caecal droppings are discharged two to four times every day.

Food is swallowed whole and is stored in the crop where a predominantly lactic acid semi-batch fermentation takes place. The Eh is fairly high so that ingested obligate anaerobes die. Other non-enteric organisms also generally do not thrive. There is thus a simple flora compared with that of the caeca. The predominant organisms are lactobacilli (Table 10.1) producing mainly lactic and acetic acids such that the crop contents pH of the healthy chicken is 4–5 with the result that less aciduric organisms do not normally grow to the same high numbers. A number of metabolic types have been isolated and characterized (Fuller, 1973) including *Lactobacillus salivarius*, *L. fermentum* and a type resembling *L. acidophilus*. Eyssen *et al.* (1965)

Table 10.1 The viable numbers of the major groups of bacteria in the alimentary tract of the chicken.

Bacterial group counted	Crop	Gizard	Small intestine Section 1	Section 2	Section 3	Section 4	Caeca	Faeces
	Log_{10} median (of 5–7 birds) viable count in contents of							
E. coli	1.7	–	2.0	1.7	1.7	2.7	5.6	6.1
Clostridia	–	–	(–)	(–)	(–)	(–)	9.0	2.0
Enterococci	4.0	3.7	4.0	4.0	3.7	4.2	6.7	6.5
Lactobacilli	8.7	7.3	8.0	8.2	8.2	8.6	8.7	8.5
Yeasts	2.7	–	1.7	–	1.7	–	2.0	1.7
Non-sporing obligate anaerobes	–	–	–	–	–	–	10.0	9.0
Anaerobic streptococci	–	–	–	–	–	–	10.0	8.7

– = log <1.0.
(–) = May or may not be present.
From Smith (1965a).

and Fuller and Turvey (1971) found large numbers of lactobacilli adhering to the squamous epithelium of the crop. Fuller (1973) found that this adherent flora became established within a day of hatching and that the numbers isolated from the homogenized crop wall were similar to those found in the lumen. Lactobacilli which adhere to crop cells were isolated from other poultry but not from mammals. The adhesion was mediated by a carbohydrate-rich capsular layer (Fuller, 1975; Brooker and Fuller, 1975). The numbers in the crop decreased with starvation but sufficient remained to inoculate fresh food entering the crop. *E. coli* is present in the crop in fairly low numbers, possibly maintained by the ingestion of faeces. *Enterococcus faecalis* subsp.

liquefaciens and subsp. *zymogenes*, *Ent. faecium*, *Ent. avium* and *Ent. gallinarum* are less aciduric than the lactobacilli and are present in lower numbers. There is no evidence for extensive colonization of the epithelium by these organisms but quantitative bacteriology indicates some colonization of the epithelium by *E. coli* in some circumstances (Barrow *et al.*, 1988).

The pH of the proventriculus and gizzard is very low (pH 1–2) and microbial survival depends on acid tolerance. Little multiplication of organisms occurs in the duodenum because of the relatively high rate of flow of the very fluid contents; however, colonization of duodenal villi by *Ent. hirae* may result in growth depression of the bird (see below). Besides the other microorganisms present in the small intestine and shown in Table 10.1, a filamentous organism, similar to *Arthromitis*, may be seen embedded in the surface of epithelial cells disrupting the brush borders. Its activities and interaction with the host are unknown but it is sensitive to dietary penicillin. *Clostridium perfringens* may occasionally be isolated from the small intestine where it splits fatty acids. Its role in growth depression is controversial (see below).

The caeca are filled with a thick viscous fluid containing no food particles. In these organs the highest viable counts (bacterial counts of 10^{11} g^{-1} of contents) and most complex microflora exist. Smith (1965a) attributed this to the slow rate of flow, the kinetics of bacterial growth resembling batch culture. Most of the microorganisms present are obligate anaerobes, there being more than 200 strains present in the highest dilutions of caecal samples from chickens of more than 4 weeks of age. Gram-positive, anaerobic cocci, including peptostreptococci, comprise up to 30% of the total viable count. Other major components include Gram-negative, non-sporing rods (20% of the total) such as the Bacteriodaceae. This important group includes *Bacteroides hypermegas*, now reclassified as *Megamonas*, *Bact. microfusus* and many other types distinguished by morphology, biochemical activity and fermentation products. Few of them can be assigned to known species. Gram-positive, non-sporing rods, including several types of *Eubacterium*, comprise up to 16% of the total count. The budding bacterium, *Gemmiger formicalis*, and the budding cocci account for 10% of the total, present at 10^9 to 10^{10} g^{-1}. *Clostridium* sp. and *Bifidobacterium* including *Bifid. gallinarum* are present at similar levels. Facultative anaerobes include Enterobacteriaceae such as *E. coli*, *Citrobacter*, *Salmonella*, *Proteus* and *Klebsiella* which are frequently present but in lower numbers. Smaller numbers of other organisms such as the aerobe, *Pseudomonas*, and yeasts may be found throughout the gut from time to time but are never present in high numbers.

The mechanisms whereby these bacteria are maintained in the caeca are poorly understood. A layer of bacteria, hundreds of cells thick, is present, embedded in the mucus lining the epithelium. Whether this is epithelial adhesion in the strictest sense is unclear. Such a thick layer which allows rapid colonization of fresh contents when withdrawn into the caecal lumen is thought to be of significance in the protection against pathogens afforded by the normal flora of the adult. Colonization by potential pathogens such as *Campylobacter*, *Clostridium* and *Salmonella* is also poorly understood. The slow rate of turnover of caecal contents is thought to contribute to colonization by the latter, but there is little evidence for adhesion being involved (Barrow *et al.*, 1988). A chromosomal colonization determinant can be recognized in *Salmonella* (Barrow *et al.*, 1988).

This is the general situation in the alimentary tract of healthy chickens. In addition, a number of factors such as disease, age and diet can affect the composition of the gut flora. These changes, important in an assessment of the need for probiotic use, are poorly understood and have been little characterized.

Detailed analysis of the intestinal flora of avian species other than chickens have yet to be made. Considerable populations of greater than 10^{10} g^{-1} have been found in the caeca of turkeys, ducks and pheasants. Microscopical observations indicate that the proportions of the predominant types vary, but whether this is a true difference or is the result of differences in age or diet is unknown. Some types such as *Megamonas hypermegas* and budding bacteria appear to be widespread.

The normal viral flora in the chicken has not been studied in depth. Reoviruses and small 50 nm viral particles can be isolated from the tissues of the alimentary tract but whether they exist in a truly healthy bird remains to be seen. Bacteriophages are undoubtedly present when susceptible host bacteria are present in the gut. They may arise from food but multiplication in the gut does not necessarily affect the numbers of the host bacterium.

10.2.1 Factors affecting the composition of the flora

A number of naturally occurring and artificial factors are able to affect the composition of the flora and give some indication as to its stability and the ease or difficulty with which it can be manipulated. These factors include age, the immune response, diet and orally administered antibiotics.

Although the alimentary tract of the healthy newly hatched chick is sterile, it rapidly becomes colonized by facultative anaerobes, particularly coliforms and streptococci. Some clostridia may also be present. Lactobacilli soon displace these types as the dominant organisms in the crop and small intestine. Obligate anaerobes may not appear for a week or more when the conditions in the caeca become favourable for their establishment. The immediate source of such strictly anaerobic organisms in the chicken is not known. The caecal flora does not stabilize until 4–6 weeks after hatching (Smith, 1965b; Mead, 1989).

The role of the immune response in controlling the components of the intestinal flora is largely unknown. Humoral responses are likely to be more effective than cell-mediated responses. Although secretory and systemic responses can be detected following infection with normal flora components, the response to some bacteria, particularly the more invasive members of the Enterobacteriaceae, is much stronger than towards other, more numerically dominant, organisms such as the lactobacilli and Gram-negative anaerobic rods. Whether invasiveness is the major determinant of immune responsiveness or whether the cell components of some organisms are able to induce a degree of tolerance is unclear. Some human strains of *E. coli* possess blood group antigens and may thus be at an advantage over other strains in a host possessing these antigens. Whether this occurs in poultry is unknown. How the immune response relates to protective members of the gut flora is unclear but some relationships have been found (Perdigón, this volume).

The most obvious changes in the flora induced by dietary change occur at the anterior end of the tract. Little change seems to occur in the caeca (Smith, 1965a). Increased carbohydrate stimulates the saccharolytic lactobacilli whereas diets artificially enriched in protein suppress the lactobacilli, while coliforms, clostridia and streptococci increase in numbers in the crop. Vitamin-producing bacterial strains may increase in number when vitamin-deficient diets are used. Appropriate dietary change may promote colonization by probiotics.

Antibiotics may be administered in the feed or water for chemotherapy, chemoprophylaxis or for growth stimulation, possibly affecting major groups of organisms. For the purposes of stability their replacement by antibiotic-resistant forms maintains the *status quo* but public health risks can follow the acquisition by potential pathogens or members of the normal flora of antibiotic-resistant plasmids (Barrow, 1987). However, the use of antibiotics over a short term, simultaneously with the administration of probiotic organisms resistant to that antibiotic, may conceivably promote their colonization and render manipulation of the flora easier.

In the absence of such major selective pressures the adult intestinal flora is difficult to change or manipulate simply by oral administration of microorganisms (Linton *et al.*, 1978). Attempts at replacing an organism occupying a particular niche are likely to be ineffective unless the indigenous organism is suppressed by one of the methods described above. Similarly, it would be easier to establish a beneficial organism soon after hatching before other organisms are able to colonize. This resistance to colonization by the adult intestinal flora is well recognized and has been exploited (see section 10.6).

10.3 HOST–MICROBIAL FLORA INTERACTIONS

10.3.1 Nutritional benefits to the host

Nutritional roles for the intestinal flora have been sought despite their questionable role in modern commercial birds administered rich diets. The crop and caecal flora may have nutritional significance in avian species in their natural environments where the diet may be poor.

The largely fermentative crop flora produce a number of organic acids which are available to the host. Whether they are extensively utilized is not known. By affecting the pH in the gut they may also affect enzyme activity (Ford, 1974). Nucleotides synthesized by lactobacilli are thought to be utilized by the host (Eyssen *et al.*, 1965). B vitamins are synthesized, but probably contribute little to the host's requirements (Coates *et al.*, 1968). Vitamin A is also synthesized and may compensate the chicken's requirements when reared on a deficient diet (Pivnyak and Konyakhin, 1973). Since the storage capacity of the gizzard is limited, the crop may act primarily as a storage organ and any nutritional benefits of microbial activity maybe incidental.

Despite the highly complex nature of the caecal microflora, surgical removal has no apparent effect on chickens reared on a commercial diet. Many caecal bacteria degrade urea which enters the caeca from the cloaca by peristalsis. The ammonia produced can be utilized by the host and the caecal flora. None of the caecal flora degrade cellulose or xylan. However, little solid food material enters the caeca. Thus, in healthy chickens reared on a complete diet administration of probiotics is unlikely to confer any substantial direct nutritional benefit.

10.3.2 Harmful nutritional effects

It is now well recognized that, in normal circumstances, the intestinal flora reduces the growth rate of the host. The increased growth rates of gnotobiotic animals and of those administered antibiotics orally indicate this. Although a considerable volume of work has been carried out in this area, it is still unclear which microorganisms depress growth rates and how they do so. One of the reasons for this is biological variation. Not only does the degree of growth depression vary between individual birds and groups of birds but the viable numbers of candidate organisms also vary.

Early antibiotics used for growth stimulation, which included benzyl penicillin and tetracycline, were eventually prohibited because they stimulated the development of resistance to these antibiotics in enteropathogens and *E. coli*. Tetracycline particularly has a wide spectrum of activity and would eliminate many potentially beneficial organisms in addition to the growth-depressing strains. The more recently developed growth-promoting antibiotics, while not affecting Gram-negative facultative anaerobes, do affect a wide range of organism types, both growth-depressing and others. In many cases it is possible that probiotics and the administration of such antibiotics would not be compatible.

The several hypotheses of microbial growth depression include reduced nutrient absorption, including glucose, reduced vitamin absorption and direct toxic effects of bacterial metabolites.

Clostridium perfringens is one of the candidate organisms (Lev and Forbes, 1959) since their numbers are reduced in antibiotic fed birds. Although the suggested method of growth depression is unclear, they produce a number of toxins and possess bile acid deconjugating ability that would effect nutrient absorption.

The effects of intestinal colonization by *E. coli* on growth in poultry has never been studied in detail. The few results available suggest that they have no effect.

Strains of *Streptococcus faecium*, now classified as *Ent. hirae*, have also been found to exert a strong growth-depressing effect either when administered on their own or, to a greater degree, when given together with a bacterial-free filtrate obtained from the faeces of conventional chickens (Eyssen and DeSomer, 1967; Fuller *et al.*, 1979). The effect is variable and is affected by diet. Duodenal colonization by adhesion of *Ent.hirae* was necessary for the manifestation of the growth depressing effect (Houghton *et al.* 1989; Fuller *et al.* 1981) These strains deconjugated bile acids but the exact mechanism of growth depression is unknown.

The work summarized above shows the paucity of information available on the harmful effects of the normal flora. For probiotics to be developed and used in anything other than an empirical way the exact mechanisms whereby such effects occur and how harmful bacteria colonize the alimentary tract must be known. The interactions of both *C. perfringens* and *Ent. hirae* with potentially beneficial members of the normal flora are also unknown.

10.3.3 Colonization resistance against pathogens

The flora of the alimentary tract of the chicken has an important role in preventing colonization by potential pathogens, primarily enteropathogens. The main areas of interest are the crop, the first major site for colonization following the ingestion of microorganisms, and the caeca, the primary colonization site for a number of pathogens including *Salmonella* and *Campylobacter*. The experimental work indicating the importance for the health of the animal of the normal flora in these sites has had profound influence on the development of probiotics for use with poultry.

Fuller (1977) established that the crop *Lactobacillus* flora was important in maintaining a beneficial microbial balance in the crop and exerted its influence on the small intestine. The gradual displacement of *E. coli* and streptococci by lactobacilli as dominant organisms in the crop could be reproduced *in vitro* using suspensions of chick diet in water. These were incubated for several hours with different *Lactobacillus* species before being reinoculated with an *E. coli* strain. Although crop contents obtained from healthy birds are bactericidal, a similar effect *in vitro* was obtained with one homofermentative strain, 74/1, which reduced the pH of the food suspension to 4.15, but not with another, *L. salivarius* strain 59, which reduced the pH to 4.4. When inoculated with *E. coli* alone the pH was 5.7. However, the inhibitory effect was not caused by pH or lactic acid alone and this suggests that additional antibiotic effects may be involved. Colonization by these strains in gnotobiotic birds was found to reduce the intestinal counts of *E. coli* by factors of 100 to 1000 in the crop and by 10 in the ileum below that found in *Lactobacillus*-free birds. Most of these studies were carried out with non-pathogenic *E. coli*. However, *Salmonella* strains are very similar to coliforms and *Campylobacter* and *Clostridium* are highly sensitive to low pH. It thus seems feasible to colonize chicks within hours of hatching with a highly bactericidal *Lactobacillus* strain to establish the most inhibitory *Lactobacillus* flora possible rather than leaving such colonization purely to chance. Extension of this work to study the effect in detail on *Salmonella*, *Campylobacter* and strains of *E. coli* involved in colibacillosis and thought to be involved in haemorrhagic enteritis would be valuable.

Other interrelationships between different types of organisms similar to that above and to the inhibition of yeasts by lactobacilli in the murine stomach have been demonstrated in the crop of poultry, albeit in a less detailed manner. Inoculation of gnotobiotic chickens with *E. coli* was found to prevent the morphogenesis of *Candida albicans* into the more virulent hyphal form. This latter form is found in germ-free birds monoassociated with *Candida*, whereas the yeast form is found in conventional birds (Balish and Phillips, 1966). This is of significance because moniliasis of the crop can be a serious disease of birds, particularly turkeys (Blaxland and Fincham, 1950). Streptococci were not effective in preventing the transformation from yeast to hyphal form and lactobacilli were not tested. The work indicates that at least some strains of *E. coli* may be beneficial to the host, even in the crop.

The inhibitory effect *in vivo* of lactobacilli against *E. coli* appears not to be due to pH alone. Organic acids such as lactic acid and acetic acid produce greater antibacterial effects at low pH than inorganic acids such as hydrochloric acid. They are thought to affect membrane structure and oxidative metabolism. Lactic acid has been found to be very inhibitory towards *Salmonella typhimurium in vitro* (Rubin and Vaughan, 1979). A number of additional antibacterial factors have been found to be produced *in vitro* by lactobacilli, but although they generate considerable interest the extent of their *in vivo* production and significance is completely unknown. Hydrogen peroxide is produced which itself contributes in part to the measurable inhibitory activity (Wheater *et al.*, 1952; Gilliland and Speck, 1977). In addition, a number of antibiotic and bacteriocin-like substances are produced *in vitro* which have potent and sometimes wide-spectrum antibacterial activity. Homofermentative lactobacilli (including *L. acidophilus*) produce several types of bacteriocin whereas heterofermentative types produce relatively few. The spectrum of activity is wider than that of bacteriocins produced by other Gram-positive bacteria and it is conceivable that they may exert activity against other genera *in vivo*. Some of the other antibiotic-like substances have been partially purified but their exact nature and *in vivo* significance is not known (Hamdan and Mikolajcik, 1974; Shahani *et al.*, 1977).

There is considerable evidence that the highly complex, normal flora of the caeca exerts protection against the establishment of microbial pathogens such as *Salmonella* and *Campylobacter* which preferentially colonize the caeca. Newly hatched chicks which have little or no intestinal flora are much more susceptible to oral infection with either of these organisms compared to adults. This can be remedied by using the competitive exclusion principle (see below). In the same way that mice treated orally with streptomycin show enhanced susceptibility to salmonellosis (Bohnhoff *et al.*, 1954) adult chickens given chemotherapeutic or growth-promoting antibiotics in the feed can show increases in faecal excretion of *Salmonella* following oral infection (Smith *et al.*, 1985). These two pieces of evidence suggest that competitive exclusion cannot be used in conjunction with some growth-promoters. The mechanism of inhibition by the flora is unknown.

10.4 APPLICATION OF PROBIOSIS TO POULTRY

Probiotics for chickens are designed either to replace beneficial organisms that are not present in the alimentary tract or to provide the chicken with the effects of beneficial bacteria. They may be absent possibly because present methods of husbandry prevent contact between the newly hatched chick and its parent, preventing rapid vertical transfer of beneficial bacteria or by management practices which may disturb intestinal microecology. There are two major groups of probiotic preparations: those which are primarily intended to be effective in the crop and the anterior regions of the alimentary tract and those whose effect is directed mainly at the caeca. However, it is likely that both types of preparation are, to some extent, effective throughout the gut.

Among the first group are the various *Lactobacillus* cultures and preparations which are thought to colonize the crop and small intestine in ways described by Fuller (1978). they are thought to exert antibacterial effects against potential pathogens (Fuller 1974, 1978) and are also considered to increase performance by unknown mechanisms. There is less rationale for the latter effect than for the former, it being based more on the long-term tradition arising from Metchnikoff's original observations and associated with the supposed beneficial effects of consumption of fermented milk products.

Although a number of empirical observations have suggested that some preparations containing dead bacteria are effective probiotics, the use of live organisms is emphasized for many products with the implication that intestinal colonization is essential for efficacy. Whether many of these organisms actually become established in the gut is questionable since a number of criteria must be fulfilled to ensure colonization ability (Morishita *et al.*, 1971; Fuller, 1978; 1986). These criteria can include adhesion to the crop epithelium, ability to grow in the nutritional environment of the gut and ability to resist innate or microbially produced inhibitory mechanisms although the latter two factors could be difficult to dissociate. The ability of strains to adhere to the crop is particularly important for organisms that have a slower rate of multiplication in the feed slurry present in the crop. From work with monocontaminated and dicontaminated gnotobiotic chickens, Morishita *et al.* (1971) found that whereas avian strains of *L. acidophilus*, *L. salivarius* and *L. fermentum* in addition to the non-intestinal *L. plantarum* and *L. casei* colonized well, a human *L. acidophilus* strain, *L. helveticus* and *L. brevis* were rapidly eliminated from the alimentary tract. This in itself indicates the importance of choosing both the right species and strain. In the presence of an *L. acidophilus* strain or an Ent. *faecalis* strain from chickens neither *L. plantarum* or *L. casei* became established indicating that the inability of a strain to colonize may result from an innate inability or may relate to the presence of the normal flora. Tolerance of a low pH is also important in allowing extensive colonization of the small intestine by survival of the gizzard environment (Fuller, 1978). Additional factors such as optimal temperature for growth (Morishita *et al.*, 1971) and resistance to unsaturated fatty acids (Niemen, 1954; Fuller and Moore, 1967) may also be significant in determining colonization.

Factors affecting the colonization by the complex flora in the caecal cultures used for competitive exclusion are almost unknown since so many different types of organisms are involved. However, all are avian strains, including lactobacilli, originally isolated generally from the species in which they are used so that problems of colonization ability, by and large, will not be encountered.

A number of technical and experimental points must be considered in assessing the value of probiotic preparations and in assessing experimental work carried out by others to do this. Published data and unpublished information and claims must be studied critically.

Statistical and biological significance must be calculated but lack of significance in one area does not necessarily imply insignificance of the other. Statistical significance must be aimed for, but even if it cannot be attained the results may nevertheless be of biological significance. For example, a small but consistent weight gain may be economically significant for a large number of birds. However, statistical significance does not necessarily imply biological significance since a reproducible and statistically significant reduction in faecal excretion of *Salmonella* from 90% to 30% of chickens means little from the point of public health because cross-contamination during and after slaughter is so extensive.

Difficulties in demonstrating statistical significance probably result largely from genetic heterogeneity in poultry or microbiological variation in the gut flora of individuals and different groups of animals. While these sources of variation can be reduced under experimental conditions it is obviously important to test probiotic preparations under the conditions in which they will be used. Under these conditions the amount of variation may be greater than the beneficial effects sought. In this case the results from several untreated control groups should be examined to assess the variability expected.

Effects on performance and productivity can be assessed with the live bird. However, microbial changes and claims for colonization of the probiotic organism can only be assessed by bacteriological analysis after *post mortem* sampling because the faecal microbial flora is generally quantitatively and qualitatively quite different from the flora of the more anterior regions of the gut. Artificial changes can be introduced inadvertently by allowing microbial multiplication following slaughter if samples, for example, are not removed immediately. Claims for colonization must be examined particularly carefully because many strains used do not possess readily selectable phenotypic markers and considerable difficulty would be encountered in differentiating administered lactobacilli from indigenous strains in the absence of such markers.

Strict adherence to a rigid experimental plan, together with perhaps a slightly overcritical appraisal of the results, will then allow the experimenter to be confident in any interesting results that may be obtained. Some of the experimental designs and claims made in the literature fall short of the high standard expected for scientific work. Additionally, some of the preparations tested successfully under experimental conditions have been found to be less effective under field conditions although the reasons for this are not clear. Other products are available commercially with little published or other information made available for the scientist or for those contemplating their use.

A critical review of the available literature on the application of probiosis to poultry performance and health is essential in assessing its value. The review is divided for convenience into (1) organisms and products that are designed primarily to modify or manipulate the anterior intestinal flora and are based primarily on lactobacilli or fermentable carbohydrates to stimulate such a flora or the administration of fermentation products and (2) cultures from the alimentary tract of adult birds designed to increase the resistance to *Salmonella* infection of the newly hatched chick/poult to the level of that of the adult (competitive exclusion). Further sections on miscellaneous techniques including the use of live, attenuated vaccines and bacteriophages are appended.

10.5 LACTIC ACID BACTERIA AS PROBIOTICS

In all the experimental work reviewed here, lactobacilli have been the organisms used for probiosis. This presumably is largely for historical reasons and because some streptococci are associated with growth depression. The problem in assessing the value of probiotics is to evaluate whether any truly beneficial effects are reported. To that end it seems appropriate to present, as part of a review, as much of the relevant experimental data as possible. Their application is divided according to their effects on nutritional and performance parameters for broiler chickens, layer chickens and other poultry and to their effects on potential intestinal pathogens. In most cases reports suggesting beneficial effects are presented first, followed by those suggesting little or no effects.

10.5.1 Effects on broiler performance

The number of reports in which oral administration of probiotics to broilers has been found to produce a substantial and reproducible effect on weight gain, feed conversion, vitamin levels or other nutritional parameters are very few. Even in some papers where considerable claims for effectiveness are made, it is clear that there is little.

Tortuero (1973) reported the results of several experiments in which a *L.acidophilus* strain, whose origin was not stated, was provided in the drinking water at 10^9 organisms ml^{-1}. In one experiment the culture was given to groups of 100 chickens for a period of 11 days after hatching. The group to which the *Lactobacillus* was administered had higher daily weight gains but the feed consumption was also increased and the feed conversion index was slightly reduced from an average between 5 and 11 days of 0.60 to 0.52. Although the authors thought that the differences were not statistically significant (no analyses were carried out), they considered the differences to be real. In a second experiment, administration in the drinking water for the first 5 days of life to groups of 30 chicks was studied. The experiment lasted for 12 days. Small reductions were observed in the caecal weight (from 0.636% of live body weight in the control birds to 0.612% at 7 days and from 0.604 to 0.489% at 12 days) and in faeces weight (average of 6.2 g in the control to 4.4 g). Increases in fat digestibility and nitrogen retention were also observed. There was no effect on weight gain which the author attributed to the use of new housing, in use for 3 months, as opposed to old housing for the first experiment, which had been in use for several years. It seems unlikely, however, that the microbial load of the two housing systems would be very different although qualitative differences in the flora may have existed. In a third experiment the author claims that *Lactobacillus* administration increased the numbers of these organisms in the gut and reduced those of enterococci. However, it is unclear whether data are presented for the small intestine or caeca. In fact the $\log_{10}$ *Lactobacillus* counts in the treated and control groups at 3, 6 and 9 days were little different, i.e. 6.45 and 6.89, 6.89 and 6.30 and 8.01 and 7.32. The reductions in enterococci at 3 and 6 days are equally small. The bigger reduction observed at 9 days, from $\log_{10}$ 7.56 in the control to 4.51 in the treated group, is not very unusual considering the natural fluctuations that occur in the numbers of these organisms (Smith, 1965b). The claims made by Tortuero were repeated by Jernigan *et al.* (1985) in their review with little critical appraisal.

Couch (1978) reported several studies in broilers carried out by others. In the first study a *L.acidophilus* strain from an unknown source was incorporated in the feed at 0.025, 0.0375, 0.05, 0.0625 and 0.075% without other additives. The birds were stressed due to abnormally cold weather. There was an increase during an unspecified period in the growth of the males of 7–10% and in the females of 5–6% when the cultures were added. In a second experiment chickens in batteries were given feed containing 0.05, 0.1 and 0.2% culture for a period of 3 weeks.

Chickens given suboptimal amino acid levels showed an accelerated growth rate when the *Lactobacillus* was administered. In a third study a non-viable *Lactobacillus* culture was fed to two houses each containing 16 000 birds at a level of 0.45 $kg^{-1}t^{-1}$. What this represented in terms of numbers of lactobacilli is unknown. An increase in the average weight in the broilers of 46 g, decreased mortality of 0.4% and improved feed conversion of 0.81 units were reported. These papers by Tortuero and Couch suggest, in line with many claims, that probiotics are of particular use when poultry are reared under stressful conditions.

Adler and DaMassa (1980) carried out small-scale laboratory trials with young chicks administered *Lactobacillus* obtained from the crop and ileal walls. Both types adhered well to the crop wall. One of the biotypes was similar to the homofermentative group D biotype described by Fuller (1973). In one experiment, groups of newly hatched chickens were administered 2.5 x 10^8 organisms orally. Wet droppings and occluded vents were observed in the control group and none was seen in the treated group although no data are given. A similar result was obtained when 10^8 lactobacilli per gram of feed were administered continuously to an unknown number of birds from an inbred line prone to vent pasting. The problem was seen in control birds only and had disappeared by 3 weeks. In a third experiment 5 x 10^7 lactobacilli per gram were administered in the feed to groups of 12 male birds for 24 days. A small statistically non-significant increase in mean weight (340.2 g in the treated birds and 331.8 g in the control group) was observed.

Arends (1981) administered a bile acid-resistant *L.acidophilus* strain, whose origin was not stated, *via* the drinking water to broilers held under field conditions. Birds were given antibiotics as normal. In the first trial, four houses of birds containing 116 000 broilers were given 10^8 lactobacilli per day for 30 days. The controls consisted of two houses of 58 242 birds. A 6% weight increase and 3% feed conversion increase were observed. In a second trial, comprising 31 000 birds in treated and control groups, the two above parameters increased by 3% and 1% respectively. In a third trial both groups of chickens were housed together. Administration of 6 x 10^7 lactobacilli daily for 52 days decreased the weight by 4% and increased feed conversion by 3%. A fourth trial produced increases in these two parameters by 2% and 3.6% respectively. The results appeared promising although no statistical analysis was carried out and the results appear only in abstract form.

To these reports indicating beneficial effects can be added others indicating variable results or little or no such effect. Many of these reports are in abstract form.

Variable effects with time were observed by Burkett *et al.* (1977)

who administered an unspecified *Lactobacillus* strain to broilers under commercial conditions, producing better feed conversion at 4 but not at 8 weeks. There was no effect on weight gain.

Dilworth and Day (1978) reported in abstract form the results of two experiments in which a commercial probiotic was evaluated. The preparation containing *L.acidophilus* and other lactobacilli was tested in groups of 60 male and 60 female birds. In the first experiment the preparation was incorporated in the feed at 0.025, 0.0375, 0.05, 0.0625 and 0.0750%, probably representing approximately 10^3 to 10^4 organisms per gram. Adversely cold weather conditions during the experiment produced a reduced growth rate and feed efficiency. These parameters were significantly improved in the groups fed the lactobacilli although no data are presented. A second experiment involving a three-week battery trial employed a different culture at 0.05, 0.1 and 0.2% in the diet. The methods were unclear since half of the diets contained reduced (10%) levels of methionine, cystine and lysine. Although in two of the three comparisons the probiotic had no effect on growth rate it is not stated which groups were involved and what the results were for the third comparison. When the diets were adequate the culture had no effect on growth rate and feed efficiency. This again might indicate the beneficial effects of probiotics under some suboptimal husbandry procedures. Fethiere and Miles (1987) also reported negative results with the same preparation.

The effect of administration of another commercial *L.acidophilus* culture supplied in the drinking water at 2×10^8 per chick per day was reported by Watkins and Kratzer (1983b). The product was administered continuously on alternate days to groups of 50 chicks for 7 weeks. The treatment had no effect on wet viscera weight, body weight or feed conversion. The numbers of lactobacilli were marginally higher ($\log_{10}$ 6.90 and 7.10) in the treated than in the control groups ($\log_{10}$ 6.50). Liver biotin levels were similar in all groups supporting previous findings by these authors (1983). There was a trend for liver biotin to be higher when broiler chicks were dosed orally with either of two avian *Lactobacillus* strains, 59 and 74/1, identified by Fuller (1977) as inhibitory for *E. coli* in the crop. This only occurred when 10^7 organisms per chick per day were administered for 21 days but not when 10^5 or 10^9 were given. The authors suggested that an optimum dose might be required below which no effect may occur and above which competition for biotin may take place. Buenrostro and Kratzer (1983) had also studied the effect of administering a commercial host 'non-specific' strain 40 on alternate days at 10^8 per chick per day on liver biotin. The source and characteristics of this organism were not mentioned. Broilers reared in battery brooders did not perform as well as control birds. The diet was marginally deficient in biotin (56 μg

kg^{-1}) but administration decreased significantly ($p<0.01$) liver biotin levels from 1596 ng g^{-1} in control birds to 931 ng g^{-1} in treated birds. The authors discussed the possibility of an association between competition between intestinal organisms for biotin and sudden death associated with low biotin levels. This correlates with the findings of Coates *et al.* (1968) that the normal gut flora does not compensate for dietary vitamin B deficiency.

10.5.2 Effect on laying hens

A number of different cultures and products have also been tested in laying hens producing equally variable results.

Krueger *et al.* (1977) reported in abstract form the results of feeding a so-called *Lactobacillus* complex to young leghorn hens at a concentration of 2.27 kg t^{-1}. No information is given about the bacteria used or the viable numbers present in the feed. Three groups each of treated and control pens housing 26 young females and 2 males were monitored for 140 days. Although actual figures were not given, the treatment produced an improvement in egg production and feed efficiency of 3.03 and 7.41%. No information is given about the statistical significance and whether there was variation with time. The effect on fertility and hatchability was confusing although considered to be favourable.

Crawford (1979) tested a mixed *Lactobacillus* preparation at 340 g t^{-1} in 101 615 commercial hens. Although there was an increase in egg production from 69.5% in control hens to 72.17% in treated birds, these production figures are extremely and unnaturally low – and this is not explained. The amount of feed required to produce a dozen eggs was reduced from 1.75 to 1.69 kg.

Miles *et al.* (1981a,) carried out a study at three sites: Florida, South Dakota and Arizona. A mixed *Lactobacillus* preparation was again tested in the feed at 0.0125, 0.0375 and 0.0625%. The viable counts of different batches of probiotic were estimated at a minimum of 4 x 10^6 organisms per gram, giving very low levels of viable organisms in the feed. Treated and untreated feed were given to seven groups of ten layers from 24 weeks of age for 280 days. The number of hens involved was rather small but an increase in egg production was seen in Florida with the three concentrations of 72.77, 72.57 and 70.88% in treated birds compared with 70.89% for control hens. Similar results were obtained at Arizona but not at South Dakota. The figures are again low by present UK commercial standards. The absence of an increase at the higher level was attributed to excessive numbers of organisms, but this is unlikely to be the case. There was no effect on feed efficiency or any statistically significant change in egg weight. Although they claim an increase in *Lactobacillus* numbers and a decrease in coliform counts, the data do not indicate this. In any case it seems incongruous to combine samples from the small and large intestine (sic) to produce such counts. A similar study by Francis *et al.* (1978) in Florida indicated improved performance and reduced intestinal coliform counts, but the information is in abstract form and no data are given.

Cerniglia *et al.* (1983) reported uniformly negative results from five trials of feeding various unspecified *Lactobacillus* products in their diets. The authors tested a liquid non-viable product and a dried non-viable product. There were no significant effects on egg production, feed consumption or mortality. When the dried product was fed at 686 g t^{-1} an increase in the number of large eggs was observed. This was also seen previously by Couch (1978) feeding a viable *L.acidophilus* strain from an unknown source. However, the significance of the results for performance is unclear.

10.5.3 Effect on other poultry

Probiotic cultures have also been administered to turkeys and other poultry. In several separate studies the mixed *Lactobacillus* preparation described in the sections on broilers (see Dilworth and Day, 1978) has been assessed.

Francis *et al.* (1978) tested the commercial preparation in groups of 48 broad-breasted large white turkey poults administering 750 mg kg^{-1} in the feed for 3 weeks. An increase in body weight from 411.8 to 424.6 g was observed but the feed efficiency fell slightly from 1.40 to 1.39. Changes in the intestinal flora included reductions in the coliform counts in the gizzard, small intestine and caeca from (log_{10}) 4.47, 7.92 and 8.61 in control birds to 1.65, 4.18 and 5.11 in treated birds respectively. The *Lactobacillus* counts in the same organ increased from 4.17, 5.07 and 5.88 to 5.17, 7.78 and 8.32 respectively. However, the coliform counts in the control birds seem abnormally high and those of lactobacilli unusually low (Table 10.1). The reproducibility of those changes would need to be further assessed.

This probiotic was also tested by Crawford (1979) who found a 6.1% weight increase at 12 weeks of age after continuous administration at 0.2 kg t^{-1}. However, the increase at 16 weeks was not significant. Very similar results were reported by Potter *et al.* (1979). Damron *et al.* (1981) tested the product in broad-breasted white turkey hens at 625 mg kg^{-1} but found no effect in weight gain, egg production or hatchability and Miles *et al.* (1981b, c) tested it in Bob White quail at the same level but found no effect on growth, feed consumption, hatchability or mortality.

10.5.4 Effects against potential enteric pathogens

Studies have been largely confined to the effects of lactobacilli on pathogenic and non-pathogenic *E.coli* and *Salmonella*.

Watkins *et al.* (1982) reported on the results of two trials testing a strain resembling *L.acidophilus* of unspecified origin. Two-day-old gnotobiotic chicks were inoculated orally with high counts (10^8–10^9) of *Lactobacillus* followed 2 days later by a similar sized inoculum of a pathogenic 02 : K1 strain of *E.coli*. Additional oral inoculations of the *Lactobacillus* produced a mortality of 3.7%, whereas the *E.coli* administered on its own at this time produced 66.7% mortality. When the *E.coli strain was inoculated followed by one or two inocula of the Lactobacillus* the mortality was reduced to 54 and 17% respectively. Watkins *et al.* (1982) reported, in abstract form, that avian *Lactobacillus* strains KTM and 59 (the *L. salivarius* described by Fuller, 1977) reduced commensal *E.coli* counts in broiler caeca whereas strains 74/1 (also described by Fuller) and A1 (described as host non-specific, although the meaning of this is not clear) did not have this effect. No information was given on the effects in the crop. Fuller (1977) considered that both 59 and 74/1 were able to reduce small intestinal *E.coli* counts.

Watkins and Miller (1983) tested the *L. acidophilus*-like strains described by Watkins *et al.* (1982) for its inhibitory activity against *Sal. typhimurium* and *Staphylococcus aureus* in gnotobiotic chicks. The numbers of *Lactobacillus*, *Salmonella* and *Staphylococcus* inoculated were 10^8, 10^9 and 10^{10} respectively. Significant reductions in mortality ($p<0.01$) and in faecal shedding ($p<0.05$) were produced when the *Lactobacillus* was administered prophylactically 2 days before the pathogens. The *Salmonella* on its own produced 33% mortality, 14% when preceded by the *Lactobacillus* and 9.6% when the *Lactobacillus* was administered therapeutically in five doses. The corresponding figures for *Staph. aureus* were 43, 31 and 28%. It must be remembered, however, that the numbers dying were very small since the original group sizes were also small.

Adler and DaMassa (1980) in a very small trial inoculated groups of only five newly-hatched chicks orally with 10^{10} lactobacilli, isolated originally from the crop and demonstrating adhesiveness. Twenty-four hours later the birds were challenged with *Sal. infantis*. After 7 or 8 days the mean $\log_{10}$ viable caecal *Salmonella* count in the birds challenged with either 10^5, 10^2 or 5×10^1 *Salmonella* organisms were ($\log_{10}$) 6.8, 6.0 and 7.4 in treated birds and 6.8, 7.6 and 8.2 in control birds respectively. Not surprisingly none of these reductions was statistically significant.

Barnes *et al.* (1980) administered homofermentative and heterofermentative lactobacilli obtained both from the caeca of chickens which had been protected against *Salmonella* infections with a competitive exclusion culture (see below) and from other sources. They were inoculated orally into newly hatched chicks followed 24 hours later by 10^3 organisms of *Salmonella typhimurium*. No protection was obtained when lactobacilli were given in the feed or via the crop or when lactobacilli were administered with *Bact. vulgatus* and *Bifidobacterium* sp. or with two *Bifidobacterium* strains together with a *Eubacterium* sp. In fact the authors considered that where lactobacilli had eliminated coliforms the *Salmonella* counts were 10- to 100-fold higher. They thought that although lactobacilli were inhibitory for *Salmonella* they were also inhibitory for other organisms and that oral administration on their own could disturb the intestinal ecology. However, Impey *et al.* (1982) indicated that lactobacilli were an important component of any effective competitive exclusion mixture and their exclusion reduced its protective ability.

Soerjadi *et al.* (1981) administered a mixture of lactobacilli obtained from the crop and caeca of a chicken, which as above produced an inhibitory microflora for competitive exclusion, together with *L. salivarius* strain 59 (Fuller, 1977). Birds were inoculated within hours of hatching and were challenged with *Sal. typhimurium* at two hours or one day post-treatment. Initially there was little effect on the *Salmonella* counts in the crop epithelial homogenate. However, some reductions ($>\log^{10}2$) were seen at 2 and 3 days after challenge. There was no significant effect on the *Salmonella* counts in the caecal tissue homogenate. Administering the lactobacilli had no effect on the number of chickens shedding *Salmonella*. Strain 59 on its own had little effect on *Salmonella* shedding. Protection was obviously more apparent in the crop than in the caeca and it took some time for maximum protection to occur.

10.5.5 Effects of administering carbohydrates and fermentation products

While it is understandable that non-viable *Lactobacillus* preparations would be less effective than live cultures there seems some logic in attempting to manipulate the normal intestinal flora by the administration of fermentable carbohydrates or by administration of major fermentation products.

Oyofo *et al.* (1989a) administered different carbohydrates at a relatively high level of 2.5% in the drinking water of two replicate groups of 15 commercial broilers. At 3 days of age they were inoculated with 10^8 organisms of *Sal. typhimurium*. Caecal swabs taken one week later indicated that although the *Salmonella* was present in 100% of control birds it was present in 16/30 (53%) and 8/20 (27%) of those administered lactose and mannose. These birds also had lower caecal *Salmonella* counts ($\log_{10}$ 3.6 and 2.9 respectively compared with the control group of 6.6). Dextrose, sucrose and maltose had no effect. It was thought that the two carbohydrates acted by different mechanisms although there was little evidence to support this. The authors thought that since the *Salmonella* possessed type 1 pili which mediate a mannose-sensitive adhesion to epithelial cells (among other cells) *in vitro* it was likely that mannose prevented colonization by this mechanism (Oyofo *et al.*, 1989b). However, there is no evidence that *Salmonella* colonize the gut by adhesion, let alone type-1 pili-mediated adhesion. *Salmonella* strains which do not possess these organelles are very frequent and colonize the alimentary tract of chickens equally well. It was considered that lactose at this concentration stimulated the development of a lactose-fermenting flora which was inhibitory for *Salmonella*. Additional studies (DeLoach *et al.*, 1990) supported the original findings by demonstrating similar inhibitory effects in the caeca with 5% lactose in drinking water or with 5% milk whey in the feed. Although the provision of lactose at this level may not be economically feasible the results are interesting and further studies on the mechanism of action are warranted since it is unclear whether the major area of activity is the caeca or crop.

Feeding high concentrations of lactose (20%) or of dried skim milk (40%) to chickens also reduced the severity of infection with *Eimeria*, the avian protozoan. This may well again have been due to the selection of a highly fermentative flora with the result that organisms such as *E. coli* and *Bacteroides*, which can exacerbate the severity of the disease, were reduced in numbers (Beach and Davies, 1925).

Feeding lactic acid and butyric acids to chickens at low concentrations had little effect on *Salmonella* excretion, whereas higher concentrations are not tolerated or consumed readily by poultry. Propionic acid and, particularly, formic acid can be incorporated in feeds at low concentrations (<1.0%) which will reduce the level of *Salmonella* contamination in feeds thereby reducing infection of poultry (Mulder, 1980; Hinton *et al.*, 1985). The mechanism of activity is not fully understood.

An unspecified fermentation product administered to 21-week-old hens at a level of 0.25% in the feed for a short period of 16 days had a small effect on egg production. However a second experiment produced no effect (Charles and Duke, 1978). Since the information was published in abstract form, and no further details were available as to the nature of the product, the significance of these findings is unclear. The understanding of the interrelationships between lactobacilli and other organisms is not yet great enough to be able to advocate the use of minor fermentation products such as bacteriocins and other antibiotic-like products produced by lactic acid bacteria as probiotics.

10.5.6 Lactic acid bacteria as probiotics – a critical résumé

This review of the published effects of probiotic cultures and preparations indicates how variable the effects can be. For critical appraisal it is essential that full details of bacterial strains used and experimental design be given. Many experiments are reported as abstracts of oral communications presented at meetings. It is unclear why these presentions were not published in full particularly since in several cases beneficial effects were claimed.

The origin of bacterial strains must be given. So much information has now accumulated indicating the microbial characteristics, particularly of lactobacilli, that are essential for colonization that it is difficult to understand the rationale of using non-avian strains. There is also rarely any assessment of the extent of implantation of the administered strains. Work in this area would be useful. Because of this much of the reviewed work has to be assessed almost empirically. This is not to detract from the clear beneficial effects that have been demonstrated in some cases.

A second criticism is that microbiological results presented in papers of primarily nutritional interest are poorly interpreted indicating poor understanding of the microecology of the avian alimentary tract.

A third comment of importance is that in some cases where a beneficial nutritional effect has been demonstrated the values of performance or production in the control groups themselves are frequently usually low. Whether this indicates purely biological variation between the control and treated groups or a real improvement under poor management or other conditions is unclear. This may be significant because one of the areas where probiotics may be of real value is where poor management or other adverse conditions result in poor performance. Perhaps more studies should be carried out in this area.

However, some of the interpretation of results is obviously over-optimistic and arises mainly from a näive and uncritical acceptance of the data or of speculations of previous workers. Attitudes such as this serve only to create a mystique of probiosis without an adequate rational assessment of its true value.

10.6 COMPETITIVE EXCLUSION

Competitive exclusion (CE) as a principle is designed primarily to reduce intestinal colonization by enteric pathogens such as *Salmonella* and *Campylobacter*. Its use is not thought to improve performance. Unlike the use of lactic acid bacteria as probiotics the demonstrable effect is profound and reproducible.

The principle of CE is that newly hatched chicks are far more susceptible to infection with *Salmonella* than are adult birds largely because of the absence of the complex inhibitory microflora present, largely in the crop and the caeca of the adult. The flora of the chick develops over several weeks mainly because of the structure of the industry combined with high standards of hygiene practised in hatcheries preventing young birds acquiring microorganisms directly from the hen. Chicks can acquire the resistance to infection possessed by the adult by oral inoculation with a suspension or anaerobic culture of faeces or caecal contents obtained from *Salmonella*-free adult birds (Nurmi and Rantala, 1973). The phenomenon has since been studied in considerable detail (Pivnick and Nurmi, 1982; Mead and Impey, 1987).

The protective effect is not affected markedly by breed or sex of the bird providing the donor material. Viable bacteria are required for the effect although the immune system is not involved in protection. The treatment is essentially prophylactic although some reduction in *Salmonella* excretion is obtained by treatment of birds already infected with *Salmonella*. Material from chickens and turkeys can protect each other unless the material is obtained from the more limited gut flora of specified pathogen-free (SPF) birds. Cultured faeces from mammals are ineffective in chickens.

The mechanism of protection is poorly understood but, essentially, treatment prevents the intra-intestinal growth of *Salmonella*, such that organisms are gradually eliminated from the caeca. The antibacterial mechanisms, although unknown, may include pH, low Eh and miscellaneous inhibitory substances such as H_2S, bacteriocins, fatty acids and deconjugated bile acids and competition for nutrients and adhesion receptor sites. There is considerable interest in this last facet of inhibition supported by the observation that the inhibition begins to take effect within one to two hours of challenge. Whether this is the reason is unclear since there is very little evidence that active adhesion of *Salmonella* to the crop or caecal epithelium occurs *in vivo* although the normal flora colonizes the thick glycocalyx covering the mucosa.

A number of factors can affect the degree of protection. In the same way that some antibiotics (Smith *et al.*, 1985) particularly those used for growth-stimulation, can increase faecal excretion by *Salmonella* in normal birds, they can also disrupt the efficacy of CE mixtures. Dietary tylosin, avoparcin and a mixture of lincomycin and spectinomycin are reported to produce deleterious effects whereas nitrovin did not. Stress, either physiological provoked by abnormally high or low temperatures or deprivation of food and water, may reduce efficacy. As discussed above, this is an area where probiotics containing lactic acid bacteria might have a role. Concurrent infections with *Eimeria*, *Mycoplasma* or avian viruses may also reduce protection. Variable results can also be produced by altering experimental design emphasizing the need for a standardized protocol in experimental work.

CE is effective in reducing excretion of *Salmonella* serotypes that produce food-poisoning but is not so effective against *Sal. gallinarum* which does not colonize the alimentary tract to the same extent. CE is effective against *Sal. arizonae* and some pathogenic *E. coli* strains, but does not affect the normal *E. coli* flora. The variation seen in the efficacy against strains of this organism indicates how little is known of the colonization characteristics of the gut flora. CE is also partially protective against *Campylobacter jejuni*, *C. perfingens*, *C. botulinum* and *Yersinia enterocolitica*, although the latter organism is rarely encountered in poultry. However, whether a reduction from 100% carriage of *Cam. jejuni* in control birds to 4% in treated birds, although it is large, would translate into a much lower carcass carriage after slaughter and processing remains to be seen.

The major practical problem of using suspensions of faeces from adult chickens is that other avian pathogens may be introduced into previously uninfected stock. Viral and parasite contamination can be eliminated by dilution by *in vitro* passage. Under anaerobic cultural conditions neither *Listeria monocytogenes* nor *Cam. jejuni* thrive. Such cultures can be tested by inoculation into SPF chickens and monitoring the birds for the production of antibodies against the agents concerned. Another alternative is to limit use of any one source of material to recipient flocks from the same farm or company. This problem can be overcome by using mixtures of cultures of individual organisms isolated from undefined faeces cultures. Such mixtures containing 10–50 strains are produced empirically since there is no information on the desired characteristics of protective organisms. Both obligate and facultative anaerobes are required although exactly which ones are essential is not known (Table 10.2). Defined cultures have some disadvantages when compared with undefined cultures. Mixtures which give consistent protection are too large to be of practical use. Obligate anaerobes also sometimes lose viability on storage. Defined mixtures are more host-specific and are less able to withstand high challenge doses of more that 10^4 *Salmonella* organisms per bird. It is possible that essential organisms present in relatively low numbers in faeces are not present in such mixtures.

Undefined cultures have been tested on a field scale on a few occasions. In one experiment in Sweden involving 2.86 million broiler chicks, the incidence of infection was 1/144 in CE treated flocks and 87/144 in untreated flocks. In most cases chickens were reared

Table 10.2 The composition of defined bacterial mixtures with a demonstrable competitive exclusion effect.

Bacterial type	Organisms present or absent in the mixtures described by Impey *et al.* (1982)	Nurmi (1985)	Stavric *et al.* (1985)
Escherichia	+	+	+
Streptococcus	+	+	+
Bacillus	+	–	–
Bacteroides	+	+	+
Fusobacterium	–	+	+
Lactobacillus	+	+	+
Eubacterium	+	+	+
Propionibacterium	–	+	+
Clostridium	+	–	+
Bifidobacterium	+	–	+
Gram-positive, anaerobic cocci	+	+	–
Other Gram-positive, anaerobic rods	+	–	+
Budding bacteria	+	–	–

on premises from which *Salmonella* had been previously isolated. Smaller reductions were observed in 284 flocks in the Netherlands. The incidence of infected flocks was 14.7% in treated and 24.1% of untreated flocks. The proportion of infected birds in these two groups was 6.4 and 14.3%. CE has also been used successfully in 20/22 trials in the United Kingdom where it has been combined with antibiotic therapy. Birds were given a period of chemotherapy to eliminate or reduce *Salmonella* infection. They were then transferred to new premises (a laying farm) and administered a CE mixture.

The problems of using large, complex mixed cultures suggest that smaller pools or individual protective organisms might be an advantage if they could be found. Soerjadi *et al.* (1978) reported that a strain of *Ent. faecalis* was protective against *Salmonella* although the effect could not be reproduced by other workers (Goren, 1982; Pivnick and Blanchfield, 1982). Rigby *et al.* (1977) reported that an unidentified strain of *Clostridium* was also effective. However, none of these studies has been pursued. Barrow and Tucker (1986) isolated three strains of *E. coli* from sewage and an abattoir drain which, when administered simultaneously to newly hatched chicks, gave partial protection against challenge by *Sal. typhimurium*. These authors were not able to isolate an individual Gram-negative facultative anaerobe which was protective on its own from more than 600 cultures examined. Other genera of the Enterobacteriaceae were also ineffective. It was considered that for a single strain to be protective it must possess the colonization characteristics of *Salmonella* without its virulence attributes. Barrow *et al.* (1987) and Berchieri and Barrow (1990) found that intestinal colonization of chicks by *Sal. typhimurium* or other serotypes could be prevented by pre-inoculation with an avirulent, rough mutant of the homologous strain. The protection was profound and lasted several weeks. The mechanism was not related to immunity, or phage or bacteriocin activity. A similar inhibition occurred between cultures *in vitro* which was mediated by a single surface protein. Although consumer resistance would prevent the adoption of such mutants because of their *Salmonella* origins, the chances of reversion to virulence can be eliminated by appropriate manipulations using transposon mutagenesis. Deletion of the chromosomal gene(s) responsible for enterotoxin should produce mutants avirulent for man, but still possess the colonization attributes of *Salmonella*. This approach of colonization inhibition by avirulent mutants of homologous organisms could also be extrapolated to *Cam. jejuni* whose colonization characteristics are only now being studied and where enterotoxin genes are thought to be important mediators of diarrhoea. Whether this approach might also apply to organisms such as clostridia and growth-depressing strains of *Ent. hirae* is unclear. For the latter organism some difficulties might be encountered since colonization of the duodenal mucosa by any organism might be sufficient to affect nutrient absorption and reduce growth rates.

Using a similar approach, the colonization of the intestine with antibiotic-resistant *E. coli* strains in mature broilers has been reduced by oral inoculation with large numbers of antibiotic-sensitive *E. coli* strains (Linton *et al.*, 1978). Although it is expected that *E. coli* would be neither detrimental nor beneficial to poultry (Siccardi and Pomeroy, 1964), Schmidt *et al.* (1988) suggested that a strain isolated from the yolk sac of stunted turkey poults, when administered orally to day-old poults, increased body weight, feed consumption and feed efficiency by 4.5, 2.1 and 2.4% respectively. This serves to illustrate how little we know about the *in vivo* characteristics of many intestinal organisms.

10.7 IMMUNITY

The stimulation of a classical immune response by oral immunization of poultry with live, attenuated bacterial vaccines is not usually considered as probiosis. Most of the work has been done in relation to *Salmonella* and this has been reviewed elsewhere (Mead and Barrow, 1990, Barrow 1991). In summary, intestinal colonization and tissue invasion by a number of *Salmonella* can be reduced by vaccination with live, attenuated vaccines prepared empirically. Live vaccines are considered to be more effective than killed vaccines but the problems of consumer resistance also apply here. The advantage of using a live mutant of *Salmonella* is that such a strain, inoculated soon after hatching, could protect chicks in the first week of so of life by colonization inhibition as described above and later by the development of an active immunity. Similar work for *Campylobacter*, *Clostridium* and *Enterococcus* is required.

10.8 BACTERIOPHAGES

Bacteriophages were sought avidly in the early years of this century in attempts to treat the major human diseases (Wilson and Miles, 1973). However, the paucity of information on microbial pathogenesis and the fact that phages active *in vitro* are not necessarily active *in vivo* meant that their early promise was not fulfilled. However, as a result of more recent work they are now being advocated as a novel method of microbial control (Report, 1988). Virulent phages have recently been used to successfully control, by treatment and prophylaxis, experimental systemic *E. coli* infections in mice (Smith and Huggins, 1982) and enterotoxigenic *E. coli* diarrhoea in pigs, calves and lambs (Smith and Huggins, 1983). They had advantages over antibiotics in that phage resistant mutants which developed were generally less virulent than the parent strain. Phages lytic for *Sal. typhimurium* can be isolated from poultry feed, faeces and sewage. Some phages isolated from the latter source can reduce mortality in chicks caused by this organism from 53 to 16% if inoculated with the phage soon after infection with the *Salmonella* (Berchieri *et al.*, 1991). Such an approach would be interesting to consider for *Campylobacter* and growth-depressing organisms. Whether they would be as effective as they are with enterotoxigenic *E. coli* is unclear since the pathogenesis of these infections are very different. Houghton and Fuller (1980) found that a bacteriophage, lytic for *Ent. hirae* SY1, was active on this organism in the chicken gut. Phage-resistant organisms developed after administration of the phage to *Ent. hirae* infected chickens and the phage was able to partially reverse the growth depression induced by the organism.

10.9 SUMMARY

The microbial flora of the alimentary tract can have a considerable effect on the health and performance of poultry. Disturbance of the flora could thus lead to detrimental effects by allowing colonization by pathogens or by growth-depressing bacteria. Many attempts to restore a stable flora either in newly hatched birds that have not yet acquired a resistant flora or in adult birds have been based more on an ill-defined conception of the role of individual components of the flora and reflect our ignorance in this area. However, the information which is available on colonization by some potentially useful organisms is often ignored by workers in this field. It is not surprising that in such instances results are poor or inconclusive. Any good results, even if discovered by accident, must be worth repetition and study but good results are more likely to be obtained if individual probiotic preparations are selected for use and administered based on sound scientific rationale using all the available information on intestinal microecology. It is lamentable, from this point of view, that so few groups of scientists are now working in this area.

REFERENCES

Adler, H.E. and DaMassa, A.J. (1980) Effects of ingested lactobacilli on *Salmonella infantis* and *Escherichia coli* and on intestinal flora, pasted vents and chick growth. *Avian Dis.*, **24**, 868–78.

Arends, L.G. (1981) Influence of *L. acidophilus* administered *via* the drinking water on broiler performance. *Poultry Sci.*, **60**, 1617 (Abstract).

Balish, E. and Phillips, A.W. (1966) Growth and virulence of *Candida albicans* after oral inoculation in the chick with a monoflora of either *Escherichia coli* or *Streptococcus faecalis*. *J. Bacteriol.*, **91**, 1744–9.

Barnes, E.M., Impey, C.S. and Cooper, D.M. (1980) Competitive exclusion of salmonellae from the newly hatched chick. *Vet. Record*, **160**, 61.

Barrow, P.A. (1987) Microbial antibiotic resistance in animals, in *Combating Resistance to Xenobiotics* (eds M.G Ford *et al.*), Ellis Horwood, Chichester, UK.

Barrow, P.A. (1991) Immunological control of *Salmonella* in poultry. *Proc. Int. Symp. on Control of Human Bacterial Pathogens in Poultry, Atlanta, Georgia, USA* pp. 199–217.

Barrow, P.A., Simpson, J.M. and Lovell, M.A. (1988) Intestinal colonization on the chicken by food-poisoning *Salmonella* serotypes; microbial characteristics associated with faecal excretion. *Avian Pathol.*, **17**, 571–88.

Barrow, P.A. and Tucker, J.F. (1986) Inhibition of colonization of the chicken with *Salmonella typhimurium* by pre-treatment with strains of *Escherichia coli*. *J. Hygiene. (Camb.)*, **96**, 161–9.

Barrow, P.A., Tucker, J.F. and Simpson, J.M. (1987) Inhibition of colonization of the chicken alimentary tract with *Salmonella typhimurium* by Gram-negative facultatively anaerobic bacteria. *J. Hygiene (Camb.)*, **98**, 311–22.

Beach, J.R. and Davis, D.E. (1925) The influence of feeding lactose or dry skim milk on artificial infection of chicks with *Eimeria avium*. *Hilgardia*, **1**, 167–81.

Berchieri, A. Jr and Barrow, P.A. (1990) Further studies on the inhibition of colonization of the chicken alimentary tract with *Salmonella typhimurium* by pre-colonization with an avirulent mutant. *Epidemiol. Infect.*, **104**, 427–41.

Berchieri, A. Jr, Lovell, M.A. and Barrow, P.A. (1991) The activity in the chicken alimentary tract of bacteriophages lytic for *Salmonella typhimurium*. *Res. Microbiol.* **142**, 541–9.

Blaxland, J.D. and Fincham, I.H. (1950) Mycosis of the crop (moniliasis) in poultry, with particular reference to serious mortality occurring in young turkeys. *Brit. Vet. J.*, **106**, 221–31.

Bohnhoff, M., Drake, B.L. and Miller, C.P. (1954) Effect of streptomycin on susceptibility of the intestinal tract to experimental salmonella infection. *Proc. Soc. Exp. Biol. Med.*, **86**, 132–7.

Brooker, B.E. and Fuller, R. (1975) Adhesion of lactobacilli to the chicken crop epithelium. *J. Ultrastruct. Res.*, **52**, 21–31.

Buenrostro, J.L. and Kratzer, F.H. (1983) Effects of *Lactobacillus* inoculation and antibiotic feeding of chickens on availability of dietary biotin. *Poultry Sci.*, **62**, 2022–9.

Burkett, R.F., Thayer, R.H. and Morrison, R.D. (1977) Supplementing market broiler rations with lactobacillus and live yeast cultures. Animal Science Agricultural Research Report, Oklahoma State University and USDA, USA.

Cerniglia, G.J., Goodling, A.C. and Herbert, J.A. (1983) The response of layers to feeding lactobacillus fermentation products. *Poultry Sci.*, **62**, 1399. (Abstract)

Charles, O.W. and Duke, S. (1978) The response of laying hens to dietary fermentation products and probiotic–antibiotic combinations. *Poultry Sci.*, **57**, 1125. (Abstract)

Coates, M.E., Ford, J.E. and Harrison, G.F. (1968) Intestinal synthesis of vitamins of the B complex in chicks. *Brit. J. Nutr.*, **22**, 493–500.

Couch, J.R. (1978) Poultry researchers outline benefits of bacteria, fungistatic compounds, other feed additives. *Feedstuffs* **50**, 6.

Crawford, J.S. (1979) 'Probiotics' in animal nutrition. *Proc. 1979 Arkansas Nutrition Conf.*, pp. 45–55.

Damron, B.L., Wilson, H.R., Voitle, R.A. and Harms, R.H. (1981) A mixed lactobacillus culture in the diet of Broad Breasted Large White turkey hens. *Poultry Sci.*, **60**, 1350–1.

DeLoach, J.R., Oyofo, B.A., Corrier, D.E., Kubena, L.F., Ziprin, R.L. and Norman, J.O. (1990) Reduction of *Salmonella typhimurium* in broiler chickens by milk or whey. *Avian Dis.*, **34**, 389–92.

Dilworth, B.C. and Day, E.J. (1978) Lactobacillus cultures in broiler diets. *Poultry Sci.*, **57**, 1101 (Abstract).

Dubos, R., Schaedler, R.W., Costello, R. and Hoet, P. (1965) Indigenous, normal and autochthonous flora of the gastrointestinal tract. *J. Exp. Med.*, **122**; 67–76.

Duval-Iflah, Y., Chappuis, J.P., Ducluzeau, R. and Raibaud, P. (1983) Intraspecific interactions between *Escherichia coli* in human newborns and in gnotobiotic mice and piglets. *Prog. Food Nutr. Sci.*, **7**, 107–16.

Eyssen, H. and DeSomer, P. (1967) Effects of *Streptococcus faecalis* and a filterable agent on growth and nutrient absorption in gnotobiotic chicks. *Brit. Poultry Sci.*, **46**, 323–33.

Eyssen, H., Swaelen, E., Kowszyk-Gindifer, Z. and Parmenteer, G. (1965) Nucleotide requirements of *Lactobacillus acidophilus* variants isolated from the crops of chicks. *Antonie van Leeuwenhoek*, **31**, 241.

Fethiere, R. and Miles, R.D. (1987) Intestinal tract weight of chicks fed an antibiotic and probiotic. *Nutr. Rept Int.*, **36**, 1305–9.

Ford, D.J. (1974) The effect of the microflora on gastrointestinal pH in the chick. *Brit. Poultry Sci.*, **15**, 131–40.

Francis, C., Janky, D.M., Arafa, A.S. and Harms, R.H. (1978) Interrelationship of lactobacillus and zinc bacitracin in the diets of turkey poults. *Poultry Sci.*, **57**, 1687–9.

Fuller, R. (1973) Ecological studies on the lactobacillus flora associated with the crop epithelium of the fowl. *J. Appl. Bacteriol.*, **36**, 131–9.

Fuller, R. (1975) Nature of the determinant responsible for the adhesion of lactobacilli to chicken crop epithelial cells. *J. Gen. Microbiol.*, **87**, 245–50.

Fuller, R. (1977) The importance of lactobacilli in maintaining normal microbial balance in the crop. *Poultry Sci.*, **18**, 85–94.

Fuller, R. (1978) Epithelial attachment and other factors controlling the colonization of the intestine of the gnotobiotic chicken by lactobacilli. *J. Appl. Bacteriol.*, **45**, 389–95.

Fuller, R. (1986) Probiotics. *J. Appl. Bacteriol. Symp.*, No. 15, pp. 1–7.

Fuller, R. and Moore, J.H. (1967) The inhibition of the growth of *Clostridium welchii* by lipids isolated from the content of the small intestine of the pig. *J. Gen. Microbiol.*, **46**, 23–41.

Fuller, R. and Turvey, A. (1971) Bacteria associated with the intestinal wall of the fowl (*Gallus domesticus*). *J. Appl. Bacteriol.*, **34**, 617–22.

Fuller, R., Coates, M.E. and Harrison, G.F. (1979) The influence of specific bacteria and a filterable agent on the growth of gnotobiotic chicks. *J. Appl. Bacteriol.*, **46**, 335–42.

Fuller, R., Houghton, S.B. and Brooker, B.E. (1981) Attachment of *Streptococcus faecium* to the duodenal epithelium of the chicken and its importance in colonization of the small intestine. *Appl. Environ. Microbiol.*, **41**, 1433–41.

Gilliland, S.E. and Speck, M.L. (1977) Antagonistic action of *Lactobacillus acidophilus* toward intestinal and foodborne pathogens in associative culture. *J. Food Protect.*, **40**, 820–3.

Goren, E. (1982) Unpublished data cited by Pivnick, H. and Nurmi, E. The Nurmi concept and its role in the control of *Salmonella* in poultry. *Devel. Food Microbiol.*, **1**, 41–70.

Hamdan, I.Y. and Mikolajcik, E.M. (1974) Acidolin: an antibiotic produced by *Lactobacillus acidophilus*. *J. Antibiot.*, **27**, 632–6.

Hinton, M., Linton, A.H. and Perry, F.G. (1985) Control of *Salmonella* by acid disinfection of chick's food. *Vet. Record*, **116**, 502.

Houghton, S.B. and Fuller, R. (1980) Ecology of *Streptococcus faecium* bacteriophage in chicken gut. *Appl. Environ. Microbiol.*, **39**, 1054–8.

Houghton, S.B., Fuller, R. and Coates, M.E. (1981) Correlation of growth

depression of chicks with the presence of *Streptococcus faecium* in the gut. *J. Appl. Bacteriol.*, **51**, 113–20.
Impey, C.S., Mead, G.C. and George, S.M. (1982). Competitive exclusion of Salmonellas from the chicken caecum using a defined mixture of bacterial isolates from the caecal microflora of the adult bird. *J. Hygiene*, (Camb.) **89**, 479–490.
Jernigan, M.A., Miles, R.D. and Arafa, A.S. (1985) Probiotics in poultry nutrition – a review. *Worlds Poultry Sci. J.*, **41**, 99–107.
Krueger, W.F., Bradley, J.W. and Patterson, R.H. (1977) The interaction of gentian violet and lactobacillus organisms in the diet of Leghorn hens. *Poultry Sci.*, **56**, 1729. (Abstract)
Lev, M. and Forbes, M. (1959) Growth response to dietary penicillin of germ-free chicks and of chicks with a defined intestinal flora. *Brit. J. Nutr.*, **13**, 78–84.
Linton, A.H., Howe, K., Richmond, M.H. *et al.* (1978) Attempts to displace the indigenous antibiotic resistant gut flora of chickens by feeding sensitive strains of *Escherichia coli* prior to slaughter. *J. Appl. Bacteriol.*, **45**, 239–47.
Mead, G.C. (1989) Microbes of the avian caecum; types present and substrates utilized. *J. Exp. Zool. Supp.*, **3**, 48–54.
Mead, G.C. and Barrow, P.A. (1990) *Salmonella* control in poultry by competitive exclusion or immunization. *Lett. Appl. Microbiol.*, **10**, 221–7.
Mead, G.C. and Impey, C.S. (1987) The present status of the Nurmi concept for reducing carriage of food-poisoning salmonellae and other pathogens in live poultry, (ed. F.J.M. Smulders) in *Elimination of Pathogenic Organisms from Meat and Poultry*, Elsevier, Amsterdam.
Metchnikoff, E. (1907) *The Prolongation of Life*. Heinemann, London.
Miles, R.D., Arafa, A.S., Harms, R.H. *et al.* (1981a) Effects of a living non-freeze dried *Lactobacillus acidophilus* culture on performance, egg quality and gut microflora in commercial layers. *Poultry Sci.*, **60**, 993–1004.
Miles, R.D., Wilson, H.R., Arafa, A.S. *et al.* (1981b) The performance of Bobwhite quail fed diets containing lactobacillus. *Poultry Sci.*, **60**, 894–6.
Miles, R.D., Wilson, H.R. and Ingram, D.R. (1981c) Productive performance of Bobwhite quail fed a diet containing a lactobacillus culture. *Poultry Sci.*, **60**, 1581–2.
Morishita, Y., Mitsuoka, T., Kaneuchi, C. *et al.* (1971) Specific establishment of lactobacilli in the digestive tract of germfree chickens. *Jap. J. Microbiol.*, **15**, 531–8.
Mulder, R.W.A.W. (1980) Control of salmonellae in poultry feed., in *Proc. 6th European Poultry Congress, Hamburg*.
Niemen, C. (1954) Influence of trace amounts of fatty acids on the growth of microorganisms. *Bacteriol. Rev.*, **18**, 147–63.
Nurmi, E. (1985). Use of competitive exclusion in prevention of Salmonellae and other enteropathogenic bacterial infections in poultry. In: (Rd. G.H. Snoeyenbos. *Proceedings of the International Symposium on Salmonella*, New Orleans. American Association of Avian Pathologists, University of Pennsylvania, pp. 64–71.
Nurmi, E. and Rantala, M. (1973) New aspects of *Salmonella* infection in broiler production. *Nature, London*, **241**, 210–11.

Oyofo, B.A., DeLoach, J.R., Corrier, D.E., *et al.* (1989a) Effect of carbohydrates on *Salmonella typhimurium* colonization in broiler chickens. *Avian Dis.*, **33**, 531–4.

Oyofo, B.A., DeLoach, J.R., Corrier, D.E. *et al.* (1989b) Prevention of *Salmonella typhimurium* colonization of broilers with D-mannose. *Poultry Sci.*, **68**, 1357–60.

Perdigón, G. (1991) Probiotics and the immune state (this volume).

Pivnick, H. and Blanchfield, B. (1982) Unpublished data cited by Pivnick, H. and Nurmi, E. The Nurmi concept and its role in the control of *Salmonella* in poultry. *Devel. Food Microbiol.*, **1**, 41–70.

Pivnick, H. and Nurmi, E. (1982) The Nurmi concept and its role in the control of *Salmonellae* in poultry, in *Developments in Food Microbiology* 1 (ed. R. Davies). Applied Science Publishers, London, pp. 41–70.

Pivnyak, I.G. and Konyakhin, A.N. (1973) Biologicheskaya effecktivnost preparatov iz zhivykh karotinoo brazuyushchikh kultur dlya tsplyat. Cited in *Nutritional Abstracts and Reviews* **43**, 330.

Potter, L.M., Newbern, L.A., Parsons, C.M., *et al.* (1979) Effects of protein, poultry by-product meal and dry *Lactobacillus acidophilus* culture additions to diets of growing turkeys. *Poultry Sci.*, **58**, 1095. (Abstract)

Report (1988) Salmonellosis control: The role of animal and product hygiene. *World Health Organisation Technical Report*, Series No. 774. World Health Organisation, Geneva.

Rigby, C., Pettit, J. and Robertson, A. (1977) The effects of normal intestinal flora on the salmonella carrier state, in *Proc. Int. Symp. on Salmonella and Prospects for Control, Guelph, Ontario, Canada*, (ed. D.A. Barnum).

Rubin, HE. & Vaughan, F. (1979). Elucidation of the inhibitory factors of yogurt against *Salmonella typhimurium*. *J. Dairy Sci.*, **62**, 1873.

Rubin, H.E., Nerad, T. and Vaughan, F. (1982) Lactic acid inhibition of *Salmonella typhimurium* in yogurt. *J. Dairy Sci.*, **65**, 197–203.

Schmidt, G.P., Domermuth, C.H. and Potter, L.M. (1988) Effect of oral *Escherichia coli* inoculation on performance of young turkeys. *Avian Dis.*, **32**, 103–7.

Shahani, K.M., Vakil, J.R. and Kilara, A. (1977) Natural antibiotic activity of *Lactobacillus acidophilus* and *bulgaricus*. II. Isolation of acidophilin from *L. acidophilus*. *Cult. Dairy Prod. J.*, **12**, 8–11.

Siccardi, F.J. and Pomeroy, B.S. (1964) Effect of *E. coli* on the growth rate, morbidity and mortality in monocontaminated and conventional white rock chickens. *Poultry Sci.*, **43**, 1361 (Abstract).

Smith, H.W. (1965a) The development of the flora of the alimentary tract in young animals. *J. Pathol. Bacteriol.*, **90**, 495–513.

Smith, H.W. (1965b) Observations on the flora of the alimentary tract of animals and factors affecting its composition. *J. Pathol. Bacteriol.*, **89**, 95–122.

Smith, H.W. and Huggins, M.B. (1982) Successful treatment of experimental *Escherichia coli* infections in mice using phage; its general superiority over antibiotics. *J. Gen. Microbiol.*, **128**, 307–18.

Smith, H.W. and Huggins, M.B. (1983) Effectiveness of phages in treating experimental *Escherichia coli* diarrhoea in calves, piglets and lambs. *J. Gen. Microbiol.*, **129**, 2659–75.

Smith, H.W., Barrow, P.A. and Tucker, J.F. (1985) The effect of oral antibiotics administration on the excretion of *Salmonellae* by chickens, in *Proc. Int. Symp. of Salmonella, New Orleans, USA* (ed. G.N. Snoeyenbos).

Soerjadi, A.S., Lloyd, A.B. and Cumming, R.B. (1978) *Streptococcus faecalis*, a bacterial isolate which protects young chickens from enteric invasion by salmonellae. *Austral. Vet. J.*, **54**, 549–50.

Soerjadi, A.S., Steham, S.M., Snoeyenbos, G.H., *et al.* (1981) The influence of lactobacilli on the competitive exclusion of paratyphoid salmonellae in chickens. *Avian Dis.*, **25**, 1027–33.

Stavric, S., Gleeson, T.M., Blanchfield, B. & Plinick, R. (1983). Competitive exclusion of Salmonella from newly hatched chicks by mixtures of pure bacterial cultures isolated from faeces and caecal contents of adult birds. *J. Food Protect.*, **48**, 778–782.

Tortuero, F. (1973) Influence of implantation of *Lactobacillus acidophilus* in chicks on the growth, feed conversion, malabsorption of fats syndrome and intestinal flora. *Poultry Sci.*, **52**, 197–203.

Watkins, B.A., Miller, B.F. and Neil, D.H. (1982) *In vivo* effects of *Lactobacillus acidophilus* against pathogenic *Escherichia coli* in gnotobiotic chicks. *Poultry Sci.*, **61**, 1298–1308.

Watkins, B.A. and Miller, B.F. (1983) Competitive gut exclusion of avian pathogens by *Lactobacillus acidophilus* in gnotobiotic chicks. *Poultry Sci.*, **62**, 1772–9.

Watkins, B.A. and Kratzer, F.H. (1983a) Effect of oral dosing of *Lactobacillus* strains on gut colonization and liver biotin in broiler chicks. *Poultry Sci.*, **62**, 2088–94.

Watkins, B.A. and Kratzer, F.H. (1983b) Drinking water treatment with a commercial preparation of a concentrated *Lactobacillus* culture for broiler chickens. *Poultry Sci.*, **63**, 1671–3.

Wheater, D.M., Hirsch, A. and Mattick, A.T.R. (1952) Possible identity of 'Lactobacillin' with hydrogen peroxide produced by lactobacilli. *Nature*, **170**, 623.

Wilson, G.S. and Miles, A.A. (1973) *Topley and Wilson's Principles of Bacteriology and Immunity*, 6th edn,. Edward Arnold, London.

Chapter Eleven

Probiotics for pigs

EVA JONSSON AND PATRICIA CONWAY

11.1. INTRODUCTION

Although the use of the term 'probiotic' to describe a feed supplement is recent (Parker, 1974), there are a few earlier reports presenting the concept of using living microbes and substances to improve piglet health. For example, in 1946 Møllgaard proposed that the phytic acid found in ungerminated seeds interfered with absorption of calcium and phosphorus, an effect which they showed could be inhibited by compounds such as lactic acid. In order to increase levels of lactic acid within the digestive tract, a lactic acid bacillus originating from the piglet intestine was introduced and subsequent improved health and skeletal formation was observed (Møllgaard, 1947).

As pig raising has become more industrialized with intensive or semi-intensive commercial units, the risk for major economic losses due to decreased performance and health has become of paramount importance and massive efforts have been made to find different ways to improve production. Intestinal disturbances, e.g. diarrhoea, contribute significantly to the piglet health and reported decreased performance. Medical products such as antibiotics and chemotherapeutics have been used very successfully against the reported decreased performance. The use of low dose antibiotics in the pig raising seems to have facilitated large-scale animal production by allowing good growth rates under sometimes suboptimal conditions. For reviews of modes of growth promotion by antibiotics see Visek (1978) and Walton (1980).

There is an increasing concern from both consumers and authorities about the health risks involved in consuming meat containing residues from feed additives as well as the potential hazards from spreading of resistance factors by indiscriminate use of antibiotics. This problem is even worse if no distinction is made between those compounds used for curing of diseases and those used as feed additives. Furthermore, there is an increase in awareness of the ethical aspects of animal production. The general use of low-dose antibiotics in animal feed has hence been prohibited in Sweden since 1986. The awareness of problems connected with antibiotic treatment of enteric conditions increased the interest in the use of prophylactic *Escherichia* coli vaccines in the early 1970s. Today the vaccination of sows in Sweden before farrowing is common and has contributed to the effective reduction in outbreaks of *E. coli* K88 induced neonatal diarrhoea during the first week of life (Söderlind *et al.*, 1988).

Other approaches that have been used to deal with intestinal problems include: acidification of feed or water (Chapman, 1988), altering dietary formulations, e.g. release of new feeds for small piglets and feeds of lowered protein concentration (Lawrence, 1983); vaccination with attenuated strains of the pathogens or with strains produced by recombinant gene technology (Greenwood and Tzipori, 1987; Trevallyn-Jones, 1987); administration of growth hormone, somatostatin immunization, repartitioning agents and enzyme supplementation (Thacker, 1988); utilization of the lactoperoxidase system (Reiter, 1985): treatment with psychopharmacological drugs (Björk *et al.*, 1987) and the stimulation of levels of hormone-like protein (antisecretory factors) that is capable of reversing intestinal hypersecretion and thus reducing the symptoms of diarrhoea (Lönnroth *et al.*, 1988). More esoteric substances such as the natural mineral zeolite which had been shown to reduce diarrhoea in piglets and increase feed efficiency by as much as 25% (Mumpton and Fishman, 1977) should also be cited.

Parker's original definition of probiotics encompassed 'organisms and substances'. Subsequent references to probiotics seem to be limited to microorganisms and/or their metabolites. To be strictly in accordance with Parker, compounds such as enzymes, growth hormones, etc., could also be referred to as probiotics. We shall limit our discussion, however, to probiotic preparations which contain viable microorganisms.

Today, treatment of diarrhoea consists largely of the use of electrolyte solutions, vaccination and administration of antibiotics but there is a need for new treatments. It appears that the concept of low protein feeds supplemented with essential amino acids is now receiving considerable interest with the release of some commercial preparations in Sweden. The lower protein levels should result in less undigested protein which can be a source of nutrients by pathogens in the small intestine. Because pathogenic *E. coli* K88 grows well in the mucus secreted by the small intestinal mucosa (Conway *et al.*, 1990), the reduction of dietary proteins does not automatically protect the host from intestinal colonization by such strains. Probiotics may have the potential to become part of the treatment and be given as prophylactic agents. However, we must understand their modes of action and when and to what extent they can be anticipated to have beneficial effects, for such preparations to be used routinely. Because the causes of digestive disturbances are multifactorial, it cannot be expected that, a single agent or treatment will be effective against all causative agents. The use of probiotics is not mutually exclusive to other alternatives and in fact combined approaches are more likely to be complementary and the most effective. For example, probiotic preparations with antibacterial activity could be used together with special dietary formulations, or with compounds which suppress symptoms of diarrhoea without directly influencing the pathogenic microbes.

11.2. SPECIAL FEATURES OF PIGS RELEVANT TO THE USE OF PROBIOTICS

An understanding of factors affecting piglet growth is required for a discussion of the action of probiotics and the scientific basis of probiotic functions. At present, probiotics are only being evaluated for oral administration to piglets. Such preparations have therefore directly to exert their effects inside the digestive tract on the activities connected with digestion, microbial competition and immunological defence. However, it is feasible that *Lactobacillus* metabolites produced in the digestive tract could be absorbed and reach many sites and organs in the body. Consequently, some aspects of the raising of pigs and their special features with reference to the state of health of the animals, the digestive tract and factors influencing piglet weight gain and survival will be discussed.

11.2.1. Rearing of pigs

In the production of meat from pigs, one of the major problems is the high mortality of around 20% especially up to weaning (Bäckström, 1973). Other countries report up to 47% and 22% incidence of diarrhoea and death, respectively (Fahy *et al.*, 1987a, b). An example of the causes of pre-weaning mortality in a piggery with a moderate diarrhoea problem is presented in Table 11.1. It is not possible to convert the entire mortality rate into profit regardless of how effective a probiotic or antibiotic may be. The mortality rate cannot only be judged from the human point of economics but the biological reasons have to be taken into account as well. The sow can bear large litters of up to 12–15 piglets. The biological cost is low for bearing an extra piglet which could survive in favourable circumstances, because the newborn piglet corresponds to 0.5–1% of the weight of the sow, as compared with about 6% for humans. The piglets are born quite immature, however, and this makes them sensitive to infections. One must emphasize the importance of the sow and her individual experience and ability to take care of and feed her young. In many cases, the raising system limits her possibilities to function optimally.

Table 11.1 Causes of pre-weaning mortality in a piggery with a moderate diarrhoea problem (compiled from Fahy *et al.*, 1987a).

Cause of death	Percentage of total
Diarrhoea	41
Overlay	17
Splayleg	9
Anaemia	6
Bacterial septicaemia	5
Necrotic enteritis	3
Cold exposure	3
Congenital defects	11
Unknown	5

It is probable that only a part of the high morbidity rate can be targeted by probiotics, namely that part induced by diarrhoea. Piglet diarrhoea can be induced by a number of agents (Table 11.2), and manifests itself by hypersecretion of fluids across the gut wall and into the lumen. This host response can be triggered by, for example, the toxins produced by enteropathogenic *E. coli* strains and functions as the host's defence. Such a rapid fluid flow into the intestine facilitates flushing the pathogen from the site. Samples submitted to a typical diagnostic laboratory in USA (Hoefling, 1989) showed that enterotoxic *E. coli* (EEC) was the primary cause of diarrhoea in 26% of cases. Other cases that were analysed showed that transmissible gastroenteritis (26%), clostridial enteritis (18%), coccidiosis (14%) and rotavirus (8%) were the causative agents, while in 8% no already characterized agent was found. Fahy and co-workers (1987a, b) reported that EEC was responsible for diarrhoea on average in 25% of the cases and even up to 80% after weaning.

The piglets are particularly susceptible to diarrhoea during three periods, i.e. during the first week, at 2 to 3 weeks of age and at weaning. During the early stages of life, the piglet is protected by maternal immunoglobulins in the colostrum (Porter, 1969). A loop-hole for infection of the piglets is the period when the piglet nuzzles the skin of the sow trying to locate the teats. If the environment is heavily contaminated as in many rearing systems, the piglets are exposed to an extensive contamination of microbes. This time period can be up to 60 minutes (Spicer *et al.*, 1986). Neonatal diarrhoea often occurs within 48 hours after birth. It can be largely attributed to enterotoxic *E. coli* strains carrying K88, K99 or 987P fimbriae (reviewed by De Graaf and

Mooi, 1986) and often producing ST enterotoxin. The disease is firstly seen in one or two pigs before spreading to the entire litter. The cause of the enhanced susceptibility to diarrhoea at about 2 to 3 weeks of age is multifactorial, including infections by the protozoan *Isospora suis* and rotavirus. One suggestion for this sensitivity to infection has been that there is a marked decline in antibodies in the sow's milk. Post-weaning diarrhoea occurs approximately 4 and 10 days after weaning and is also multifactorial with enteropathogenic *E. coli* strains as the major infection agent (Fahy *et al.*, 1987b). Recently *Treponema hyodysenteriae* has been more often isolated from younger pigs (O. Söderlind pers. comm.).

Table 11.2 Causative agents of diarrhoea in pre-weaning piglets (prepared from Tzipori, 1985; Fahy *et al.*, 1987a).

	Percentage of piglets		
Agent	Canada	USA	Australia
Enterotoxigenic *Escherichia coli*, e.g.			
K88 (most frequent), K99, 987P and F41	22	31	82
Salmonella spp. *Campylobacter* spp.			
Clostridium perfringens type A and C	ND	ND	ND
Cryptosporidium	15	18	1
Transmissible gastroenteritis virus	52	16	0
Rotavirus	9	14	4
Porcine adenovirus, coronavirus	ND	ND	ND

ND = No data.

Piglets prior to weaning and newly weaned pigs may carry enteropathogenic *E. coli* strains but remain healthy. Many theories have been proposed as to why disease occurs at weaning and these include: (1) sudden deprivation of maternal antibodies and other protective factors in the sow's milk; (2) the altered diet could result in undigested material being available for growth of the pathogen until the host digestive enzymes are induced; (3) extremes of temperature and humidity can be detrimental to lymphocytes which secrete protective antibodies into the mucus layer of the intestines (Carghill, 1982); (4) dietary stress because piglets often do not eat for 24–48 hours (Fahy *et al.*, 1987b); (5) social stress increases the frequency of diarrhoea (Björk *et al.*, 1984). Any of these factors may enhance sensitivity to infection and if they are combined, the risk for infection is amplified.

The process of weaning piglets exposes them to a multitude of physiological shocks which make the piglets more sensitive to infections. These shocks are sometimes referred to collectively as stress.

Traditionally, weaning was done at 7–10 weeks of age; however, today piglets are often weaned earlier. In Sweden weaning generally is done at 5–6 weeks with some incidences at 3–4 weeks or even earlier, as as is more common in other countries. The weaned pig requires a relatively larger digestive system than a sucking pig if it is to satisfactorily digest and absorb the less digestible post-weaning diets (Cranwell and Moughan, 1989). At 3 weeks the digestive and immunological systems are not fully developed and it takes some time for the piglets to adapt their digestive system to the necessary size and enzyme production when changing from sow milk to pig feed (for review see Cranwell and Moughan, 1989). Also, abnormal behaviour might occur with inferior health as a result (Algers, 1980). Some of the problems in connection with early weaning can be overcome by using suitable food, good hygiene and careful management (English, 1984; Campbell, 1989; Partridge, 1989).

During the fattening stage, swine dysentery is a considerable problem. To achieve desirable performance at this stage, it is necessary for the pigs to be in good health, and for digestion to function optimally. During this period, probiotics are most likely to be used as a means of obtaining the desired high growth rate rather than to prevent infections, as is the case for younger pigs. During this stage diseases unrelated to the digestive tract and its microflora cannot be expected to be influenced by the use of probiotics.

Many factors will add to the stress of the animals. If pigs are raised in a suboptimal environment with a poor atmosphere (manure gases, particles in the air), overcrowding, lack of stimulation and unsuitable temperatures, the stress on the animals will increase and as a result they become more susceptible to infections. Stress, including dietary stress, has been shown to affect the microbial flora in the digestive tract of mice, humans and piglets (Tannock and Savage, 1974; Pollman *et al.*, 1980c; Lencner *et al.*, 1984; Tannock, 1984).

11.2.2 Digestive tract

If probiotic preparations are to survive and be active in the digestive tract, they have to be suitable for that environment and resist the host's protective mechanisms which are inhibitory to microbes. For example, there are powerful stomach defences such as low pH and proteolytic enzymes. The retention time as well as the degree of mixing of the ingested material with the gastric juices and previous digesta also influences survival of administered strains. In the anterior small intestine, the most important defence is the very fast flow rate which prevents microbial overgrowth unless the microorganisms can be attached to the epithelium in this site. Among other factors, the presence of bile in this region also negatively influences survival and activity of the microbes. A relatively rapid transit time in the posterior small intestine also protects the host unless invading microbes can adhere to the epithelial mucosa. In the caecum and large intestine the passage rate is lower and the microbes can establish, however, they must compete with a stable indigenous microflora in the healthy host. The extent of survival in the stomach, together with the volume of the digest found in the different parts of the digestive tract, influence the numbers of the probiotic organisms required for dosage. For reviews of digestion in the pig see Kidder and Manners (1978) and Rerat *et al.* (1976).

(a) Stomach

At the entrance to the stomach there is an area, *pars oesophagea*, which has the same type of keratinized squamous non-secreting epithelium as the oesophagus and constitutes approximately 5% of the surface area of the stomach (Noakes, 1971). (See Figure 11.1.) The size of the entire *pars oesophagea* area varies from 1–2 cm^2 in the neonatal pig to about 10–15 cm^2 in the adult pig. The squamous cells in this region undergo continuous desquamation, thereby releasing cells covered with lactobacilli and exposing new epithelial cells which are very rapidly colonized by lactobacilli. The rate of this is not known for pigs, but as a comparison it can be mentioned that in rodents the oesophageal cells undergo a proliferative cycle of 54 h with a life span of the cells of 80 h (Lipkin, 1987). It is believed that the released squamous cells that are colonized by lactobacilli may help to regulate the composition of the digestive microflora by ensuring a dominance of the lactic acid bacteria (Fuller *et al.*, 1978; Barrow *et al.*, 1980).

In the stomach, gastric juice which contains mucus, HCl and proteolytic enzymes are secreted. The composition of the gastric juice is affected by many factors, for example, the type of diet. For a review of this topic see Cranwell and Moughan (1989). Gastric pH is influenced by the age of the pig, the degree of mixing of gastric juice with the stomach contents and the rate of passage of the contents through the stomach. The mixing of the digesta depends, among other things, on its dry matter content and particle size. Liquid feed and finely ground feed can mix more easily than drier or coarsely ground cereal diets (Kvasnitskii, 1951; Maxwell *et al.*, 1970). In suckled piglets, the pH of the stomach digesta ought to be homogeneous as the curd of sows milk is soft (White *et al.*, 1969).

The pH in the stomach varies according to the site. The lowest values are generally found in the pyloric antrum near the exit to the small intestine. The contents near the *pars oesophagea*, as well as the region itself, have the highest pH. This depends on the slow mixing in the proximal part of stomach and the fact that gastric juice is only secreted in the distal part of the stomach. In neonatal piglets, the pH is relatively high in the pyloric antrum, pH 5.2–5.9 (Smith, 1965) but at 4–10 days of age it reaches 4.1–4.4 (Barrow *et al.*, 1977) and drops successively with age. Lawrence (1970) showed in cannulated pigs that the *in vivo* gastric pH varied according to type of diet and time after feeding from a maximum pH of 3.8–4.8 at 30 min after feeding through a gradual decrease down to about pH 2 after 7.5 h. He also showed that *ad lib* feeding gave higher pH values. In the very young piglet, the pH remains close to that of the diet even in the outer layer of the digesta, whereas in the adult, pH falls rapidly to approximately pH 2 in the outer layer (Manners, 1976). See Table 11.3.

Stomach emptying is affected by contractions passing down the pyloric region into the duodenum. Nervous and hormonal feedback mechanisms mainly in the duodenum regulate the passage. Factors triggering the passage include volume of gastric contents and its composition. Parameters of the duodenal digesta which decrease the stomach emptying are low pH, high contents of fat and fatty acids and amino acids and high osmolality (Kidder and Manners, 1978). These factors are generally the same in the pig as for other animal species. Stomach emptying in the pig seems, however, to be somewhat different in two aspects. Firstly, the pH of the emptied gastric contents only plays a minor role for the passage out from the stomach. Secondly, it appears that not only influences from the duodenum but from a larger area in the small intestine affects the gastric emptying (Auffray, 1965).

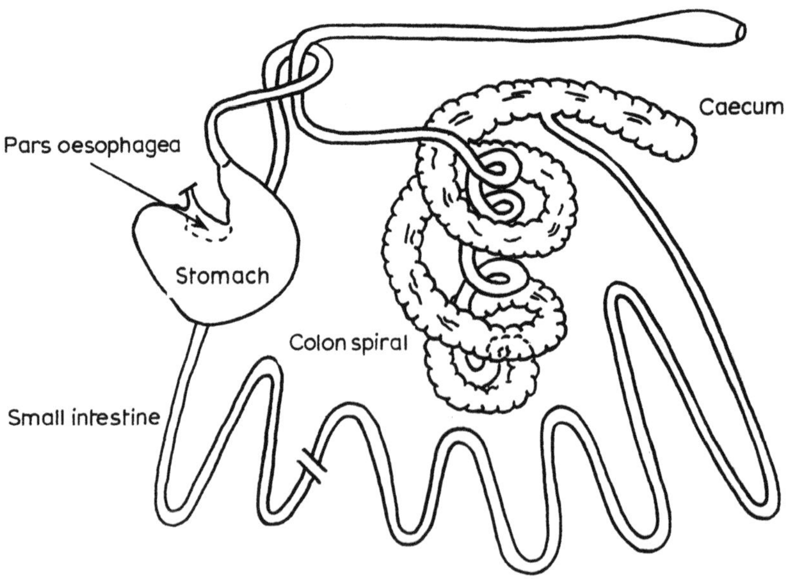

Figure 11.1 The gastro-intestinal tract of pigs, from stomach to rectum.

After a small feed or initially after a large feed, the stomach emptying is approximately exponential. The stomach does not, however, empty completely between meals (Braude *et al.*, 1976; Kidder and Manners, 1978). Food is likely to remain longer in the stomach if the pig is fed at infrequent intervals (Kidder and Manners, 1978). In piglets who suckle frequently (Jensen and Recén, 1989), the stomach empties almost completely between meals (Kvasnitskii, 1951). Reiter (1985) reported that undiluted cow's milk fed to artificially reared piglets may accumulate in the stomach and cause death, while this is not the case for diluted cow colostrum. However, in our experience UHT-treated cow milk with 7% fat has proved satisfactory for the raising of gnotobiotic piglets. After a meal, the liquid part of the feed will rapidly start passing out from the stomach and reach the colon about 12h after feeding (Clemens *et al.*, 1975). In very young piglets, scours was always preceded by gastric stasis (White *et al.*, 1969). These authors considered that gastric stasis with the following lack of nutrients for the body gave general unthriftiness which predisposed the piglets to bacterial infections. Although not discussed by these workers, gastric stasis may result in decreased transit time in the small intestine. In turn, this may reduce the hosts' defence against pathogen colonization of the epithelia, because the pathogens will not be removed as rapidly and therefore could proliferate.

Table 11.3 pH values in the digestive tract of pigs.

Age	Stomach	Small intestine		Caecum	Colon
		Anterior part	Posterior part		
Neonatal	4.0–5.9	6.4–6.8	6.3–6.7	6.7–7.7	6.6–7.2
Unweaned	3.0–4.4	6.0–6.9	6.0–6.8	6.8–7.5	6.5–7.4
Weaned	2.6–4.9	4.7–7.3	6.3–7.9	6.1–7.7	6.6–7.7
Adult	2.3–4.5	3.5–6.5	6.0–6.7	5.8–6.4	5.8–6.8

Compiled from: Barrow *et al.* (1977); Boucourt and Ly, (1975); Braude *et al.* (1976); Clemens *et al.* (1975); Cranwell *et al.* (1976); Schulze (1978); Schulze (1987); Schulze and Bathke (1977); Smith and Jones (1963); Smith (1965).

Volumes measured for different parts of the digestive tract can be found in the literature. Some of these values have to be regarded as unrealistic judging from the volumes available within the animal. Also, it should be remembered that not all organs are filled with contents at all times. In very young piglets, the contents in the stomach might be up to 50 ml, while in weaned pigs it may be up to 1l. In adult pigs the stomach may contain 3–7 kg of digesta (Smith and Jones, 1963; Kidder and Manners, 1978).

(b) Small intestine

The acidified small portions of digesta let out from the stomach into duodenum will be mixed with bile and pancreatic juice containing enzymes and buffer substances. The pH will increase along the passage through the small intestine. The variations in pH values in the small intestine is less than for the stomach and also the difference between piglets and adults is less pronounced (Kidder and Manners, 1978). The variations are large in the duodenum (pH 2–6) and progressively smaller down to the ileum (pH 7.0–7.5). There is also a considerable influence of the type of diet on the lumenal pH (Braude *et al.*, 1976). The activity of the microflora in the distal part of the small intestine (Friend *et al.*, 1963) will lower the pH in this region.

The passage rate through the small intestine is very fast in the anterior part but slows down successively as the digesta progresses towards the anus. Normally, it takes about 2.5 h for a digesta particle to pass through the small intestine (Kidder and Manners, 1978). At this flow rate, it would be difficult for bacteria to multiply sufficiently fast to avoid being washed out. Attachment to the epithelial cells is therefore virtually a prerequisite for microorganisms to colonize this region. Because the epithelium is continuously regenerating and sloughing off cells and overlying mucus, bacteria can colonize this region only if their generation time is faster than the sloughing rate. For conventional mice, cells in the small intestine migrate from the crypts to the top of the villi in about 2 days (Abrams *et al.*, 1963). This is much slower than the anticipated generation time for many bacteria. In addition to bacteria adhering to epithelial cells *via* specific adhesins, it has recently been proposed that bacteria may also colonize the mucus overlaying the epithelial cells (Conway *et al.*, 1990, 1991). This mucus supports rapid growth of an *E. coli* strain with an *in vitro* generation time of 26 min.

Volumes measured for the small intestine can be up to 0.1, 0.6 and 20 litres for very young piglets, weaned and adult pigs respectively (Vodovar *et al.*, 1964). In the newborn piglets the length of the small intestine is 2–3 m (review Cranwell and Moughan, 1989) and the surface (not taking the enlargement by microvilli into account) can be estimated to be *c.* 1 m^2 at birth and by day 10 to be 2 m^2 (Smith and Jarvis, 1978). In a 25 kg pig the length of the small intestine is approximately 16 m (Pettersson, 1990), which approaches that of the adult pig (Nickel *et al.*, 1967). Based on the assumptions given above and our experiences, the area in the 25 kg pig could be estimated to be 15–20 m^2.

(c) Large intestine

The large intestine consists of the caecum, the spiral colon and the distal colon. The caecum and centripetal part of the spiral are wide and the rate of passage of digesta is low, thus allowing activity of a dense and complex anaerobic microflora. The passage rate of a meal is not very easily determined as particles from different meals become randomly distributed. Generally, the first part of a meal will reach the anus after 10–24 hours. The mean retention time is much more variable and can be 2–4 days (Kidder and Manners, 1978).

Volumes measured for the large intestine can be up to about 0.04, 1 and 25 litres for very young piglets, weaned and adult pigs respectively (Kvasnitskii, 1951). The pH in the large intestine (around pH 6.0) does not fluctuate as much as in the foregut as the contents are more homogeneous and microbial fermentation is dominant (Kidder and Manners, 1978).

11.2.3 Indigenous lactic acid bacteria

The pig is a monogastric animal in which the foregut (stomach and small intestine) is colonized by a relatively rich microflora. The flora is not as rich as in ruminants which have a specialized foregut fermentation system, but stomach contents still contains about 110^7g 10^8 bacteria per gram of digesta. As the killing by low pH is not so great bacterial numbers found in the small intestine are generally also high, 10^7–10^9. See Table 11.4. The microflora of the pig foregut is dominated by lactic acid bacteria (LAB), mostly *Lactobacillus* and *Streptococcus* spp. They are found both in the digesta and attached to the epithelia. The non-secreting *pars oesophagea* area in the stomach is densely colonized with layers of LAB (Fuller *et al.*, 1978), as illustrated in Figure 11.2. Attachment to the epithelial cells in the small intestine has also been demonstrated (McAllister *et al.*, 1979). There is a difference in the composition of the microflora in the caecum and colon with Gram-negative organisms dominating the caecum (Robinson *et al.*, 1981) and Gram-positive bacteria dominating the colon (Salanitro *et al.*, 1977).

Several functions within the pig digestive tract enable the LAB microflora to be dominant. These include the fact that lactic fermentation in the stomach is facilitated by the relatively high stomach pH. Furthermore, food entering the stomach is inoculated with the indigenous LAB as the ingesta is mixed with gastric contents and continuous inoculation of LAB is ensured by the sloughing of *pars oesophagea* cells with attached LAB. The relatively high pH in the greater part of the stomach will also mean that the killing of LAB in the gastric content is not so great, hence they will become a major component of the microflora in the small intestine.

The importance of LAB microflora in the foregut relates to physiological, microbiological and digestive functions. It helps the young pig to lower the pH in the stomach by the production of lactic acid and other organic acids formed mainly from lactose in the sow's milk (Cranwell *et al.*, 1976; Barrow *et al.*, 1977). Both the organic acids and the low pH value in the pyloric antrum are important in decreasing the numbers of bacteria passing into the small intestine (Smith, 1965; Schulze, 1978). LAB may regulate the microflora of the small intestine by inoculating

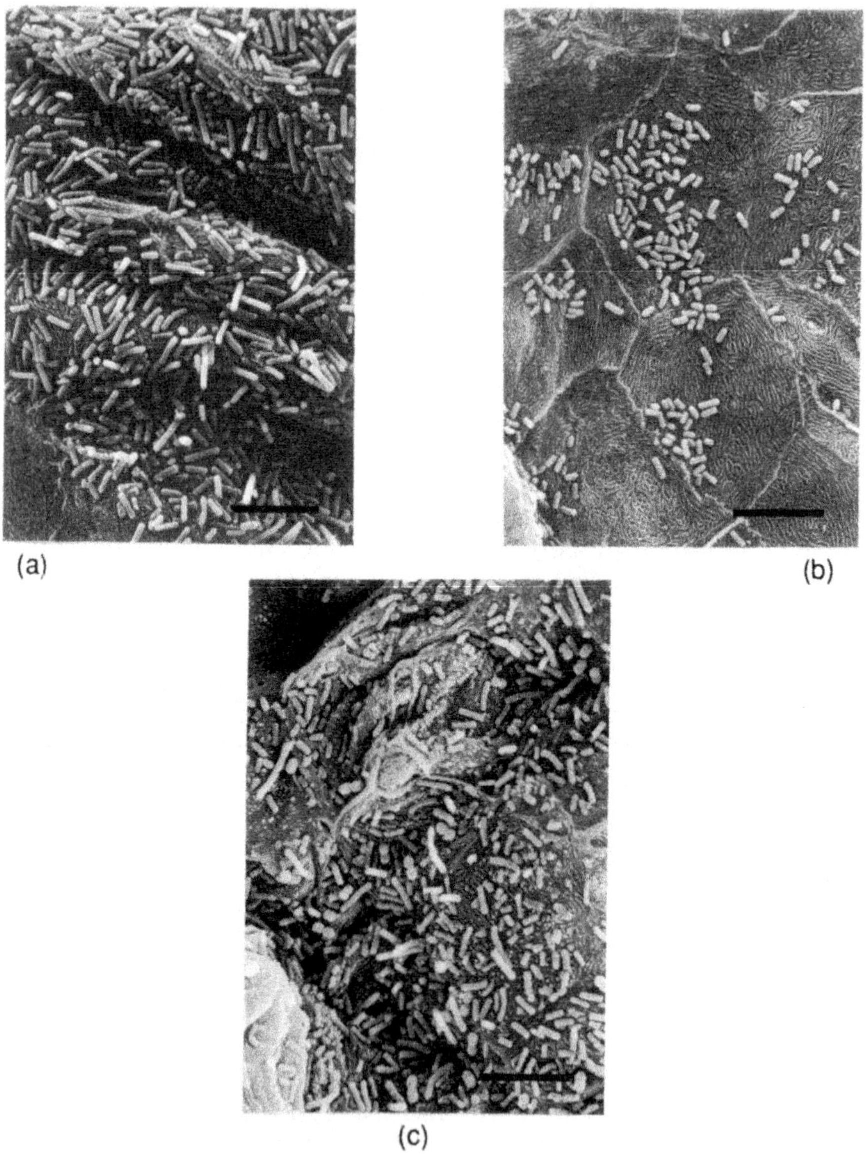

Figure 11.2 Scanning electron micrographs of the squamous stomach epithelium from piglets aged (a) 5 days, (b) 26 days and (c) 47 days. Bar markers = 10 μm. (Blomberg and Conway, 1989; reprinted by permission of John Wiley & Sons, Ltd.)

the digesta passing down the digestive tract (Fuller *et al.*, 1978). In addition to the pH effect, LAB can compete with other bacteria and in this way reduce the contamination in the lower digestive tract (Barrow *et al.*, 1980). It has also been shown that LAB can improve the digestion of mixed-linked (1→3, 1→4) β-D-glucans in fibres (Graham *et al.*, 1986). For reviews of the pig microflora see Cranwell (1968), Schulze (1987) and for its activities see Kenworthy (1973) and Ratcliffe (1985). For a review of the ecology of the lactobacilli of the digestive tract see Tannock (1990).

Species often found are *Lactobacillus acidophilus, L. delbrueckii, L. fermentum, L. reuteri, L. salivarius* (Fewins *et al.*, 1957; Fuller *et al.*, 1978; Mäyrä-Mäkinen *et al.*, 1983; Axelsson and Lindgren, 1987) and *Enterococcus bovis, Ent. durans, Ent. faecalis, Ent. faecium, Streptococcus intestinalis, S. porcinus* and *S. salivarius* (Raibaud *et al.*, 1961; Barrow *et al.*, 1977; Fuller *et al.*, 1978; Collins *et al.*, 1984; Robinson *et al.*, 1988). Bifidobacteria detected from the digesta are *Bifidobacterium adolescentis* and *Bif. suis* (Zani *et al.*, 1974; Robinson *et al.*, 1984).

Table 11.4 Occurrence of bacteria in the digestive tract of healthy pigs older than 2 weeks. Numbers given as colony-forming units per gram digesta or per square centimetre of epithelium (wet weight).

	Digesta						Epithelia	
Bacteria	1	2	3	4	5	6	3	5
Total count	5.5–9.3	5.5–8.4	5.5–9.5	8.4–9.7	7.8–9.8	5.5–5.9		9.4–10.0*
Lactobacilli	5.2–8.9	5.4–8.5	6.1–9.6	7.8–9.3	7.6–9.6	5.1–7.9	6.9	7.3– 7.6*
Streptococci	4.0–6.8	4.0–6.7	5.1–7.4	7.6	5.9–8.0	5.0	4.3–5.8 *	6.8– 7.4*
Bifidobacteria	4.3–6.7	3.9–5.3	5.6–7.3	5.1–7.9	5.5–8.7			

1, stomach; 2, anterior small intestine; 3, posterior small intestine; 4, caecum; 5, colon; 6, non-secreting epithelium in the stomach.

*Per gram of mucosa.

Compiled from: Conway (1989); Horvath *et al.* (1958); Kenworthy and Crabb (1963); Kovacs *et al.* (1972); McAllister *et al.* (1979); McGillivery *et al.* (1984); Ogata *et al.* (1968); Russel (1979); Schulze (1977); Smith (1961); Smith and Crabb (1961); Tannock and Smith (1970).

Although both McGillivery (1984) and Cain (1989) reported that lactobacilli were the dominant group of bacteria associated with the duodenal mucosa, and McAllister and co-workers (1979) reported that lactobacilli and *Bacteroides* and *Clostridium* were predominant throughout the small intestine, no characterization of the various species has been reported for the small or large intestinal mucosa. However, Fuller and co-workers (1978) found *L. fermentum* to be the most common species (93%) of isolates from 2–10 day old piglet *pars oesophagea*. Subsequently, it has been shown that there is a succession of *Lactobacillus* spp. in this region in neonatal pigs (Tannock *et al.*, 1990) and that these populations vary between 5, 26 and 47 day old piglets as illustrated in Figure 11.2 (Blomberg and Conway, 1989).

Piglets should be born in a conventional environment to allow the development of a normal adult microflora. If they are moved at a later stage to an environment devoid of indigenous microorganisms, the microflora will continue to develop normally (Gouet *et al.*, 1984). The selection and establishment of the indigenous LAB microflora in the neonate occurs progressively from birth (Sinkovics and Juhasz, 1974; Schulze, 1978). Ducluzeau (1985) concludes that several investigations point to the fact that the effect of sow milk on the digestive microflora is not so important because no major change in composition of the digestive microflora occurs at weaning, while sow colostrum does indisputably have a protective effect against diarrhoea caused by pathogenic *E. coli*. This conclusion of Ducluzeau may only apply to the indigenous microflora at weaning, because lysozyme in sow's milk has a significant effect on bacterial colonization of the unweaned piglet (Schulze and Müller, 1980) and factors such as lactoferrin, transferrin and vitamin B_{12} binding protein will also play a role (Cranwell and Moughan, 1989). For reviews of the development of the digestive microflora in the milkfed pig and milk-fed human infant, see Ducluzeau (1985) and Cranwell (1990).

Type and amount of diet have been shown to affect the lactobacilli in the digestive tract. Brockett and Tannock (1981) concluded that the relative amounts of palmitic and oleic acids in the food can influence the number of tissue-associated lactobacilli in the mouse stomach. Jonsson and Henningsson (1991) showed a correlation between the diet of the piglet and the occurrence of faecal lactobacilli with ability to degrade mixed-linked (1→3, 1→4) β-D-glucans. These bacteria that were commonly found in the neonatal pig probably originated from the faeces of the sow. During the sucking period, they could not be detected but they reappeared at 7 weeks when the piglets had consumed creep feed containing such fibre for some time.

Lactobacilli in the digestive tract appear to be more affected than, for example, *E. coli*, by lack of nutrients. In rats fed restrictively, Brownlee and Moss (1961) found decreased numbers of lactobacilli adhering to the non-secreting part of the stomach. In mice deprived of nutrients for up to 4 days, the surface associated lactobacilli of the stomach were lost; however, colon lactobacilli levels remained unchanged and coliform numbers increased (Conway *et al.*, 1986). Ogata and Morishita (1969) and Morishita and Ogata (1970) showed markedly decreased numbers of lactobacilli and bifidobacteria in the foregut when pigs were starved for 24 h, and in the ileum when they were deprived of food and water for 72 h, while numbers of *E. coli* and *Bacteroides* increased in the ileum. Ducluzeau and co-workers (1986) used pigs with ileo-rectal anastomosis. They found that in the contents of the large intestine 14–54 days after the operation, the strictly anaerobic bacteria still remained at the same level, while the numbers of lactobacilli decreased and clostridia could no longer be detected. Morishita and Ogata (1970) suggested that *E. coli* and *Bacteroides* rely more on endogenous nutrients than do the lactobacilli, which have complex nutrient requirements (Sharpe, 1981). Another reason could also be that these other groups of bacteria can grow more quickly and hence are able to compete more efficiently for limiting nutrients (Freter, 1983). Survival during nutrient deprivation has been extensively studied *in vitro* for many species of *Vibrio* and *Pseudomonas* as well as strains of *E. coli*. A very elaborate survival mechanism involving the synthesis of new proteins has been elucidated (reviewed in Kjelleberg *et al.*, 1990). Preliminary studies of the starvation survival capacity of *Lactobacillus* strains of digestive origin have shown excellent survival in the total absence of nutrients; however, in the presence of low levels of a single amino acid, the cells started to grow and then died (Conway, unpublished data). It was hypothesized that the starvation survival mechanisms of *Lactobacillus* have a very low threshold for nutrients, above which growth is initiated. We are not aware of *in vitro* starvation survival studies of any *Bacteroides* spp.

11.2.4 Immunology

In the healthy adult pig, immunoglobulins are released into the digestive tract and thereby contribute to the host's defence against infection. This immune defence does not begin to function in the newborn piglet until about 3 weeks of age. During the first weeks of life, the piglet is protected by maternal antibodies and non-immunological defence factors (Porter, 1969). After 3 weeks of age, IgA is secreted. Sow milk immunoglobulins have been shown to inhibit *E. coli* growth (Wilson and Svendsen, 1971), adhesion to enterocytes (Nagy *et al.*, 1979) and to neutralize the toxins (Brandenburg and Wilson, 1973). The milk immunoglobulins are produced by plasma cells which are derived from lymphocytes sensitized in the gut against intestinal pathogens (Hartmann *et al.*, 1984). Activation of the gut–mammary gland link by feeding sows with either live cultures of *E. coli* (Stevens and Blackburn, 1967) or killed *E. coli* cells (Porter and Linggood, 1983), induced IgM-mediated protection against neonatal scours. A further example of the phenomenon is the fact that vaccinating the sow against *E. coli* K88 has led to protection of the neonatal piglets from K88-induced diarrhoea (Söderlind *et al.*, 1982, 1988). Following the introduction of such a vaccine, the main pre-weaning diarrhoeal period has been transposed from the neonatal period to the 14–21 days period.

The success of protection *via* ingested maternal lactogenic antibodies has led to applications such as concentrating porcine serum from abattoir blood and adding it to the diet of early weaned pigs (Drew and Owen, 1988). Several non-immunological host defence factors, e.g. lactoferrin, transferrin, vitamin B_{12} binding protein and the bifidus factor, have been identified in sow milk (reviewed by Cranwell and Moughan, 1989). At weaning, the piglet is suddenly deprived of both the milk antibodies as well as these non-immunological factors. Although the immune system of the piglets is generally functioning at the time of weaning, it may need to be stimulated to overcome the cessation of the protective factors in the sow milk. Perdigón and co-workers (1987) report stimulation of the immune response in mice fed with LAB. Similarly, Shahani and co-workers (1989) also proposed that one of the functions of an administered probiotic at this time could be to stimulate the immune system.

The piglet immune system could also contribute to digestive disturbances should dietary components induce hypersensitivity responses in the early weaned piglet (Newby *et al.*, 1984). These workers suggested that the intake of small amounts of certain proteins, particularly soya, before weaning sensitizes the immune system and consequently when the post-weaning piglet consumes larger quantities of the same protein, a hypersensitivity reaction is induced. Mild diarrhoea and intestinal damage could result and thereby increase susceptibility to infections by pathogens. As the risk of hypersensitivity is a difficult parameter to assess, there appears to be considerable confusion about the significance of nominal antigenic ratings of some dietary ingredients (Campbell, 1989).

Various types of stress have been shown to exert negative effects on the immune system. For example, extremes of temperature and humidity have a detrimental effect on lymphocytes which secrete protective antibodies into the gut (Carghill, 1982). Furthermore, an association between a depressed immune response and stress-induced elevated corticosteroid concentration in the blood has been shown for sows (Barnett *et al.*, 1987).

11.3. CURRENT USE OF PROBIOTICS

The interest in probiotics has fluctuated during the twentieth century, with a peak in the 1940s. Thereafter the interest waned, but it is increasing again as can be seen from the amount of published literature containing more positive interpretations and predictions. Emphasis has now shifted from using milk fermented with microbes selected for their capacity to yield organoleptically desirable fermented dairy products, to selecting indigenous microbes. The pioneering work in this area was carried out in chickens (as presented in the chapter by Barrow). For reviews see Kinsey (1980), Wolter and Henry (1982), Søgaard (1987), Fuller (1986, 1989), Sissons (1988, 1989), Conway (1989) and Vanbelle *et al.* (1989).

Currently, probiotics are being sold as solitary microbial additives or mixed with different substances. Reported dosages range from 10^9 to 10^{12} microbes per animal per day for administration at different times and in different ways. One common problem for LAB preparations used earlier was that the numbers found in some of the preparations was considerably lower than claimed.

Table 11.5 Microorganisms used as probiotics for pigs.

Organism	Form	Reference(s)
Lactobacillus acidophilus	Cultured milk	11, 16, 17, 38, 40
	Frozen culture	33, 34, 36
	Freeze-dried	29
	Dried	10
	FP	35
L. lactis	Frozen	26
L. reuteri	Cultured milk	37
Lactobacillus spp.	Cultured milk	12, 24, 25
	Dried	32
	FP	2, 8
Enterococcus faecalis	Spray-dried	31
Ent. faecium	Freeze-dried	3,4,6,7,15,17,18,22,23,27,39,42
Bacillus licheniformis	Spores	19
B. subtilis	Spores	30
B. subtilis var. *toyoi*	Spores + FP	13, 28, 40, 41
Bifidobacterium bifidum	Dried	5
Bif. pseudolongum	Dried	14
Bif. thermophilus	Dried	14
Clostridium butyricum	Dried	9
Saccharomyces spp.	Dried	1
Yeasts	Dried	17
Mixed organisms:		
Pediococcus acidilactici	Dried	19
L. plantarum		
L. casei		
L. fermentum		
L. brevis		

L. delbrueckii subsp. *bulgaricus*	FP	21
L. casei		
Streptococcus salivarius subsp. *thermophilus*		
L. plantarum	Dried	32
L. acidophilus		
Ent. faecium		
Yoghurt organisms	Yoghurt	37
Lactic culture	Freeze-dried ± FP	28

FP = Fermentation product.

References: 1 Burnett and Neil (1977); 2 Cowman *et al.* (1978); 3 Danek (1986); 4 Deprez *et al.* (1989); 5 Ervolder *et al.* (1984); 6 Gualtieri and Betti (1984); 7 Gudding and Larssen (1985); 8 Hale and Newton (1979); 9 Han *et al.* (1984); 10 Ingram *et al.* (1973); 11 Jensen (1974); 12 Jonsson (1986); 13 Jørgensen (1988); 14 Kimura *et al.* (1983); 15 Kluber *et al.* (1985); 16 Kornegay (1985/86); 17 Kornegay and Thomas (1973); 18 Krarup (1987); 19 Landsudvalget (1988); 20 Landsudvalget (1989); 21 Lessard and Brisson (1987); 22 Maeng *et al.* (1989); 23 Moen (1982); 24 Møllgaard (1946); 25 Møllgaard (1947); 26 Muralidhara *et al.* (1977); 27 Neupert (1988); 28 Ogle and Inborr (1987); 29 Olsson (1966); 30 Ozawa *et al.* (1981); 31 Ozawa *et al.* (1983); 32 Pollman *et al.* (1980a); 33 Pollman *et al.* (1980b); 34 Pollman *et al.* (1980c); 35 Pollman *et al.* (1984); 36 Premi and Bottazzi (1974); 37 Ratliffe *et al.* (1986); 38 Redmond and Moore (1965); 39 Roth and Kirchgessner (1986); 40 Roth and Kirchgessner (1988); 41 Spriet *et al.* (1987); 42. Underdahl *et al* (1983).

11.3.1 Organisms used

Several organisms are used as probiotics for pigs while others are under investigation as potential probiotics (Table 11.5). For a review, see also Tuschy (1986), who gives details about different commercial brands, and Tournut (1989). Lactobacilli are strong acid producers and they are seldom pathogenic (Sharpe *et al.*, 1973; Sharpe 1981). Some strains of *Ent. faecalis* and *Ent. faecium* have been found to be pathogenic (Hardie, 1986; Mundt, 1986), however, streptococci are more tolerant to harsh conditions, and therefore would yield more stable preparations than those containing lactobacilli. *Bifidobacteria* are more oxygen sensitive than most lactobacilli, and some strains are strictly anaerobic. Consequently, their survival in air will be more of a problem than for the other LAB used.

The bacteria can be largely divided into two main types: those which originate from the indigenous digestive microflora and those which do not. Species normally found in the pig indigenous microflora are *L. acidophilus*, *L. fermentum*, *L. reuteri*, *Ent. faecium* and *Ent faecalis*. In the preservation of food or feed, *L. plantarum*, *L. helveticus*, *L. delbrueckii* subsp. *bulgaricus* and *S. salivarius* subsp. *thermophilus* have been employed as starter cultures.

The tradition to use bifidobacteria has developed in Japan, where three preparations of bifidobacteria, namely *Bif. bifidum*, *Bif. pseudolongum* and *Bif. thermophilum* have been used for animals since 1968 (Kurman, 1983). Other bacterial genera used as probiotics include *Bacillus* and *Clostridium*, for example, *Bacillus licheniformis* and *B. subtilis* and the related *B. toyoi* and *Clostridium butyricum* (Han *et al.*, 1984; Tournut, 1989). *Saccharomyces* sp. and unspecified yeast preparations were used by Kornegay and Thomas (1973) and by Burnett and Neil (1977).

Bacillus spp. are mostly soil organisms. Some species are used for the production of antibiotic substances. It is not known if *Bacillus* spp. constitute a component of the indigenous microflora or whether they are transient in the digestive tract (Savage, 1977). It has been claimed (Ozawa *et al.*, 1981) that the spores would germinate in the anterior part of the digestive tract and compete with enterotoxigenic *E. coli*. They may increase the *Lactobacillus* flora (Roth and Kirchgessner, 1988) or stimulate the immune system against *E. coli* (Pollman, 1986). It is feasible that their action could be mediated by preformed antibiotic substances especially if they are given as spores dried in their propagation medium. It has been proposed that *B. licheniformis* will produce antibiotic substances in the upper digestive tract of mice (Ducluzeau *et al.*, 1978a) and that a mixture of *B. subtilis* and *B. licheniformis* may be effective because of enzymatic activity and/or by suppressing pathogens and production of compounds such as NH_3 and amines (Landsudvalget, 1989).

Another approach is the use of competitive colonization where non-pathogenic strains are given in order to prevent subsequent colonization by pathogenic strains. For example, Davidson and Hirsch (1976) and Duval-Iflah *et al.* (1983) administered non-pathogenic *E. coli* to pigs and Borriello and Barclay (1985) non-toxic strains of *C. difficile* to hamsters. As with vaccinations, this approach will only protect the host from very closely related strains to that administered since its action is solely to compete for the same site. A related approach was that of Nakamura and co-workers (1983) who showed that drug-sensitive *E. coli* strains fed to young animals prevented establishment of resistant strains. Another way to try to improve animal health is that tested by Smith and Huggins (1983) using phages which attach specifically to K-antigens. This method also has the disadvantage of being too specific, however. Fuller (1986) proposed that phage probiotics may have a potential if the phage could attach to receptors to which the pathogens attach. Alternatively, this phage could attach to the adhesive determinants of the pathogens, e.g, K88, K99 fimbriae (Smith and Huggins, 1983).

11.3.2 Usage

Probiotic preparations can be given soon after birth, at times when the farmer expects diseases (preventive or curative), or mixed into the food for continuous supply. The microorganisms can be injected orally or distributed in the water or in the feed, which can be pelleted or ground. They can be given as viable organisms in wet, frozen or freeze-dried preparations or as pastes (Tournut, 1989), or as fermentation products with or without dead organisms (Pollman *et al.*, 1984).

The form of presentation of the microorganisms as probiotics is largely dependent on convenience for distribution and administration, provided essential characteristics are maintained. For example, if the microbes must be metabolically active for effective action, the preparation must contain viable microbes. Although pelleted food is often preferred, the pelleting process involves high temperatures and high pressures which can be lethal to many microorganisms. Some streptococci and the spore-forming bacteria such as *Bacillus* ssp. are less affected and survive quite well. Lactobacilli are more sensitive, but Lyons (1988) reported a process of microencapsulation which gives better survival ability, with about 40% of the microencapsulated bacteria remaining viable after pelleting. Growth conditions of the microorganism, methods used for harvesting and protection prior to freeze-drying will also influence survival of the bacteria. Storage stability of viable *Lactobacillus* preparations is influenced by the storage environment, length of storage and type of medication in the feed. Storage at refrigeration or ambient temperatures is to be preferred to storage in nurseries at higher temperature (32°C) (Pollman and Bandyk, 1984). *L. acidophilus* cells are quite sensitive to both freeze-drying and air-drying, and the drying medium and the residual moisture content influences their survival. If too much of the bound water is lost, damage to the cell wall and the cytoplasmic membrane appears to occur with increased sensitivity of *L. acidophilus* cells to bile and lysozyme (De Valdez *et al.*, 1985; Brennan *et al.*, 1986). Gilliland and Rich (1990) found that two strains of *L. acidophilus* could be stored at 5°C for 3 weeks if cultured and maintained in skim milk at pH 5. At higher pH the viability of these strains declined for the same storage conditions. Freezing and frozen storage for 4 weeks at −196°C had no significant effect on the viability of the two strains.

(a) Time of administration

The probiotics may be given at different ages of the pig depending on their proposed mechanism of function. If there is reason to believe that the normal indigenous microflora of a healthy pig will not be established, preparations containing solely LAB are probably most desirable because the natural sequential colonization of the digestive tract can be initiated. Such situations might occur when the piglets are moved directly after birth into a scrupulously clean environment, as described by Cranwell *et al.*, (1976), or for example after antibiotic treatment. For normal pig rearing, the piglets stay in close contact with the sow for the first weeks of life and will therefore be colonized by LAB. It should be stated, however, that some LAB strains have a greater capacity than other LAB strains to inhibit the establishment of pathogens, for example, by the production of antagonistic metabolites (see section 11.5.2(a)). Consequently, in farms with a high incidence of diarrhoeal disease, it may be appropriate to introduce a probiotic strain as early as possible and thereby colonize the digestive tract with the probiotic strains that have capacity to inhibit pathogens.

If the reason for using probiotic preparations is to counteract a high level of potentially pathogenic organisms such as *E. coli*, the preparations should be given prior to and during periods of risk for disease unless colonization can be ensured with limited dosages. The question is often raised as to whether single doses or continuous dosage should be used. The characteristics and mechanisms of action of the strains to be used as probiotics will influence this point. For example, if the desired strain functions by producing desirable metabolites in the digestive tract during passage through the system, but lacks the capacity to colonize the tract, then continuous or daily dosage would be required. If the probiotic strain has characteristics which facilitate colonization of the tract, e.g. adhesins, then single dosages may suffice until such time as the pig is exposed to some form of stress or treatment that could result in loss of the probiotic strain.

(b) Dosage levels

The numbers of organisms given as therapeutic doses are often 10^9–10^{12} per animal per day or 10^6–10^7 g^{-1} feed. These levels are mostly arbitrary as no dosage studies for different strains have been reported. The numbers given should be sufficient to elicit a beneficial response in the host and therefore be given in significant numbers in relation to the indigenous flora or achieve such numbers by growth within the digestive tract. Conversely, levels should not be so high that they induce digestive problems. It has been reported that overdosing humans with LAB could have a laxative effect (Gordon *et al.*, 1957). These workers report cases of diarrhoea in patients dosed with milk fermented with *L. acidophilus* such that individuals received 10^9 cells per day. For pigs, a dose of 10^9 per day corresponds to the LAB already present in only 10–100 g of digesta. This is similar to the numbers of lactobacilli found in the stomach of sucking piglets. A comparable dose in weaned pigs would have to be $10^{9.7}$–$10^{10.5}$. To our knowledge, no reports have shown that too high doses of LAB given to pigs would bring about diarrhoea. Conway and co-workers (manuscript) observed no detrimental effects of dosing 4 weeks old piglets with 10^9–10^{10} *L. murinus* cells twice daily for 10 days as compared to treatment for only 1 day.

When discussing dosage levels required to achieve a certain level of bacteria in the anterior small intestine, one has to take into account killing in the stomach and dilution with digestive secretions to 3–6 times the volume. Studies by Jonsson and co-workers (1985) in pigs with cannulas in the anterior and posterior part of the small intestine show that the numbers of lactobacilli per gram of digesta are between 4 and 15 times higher in the ileum compared to the numbers found in the anterior part of the small intestine. For the same sites, a *Lactobacillus* strain with good survival ability fed once to these pigs increased in numbers 3–5 times. Passage time between these sites in the intestine is only a few hours, hence multiplication of the organisms would not be extensive. The origin of *Lactobacillus* strains and their bile tolerance were reported to affect survival in the digestive tract of pigs (Jonsson *et al.*, 1985). Consequently the capacity of the potential probiotic strain to survive *in vivo* ought to be tested for individual strains. This aspect is further discussed in section 11.5.

The issue of whether administered probiotic microorganisms will be transient or whether they will colonize the digestive tract also influences the dosage required. Transient strains will probably need to be administered in higher levels than those strains that will multiply within the tract if similar numbers are required to achieve the desired function. For bacteria to colonize, the strains will probably need to associate with the epithelial mucosa (discussed in Conway, 1989). Required dosage levels of strains with colonizing potential have not been reported to date. As cited previously, the *pars oesophagea* of the healthy stomach is colonized by indigenous lactic acid bacteria. Approximately 10^8 bacteria cm^{-2} would be needed to cover the mucosal surface and thus one could anticipate that the entire area would thus have about 10^8 LAB in the neonatal pig and about 10^9 LAB in the 25 kg pig. If a LAB strain could multiply and colonize within the stomach it would require administration levels considerably less than 10^8 and 10^9 per neonatal pig and 25 kg pig, respectively.

If the small intestine of a newborn pig is viewed as a straight tube, it should take *c.* 10^{11} LAB to cover the surface area totally. In the 25 kg pig corresponding values would be up to 100 times larger (10^{13}). Although these values are much higher than for the stomach, dosage levels may not need to be very great for colonization to occur because those released from the stomach due to multiplication at this site could serve as an inoculum for the intestine. While such calculations yield bacterial numbers consistent with the *in vivo* reports for the *pars oesophagea* region, it is more difficult to confirm values for the small intestine. Fossum and Liven (1974) pointed to the fact that the small intestine contained parts which were almost sterile with little contents other than intestinal mucus and other dilated regions which were full of digesta, bacteria and enzymatic activity. However, the mucosa was not studied, and hence no conclusions can be drawn regarding LAB colonizing the epithelium.

11.4. EFFICACY

Use of probiotics often gives inconclusive or conflicting results (reviewed by Jonsson 1985; Tuschy 1986; Conway 1989), and there is no direct evidence to explain these differences. Strains used as probiotics can vary in their characteristics and thus their actions; for example, not all *L. acidophilus* strains have exactly the same characteristics. However, one can hypothesize that the host susceptibility may vary from one study to the next. Altered *in vivo* conditions could also result in altered activities of the probiotic cells. The viability or survival potential of the probiotic cells would also be important facts.

If the host's body is already functioning optimally, animals will most likely be growing at their maximum capacity and therefore it may not be possible to achieve any improvement by administering beneficial bacteria. It is interesting to note, however, that in SPF pigs kept under strictly hygienic conditions, advantages in health were seen by administering of antibiotics (Jucker *et al.*, 1973). This may mean that non-pathogenic microbes may exert some stress on the animal. A lot of research and trials using probiotics are being done, both by commercial companies, experimental stations and universities. As there are strong commercial interests, some of the work in this area is not accessible. In addition, the practice of not publishing inconclusive or negative results contributes to the difficulty of obtaining the overall picture.

Evaluations of the use of probiotics *in vivo* include studies of

- influence on the digestive microflora, including pathogenic bacteria
- influence on the digestive tract, its function and morphology
- performance and health of animals

- various effects where animals are used as models for humans, for example, anticholesterolaemic properties, stimulation of the immune system and anticarcinogenic activity.

For further details see review of the nutritional and therapeutic role of dietary LAB (Fernandes *et al.*, 1987, Bourlioux and Pochart, 1988)

11.4.1 Influence on the digestive microflora

To exert any beneficial effects, the bacteria or possibly the active substances have to reach the site of function. In order to do so, first of all they have to be consumed. This means that the preparations have to be appetizing or at least not repulsive to the pig. Pigs are sensitive to the long-term effects as well as the flavour and taste of their feed (Houpt *et al.*, 1979). This fact is also stressed by Premi and Bottazzi (1974). The next important step is that the preparation should contain viable and active organisms or substances in sufficiently large amounts. In the following discussion, emphasis will be put on viable organisms.

One of the reasons for using lactic acid bacteria as probiotics is that they are said to stabilize the digestive microflora and compete with pathogenic microbes. These statements probably arise because LAB are strong acid producers; they contribute to the lowered pH and decrease in numbers of bacteria entering the small intestine in neonatal pigs (Cranwell *et al.*, 1976, Barrow *et at.*, 1977). In fact, half of the lactose in sow's milk is metabolized to yield organic acids in the stomach (Cranwell *et al.*, 1976). Once the complex indigenous microflora in the digestive tract develops, the system is relatively stable. Addition of further microbes to the stable indigenous microflora should not give any change in numbers if the other conditions remain the same (Hungate, 1984). However, although the microflora is a stable system, it is also dynamic. Strains will replace each other and the metabolism will adapt itself to the substrates available. In his discussion about the large intestine, Freter (1983) states that competition for limiting nutrients is one of the determining factors that has received the most scientific support. If this is the case, the addition of small numbers of bacteria or their metabolites could have substantial influence on the microflora. Of course, the composition of the diet would also be of major importance.

The composition of the microflora in the digestive tract, except for faeces, is difficult to study in the living animal. Unfortunately, the faecal flora cannot be used as an indicator of microbial populations in the anterior digestive tract as these populations vary enormously (Pollman *et al.*, 1980b).

Several reports of administering different *Lactobacillus* strains to neonatal or weaned piglets state that the numbers of *E. coli* and lactobacilli are influenced. When feeding *L. acidophilus* to piglets, Pollman and co-workers (1980b) found no influence on the succession of the microflora in the digestive tract, except an increase in the numbers of lactobacilli and coliform bacteria in the non-secreting part of the stomach (cardia region). Higher ratios of lactobacilli to coliforms in the colon microflora were obtained after administration of *L. murinus*, but not *L. fermentum*, to piglets up to 4 weeks old (Conway *et al.*, manuscript) and in faeces of 3 weeks old piglets fed *L. acidophilus* (Premi and Bottazzi, 1974). Olsson (1966) showed decreased levels of *E. coli* and haemolytic *E. coli* in the faeces of weaned pigs dosed with *L. acidophilus*. A lactobacillus fermentation product fed to artificially raised piglets challenged with *E. coli*, suppressed the counts of *E. coli* but not the lactobacilli in the stomach (Pollman *et al.*, 1984). Ratcliffe and co-workers (1986) found that both yoghurt and milk fermented with *L. reuteri* given to 2-day-old piglets increased the numbers of lactobacilli and decreased numbers of *E. coli* throughout the gut. When milk was acidified by lactic acid to the same pH both numbers of coliform bacteria and lactobacilli decreased. *L. delbrueckii* subsp. *bulgaricus* could be detected in the stomach and duodenum but not in the colon.

For streptococci, *Ent. faecium* Cernelle 68 fed to piglets in herds, with or without *E. coli* enterotoxaemia, decreased the faecal excretion of coliforms and haemolytic *E. coli*. However, no prevention of disease occurred (Deprez *et al.*, 1989). Using a preparation containing *Ent. faecium*, Danek (1986) influenced the faecal microflora by decreasing numbers of coliform bacteria, and increasing lactic fermentation and streptococci counts. Ozawa *et al.* (1983) fed *Ent. faecalis* to weaning pigs and reported that the microflora (bifidobacteria, streptococci and lactobacilli) was stabilized and that *Ent. faecalis* was antagonistic to *Salmonella* and yeasts.

Kimura *et al.* (1983) observed that animals with diarrhoea had disturbed microflora with decreased numbers of bifidobacteria and lactobacilli and increased numbers of enterobacteria. Oral administration of *Bif. thermophilum* and *Bif. pseudolongum* reinforced the normal intestinal flora and alleviated clinical symptoms in scouring animals and had a prophylactic effect on diarrhoea in suckling piglets.

Bacillus spp. given to slaughter pigs did not influence the intestinal microflora (Spriet *et al.*, 1987). However, Ozawa *et al.* (1981) showed significantly increased numbers of streptococci and bifidobacteria in the proximal small intestine and decreased levels and detection frequencies of *Bacteroides*, by administration of *B. subtilis* strain BN.

A marked change in microbial flora after 3 weeks of administration of *C. butyricum* ID was observed with increased numbers of *C. butyricum* and lactobacilli and reduced counts of staphylococci and coliforms (Han *et al.*, 1984).

To summarize, it seems that administering some probiotics might, in many instances, influence the microbial flora in the digestive tract, especially when the indigenous microflora has been disturbed. This so-called balancing of the flora is difficult to analyse and may not always be clearly connected with improvements in health or performance. Should this stabilizing effect not occur, it is possible that detrimental effects on health and performance could well be demonstrable.

11.4.2 Influence on the digestive tract, its function and morphology

(a) Digestion

Møllgaard (1946, 1947) observed improved skeleton formation in pigs given skim milk cultures of intestinal lactobacilli. He proposed that this was due to improved absorption of calcium by increased lactate production and lower pH in the small intestine. Administration of *Bacillus* spp to cannulated pigs did not affect the pH, the concentration of bacterial metabolites (lactic acid, ammonia, volatile fatty acids) or the apparent digestibility of protein or organic matter (Spriet *et al.*, 1987). Hale and Newton (1979) fed *a Lactobacillus* fermentation product to weaned piglets. Although scouring was reduced, the average daily gain was not significantly improved and the only improvement in digestibility was noted for crude fibre. The authors proposed that this was may be attributable to a decreased transit time.

(b) Morphology

Germ-free animals are frequently used to study the effect of microorganisms on the host. The presence of a microflora causes many changes in the physiology and morphology of the animal. The digestive tract in germ-free animals has more slender and finger-like villi, thinner small intestines with reduced *lamina propria*, longer cell renewal times and higher activities of digestive enzymes (for review see Coates, 1973). Savage *et al.* (1981) found that monoassociation of germ-free mice with a *Lactobacillus* strain did not affect the renewal time of cells in the small intestine. The authors, however, expressed a caution that more microbial species may be needed to cause the changes seen in conventional animals. It has recently been shown using germ-free rodents that the presence of LAB does not stimulate any of the host characteristics referred to by these workers as microflora-associated characteristics (Norin *et al.*, 1991).

Pollman *et al.* (1984) found that artificially reared pigs fed a *Lactobacillus* fermentation product and challenged with *E. coli* had no difference in morphology of the small intestine. By contrast, notable differences in the structure of the small intestine in conventional piglets were shown by Jonsson and Henningsson (1991) using two different herds. In this study, the villi of the control piglets were more or less damaged whereas the pigs receiving *L. reuteri* had more normal villi about 1 week after weaning. Adding *L. acidophilus* to the feed of pigs to be slaughtered was suggested by Premi and Bottazzi (1974) to yield intestines of a better quality for use in the sausage industry.

11.4.3 Performance

A major aim with using probiotics is to improve the performance and health of the animals. Growth rates, feed utilization, number of deaths and days with diarrhoea, sometimes irrespective of cause, are most commonly measured. These studies do not, however, yield information about the mode of action. Depending on the experimental design and the types of controls, it can be difficult to specify with certainty which factors have contributed to the reported changes. It should not be overlooked that the probiotic preparations may have a nutritive effect in addition to other reported functions. It is very laborious to carry out studies in such a way that there should be as little doubt as possible over the results. Many factors influence the performance and health of the animals. In order to conduct experiments in such a way that the addition of the probiotics is the only factor being varied between the groups requires control over a multitude of factors, for example, the environment, handling of the animals, genetic background of the animals, different stress factors and chances for cross-contamination. Variations in these factors could lead to inconclusive or contradictory results. It may be difficult to judge from published results if controlled and comparable conditions were used unless specifically stated. To reduce interference of the factors which are difficult to control and to obtain statistically meaningful results a large number of animals need to be tested. It should also be acknowledged that results can seem more favourable than in reality by incorrect usage or understanding of the statistical analyses. Ideally, a standardized method in which conditions could be exactly specified is clearly needed for probiotic testing. However, because of the different types of microbes and preparations available as probiotics, one standardized method may be too general to be really valuable.

It has not been possible to obtain the experimental details from all reports of probiotic testing, hence the experimental design cannot be extensively judged. Many reports deal with young piglets and it seems that streptococci have the best effect on piglets exposed to suboptimal conditions. For growing pigs, the reports are fewer and the effects are more variable.

(a) Lactobacilli

L. acidophilus has also been reported in many trials to improve performance and health. Sucking piglets with chronic diarrhoea problems showed improvements after administration of *L. acidophilus* strains (Redmond and Moore, 1965; Jensen, 1974; Premi and Bottazzi, 1974). Similarly, Olsson (1966) improved the health of weaning pigs with an *L. acidophilus* strain. However, Kornegay (1985/86) did not find any improvement in growth rate of nursing pigs. This points to the fact that there might be situations when the pig is sensitive to digestive tract disturbances and therefore benefits from enrichment by lactobacilli.

In some trials, starter pigs given *L. acidophilus* have shown improved average daily gain and feed conversion; however, growing-finishing pigs have not been influenced (Kornegay and Thomas, 1973; Pollman *et al.*, 1980a, b, c). Dietary carbohydrates were found to be of importance and the addition of lactose improved the observed beneficial effects. In other trials, no consistent beneficial effects were found (Ingram *et al.*, 1973). It should be emphasized that not all *L. acidophilus* strains are identical and hence strain variation will give rise to conflicting results. A significantly improved weight gain and decreased incidence of diarrhoea was noted for piglets receiving an *L. fermentum* strain (Conway *et al.*, manuscript). Ratcliffe *et al.* (1986) used the closely related *L. reuteri* and found significantly decreased weight gain for very young piglets and tendencies to lower feed conversion.

(b) Streptococci

Almost all reported studies using streptoccocci have utilized various strains of *Ent. faecium*. The greater part of the experiments reported for *Ent. faecium* M 74 show positive effects on health and growth of neonatal piglets (Moen, 1982; Gualtieri and Betti, 1984; Gudding and Larssen, 1985; Danek, 1986; Roth and Kirchgessner, 1986 and Maeng *et al.*, 1989). In contrast, Kluber *et al.* (1985) failed to show these effects and in fact reported higher mortality figures for the groups receiving the strain M 74. For *Ent. faecium* C68 the effects depended on the conditions under which the animals were reared (Jørgensen, 1988). Under suboptimal conditions, but not for pigs raised under good hygienic conditions, improvements of performance and health were shown. Increased weight gain, feed consumption and feed conversion as well as reduced scouring were found by Maeng *et al.* (1989) when dosing piglets up to 4 months with *Ent. faecium* C68. Decreased incidence of diarrhoea was found by Krarup (1987) for neonatal piglets dosed with this strain. Pollman *et al.* (1980a) showed an additive effect of the strain C68 and an antibiotic for piglets with an increased average daily weight gain and feed conversion while for older growing and fattening pigs the effects of the probiotic were not significant. For starter pigs, the C68 strain was ineffective in promoting enhanced performance (Kornegay and Thomas, 1973; Pollman *et al.*, 1980a), whereas Neupert (1988) showed better feed conversion and shorter fattening times and a tendency towards more lean meat for growing pigs by administration of the C68 strain.

(c) Bacillus spp.

The use of *Bacillus* spp. as probiotics is more recent. Jørgensen (1988) showed that sows treated with *B. toyoi* had lower frequencies and milder symptoms of the mastitis-metritis-agalactiae syndrome. For piglets, Roth and Kirchgessner (1988) and Ogle and Inborr (1987) reported tendencies to increased daily gain, feed consumption and feed conversion ratio as well as fewer deaths and fewer treatments with antibiotics when administering *B. toyoi*. Pollman (1986) summarized some studies in which *B. subtilis* was given to pigs. When sows were dosed with *B. subtilis* before farrowing, the piglet survival rate was improved. When the sows had previously received an *E. coli* vaccine there was no effect on performance of either the sow or the piglet. For starter pigs no consistent benefit from *B. subtilis* addition to diets containing carbadox was noted when no major post-weaning diarrhoea was evident. He concluded that administration of *Bacillus* spp. to pigs would be most promising for sows which have not received an *E. coli* vaccine, and that this treatment had little effect on starter pigs.

(d) Various organisms

A mixture of *L. plantarum, L. casei, L. brevis. L. fermentum* and *Pediococcus acidilactici* (Landsudvalget, 1988) was tested in herds with problems with weaning diarrhoea. A non-significant effect towards decreased diarrhoea was seen. However, the numbers of viable bacteria in the preparation was only 0.1% of the desired value. One can speculate that the effect could have been more pronounced if the mixture had contained higher counts. Ogle and Inborr (1987) found significant effects on daily gain when including LAB in the feed of starter pigs.

When yoghurt was administered to 2–12 days old piglets Ratcliffe *et al.* (1986) found decreased feed to weight gain ratio. Using a human *Bif. bifidum* strain Ervolder *et al.* (1984) reported less incidence of diseases in suckling and weaned piglets after dosage. The dosage of *C. butyricum* ID to growing pigs showed a tendency towards better daily weight gain and greater feed consumption as well as significantly improved feed conversion. By varying the dosage, the digestibility of the diet and nitrogen retention rate could be slightly increased (Han *et al.*, 1984). Yeast preparations have also been used but neither Kornegay and

Thomas (1973) nor Burnett and Neil (1977) detected any improvement in performance.

(e) Fermentation products

Fermentation products of lactic acid bacteria (LFP) have been administered separately or in combination with freeze-dried cultures. The organisms involved are not always stated. Generally these investigation show some positive results for weaned pigs, but it must be emphasized that the success of LFP is dependent on the bacterial strain(s) being cultured. Cowman *et al.* (1978) reported briefly that in an extensive investigation they obtained positive responses to different levels of LFP on average daily gain and feed consumption in post-weaned pigs. In contrast, Hale and Newton (1979) found reduced scouring but no significant difference in average daily gain. Pollman and his group (1984) used artificially raised piglets which were housed individually. These authors found increased weight gain and decreased numbers of *E. coli*. They related the weight gain to the reduction of *E. coli* in the stomach without giving any plausible explanation as to how this connection could occur. Lessard and Brisson (1987) used a LFP made from *L.delbrueckü bulgaricus*, *L. casei* and *S. salivarius* subsp. *thermophilus* and demonstrated stimulated growth and feed intake. When feeding a fermentation product together with the mixed LAB culture, Ogle and Inborr (1987) showed a tendency to increased daily gain up to 9 weeks of age.

In summary, it seems as if there might be situations where the total microflora of the piglets is not optimal and the animals therefore benefit from the use of probiotics. Many of the reports deal with young piglets and it seems generally as if the most positive effects are reported for the young animals held under suboptimal conditions. Lactobacilli seem to be the most efficient in influencing the microflora in the digestive tract while streptococci seem to have the best effect on piglet performance. For pigs after weaning the reports are fewer and the effects are more variable.

11.4.4 Immunology

Kluber *et al.* (1985) administered Ent. *faecium* M74 to artificially reared pigs. The *in vivo* cell-mediated immunity was not affected. It is noteworthy that the mortality was greater for pigs treated with the LAB. These pigs also had an increase in mature and immature neutrophils with suppressed lymphocyte counts. Pollman and his group (1980b) report stable levels of *L. acidophilus* 9 days after dosage to gnotobiotic pigs. The animals had elevated counts of white blood cells but similar levels of serum metabolites as compared with gnotobiotic piglets, which however were shown also to be colonized by some other lactobacilli. In a later investigation, this group found an increased white blood cell count in artificially raised piglets challenged with *E. coli* after dosage with LAB metabolites (Pollman *et al.*, 1984). Lessard and Brisson (1987) fed metabolites from dairy LAB (see above) to weaned pigs and found slightly increased serum levels of IgG.

11.5. FUNCTIONAL CHARACTERISTICS OF POTENTIAL PROBIOTIC STRAINS

Lists of criteria to be fulfilled for probiotic organisms have been compiled and new items have been added as the knowledge about the interrelationships between the host and the microbes increased (see, for example, Tannock, 1984). Procedures for selection of new strains are treated elsewhere in this book and hence are only discussed briefly here in relation to the experiences obtained in isolating and selecting *Lactobacillus* strains from the indigenous microflora of pigs (cf. also Jonsson, 1985). Factors to consider when developing dietary adjuncts containing LAB and bifidobacteria have also been described by Klaenhammer (1982), Kurman (1983) and Gilliland and Walker (1990).

11.5.1 Isolation techniques

The common practice of using standard selective media (e.g. Rogosa and MRS agar) for isolating LABs from the indigenous gut microflora yields the most dominant species. Consequently, to reduce the isolates needing screening as described below, care must be taken in the selection of the source of the inoculum. In addition, the target for the intended use of the probiotic strain must be clearly defined. For example, if post-weaning bacterial-induced diarrhoea is the target, it would be reasonable to select the digestive tract microflora of an extremely healthy post-weaning pig as the inoculum source. While it is feasible to propose that the isolation medium could be specially designed to enrich for LABs with specific characteristics, in practice this could be difficult to implement. Most of the reported desirable characteristics of LABs for porcine use cannot be selected for by compositional changes of the medium, e.g. capacity to adhere to epithelial mucosa or to produce metabolites inhibitory to pathogen colonization. It can be noted, however, that Jonsson and co-workers (unpublished data) prepared media from mucosal homogenates from various regions of the gut and showed a tendency for these media to yield a greater number of isolates which were adhesive to the mucosal region used in the medium on which they had been isolated.

It is often specified that anaerobic conditions are required for culturing isolates from the digestive tract. The redox-potential of the digestive tract of the piglet immediately after birth is high, *c.*+400 mV in the stomach, +150 to −150 mV in the small intestine and +100 to −200 mV in the large intestine. After 5 days the Eh decreases to +50 mV, 0 to −250 mV and −150 to −250 for the respective sites. There is no great change in Eh during weaning, despite the changed feeding and rearing systems. For growing pigs, the Eh found was −100 to −250 mV, with the highest values found in the stomach and anterior small intestine (Schulze and Jacob, 1981). Consistent with other reports, the authors observed that these values were reflected in the values obtained for the total counts of bacteria. Lactobacilli grow best *in vitro* in microaerophilic and anaerobic conditions. Although lacking a cytochrome system and therefore not able to utilize oxygen for aerobic growth, lactobacilli can grow in the presence of oxygen using fermentative metabolism. This taxonomic description may not be entirely correct for lactobacilli of digestive origin because *L. acidophilus* Po13 isolated from the *pars oesophagea* area of a pig only grows in anaerobic conditions (Jonsson *et al.*, 1985).

Lactobacillus strains can often have distinct colony morphology and Klaenhammer and Kleeman (1981) raised the issue that colony morphology was important because isolates producing smooth colonies were more resistant to bile acid and freeze damage than isolates producing rough colonies. In contrast, recent studies have shown that the rough variant of an *L. fermentum* strain of porcine origin was generally more resistant to bile acids and low pH than a smooth colony variant of the same strain (Szewzyk *et al.*, manuscript).

To date, the *Lactobacillus* flora from the digestive tract has been poorly studied taxonomically and many new species from this habitat are now being described (reviewed by Tannock, 1990). The morphology and physiology of the isolated strains might be different to what has previously been considered normal for lactobacilli. The morphology of these wild types may sometimes cause some confusion as coccoid rods are often found and the morphology also might change as the strain is propagated, (see Kandler and Weiss, 1986). On isolation, such strains may also have an increased sensitivity to oxygen.

11.5.2 Evaluation of strains

Probiotic preparations have been said to have several beneficial roles in pigs. Although their modes of action are not known, it is assumed that strains intended to be used as probiotics should be isolated from the indigenous microflora and have characteristics such as colonization capacity, and capacity to inhibit the activity as well as colonization ability of pathogenic microbes. Initially, such screening relies on *in vitro* test systems, the result of which can be confirmed *in vivo*. Subsequently, extensive field testing is needed to confirm that the strain(s) do in fact have a beneficial influence on performance and health of the pig.

The success of a probiotic preparation will depend not only on its biological effect under controlled experimental conditions but also relies on its dependability when produced on a larger scale. It has to be possible to propagate the strain or strains on a commercial scale such that essential strain characteristics are maintained without excessive production costs. The commercial preparation should also withstand storage under conditions likely to be encountered in the distribution and farm situations without great loss of viability or strain characteristics, (see Pollman, 1986).

To exert beneficial effects, the bacteria or possibly the active substances have to reach the site of function. In order to do so, first of all they have to be consumed. This means that the preparations have to be appetizing or at least not repulsive to the pig. Pigs are sensitive to the long-term effects as well as the flavour and taste of their feed (Houpt *et al.*, 1979). This fact is also stressed by Premi and Bottazzi (1974). The next important step is that the preparation should contain viable and active organisms or substances in sufficiently large amounts.

(a) **In vitro *versus* in vivo *tests***

Ideally, strain selection should be performed under *in vivo* conditions but logistics require the development of *in vitro* assays. Consequently, the probiotic characteristics which mediate the beneficial effects on health and performance of the pig need to be defined such that specific parameters can be measured *in vitro*, e.g. colonization capacity is measured as the *in vitro* adhesive capacity. Designing *in vitro* and *in vivo* studies to evaluate the potential of a probiotic strain raises more questions: for example, should conventional or gnotobiotic pigs be used? These *in vitro* and *in vivo* issues are discussed below in terms of survival, colonization and inhibition of pathogen colonization.

Many *in vitro* tests for studying strain characteristics of probiotic organisms have been reported, however, not all studies have included comparisons of the results with results obtained under *in vivo* conditions. As discussed below for the various characteristics, this comparison of the *in vivo* and *in vitro* results is feasible for factors such as survival and colonization of the strain(s), because these can be measured directly *in vivo*. Other factors such as the production of antagonistic substances can only be measured indirectly unless the active substances are characterized and probes for the component are produced. Unfortunately, measuring survival and colonization *in vivo* is also limited because of difficulties in detecting administered strains in the complex indigenous microflora, especially if the strain originates from the digestive tract. This aspect is discussed in detail in section (b) below on gnotobiotic and conventional animals.

Survival

Survival of administered bacteria in the anterior digestive tract is mainly discussed and studied *in vitro* in terms of bile acid and low pH tolerance. The effects of proteolytic enzymes in the stomach and small intestine on survival of bacteria have been seldom investigated but it cannot be excluded that these enzymes also may have some influence on the survival of probiotic strains. Lactobacilli of pig stomach origin have been shown to survive passage of the pig stomach and small intestine (Jonsson *et al.*, 1985). In addition, these workers showed that strains isolated from the *pars oesophagea* area varied in their bile tolerance *in vitro* and a similar variation could be observed *in vivo* in the small intestinal contents of cannulated pigs. Even a *L. delbrueckii* subsp. *bulgaricus* strain from yoghurt was found to survive in the small intestine of pigs after ingestion (Ratcliffe *et al.*, 1986). Using lactobacilli of human origin and low pH buffer or human gastric juice, survival in gastric juice *in vitro* and *in vivo* were almost comparable, with *in vivo* studies showing slightly better survival than those using low pH buffer (Conway *et al.*, 1987).

In an attempt to develop an *in vitro* survival assay which simulated *in vivo* stomach conditions, sterile pig feed was slowly mixed with artificial saliva, bacterial culture and acid mixture and incubated anaerobically at 40°C for 24 h (Jonsson, unpublished). The results obtained with this test did not correspond very well to what was found *in situ* in the pig. One explanation could be that the influence of the gastric LAB microflora was missing; this is very relevant for the conditions in the pig. In other experiments where the pH of MRS-broth was decreased stepwise by 0.5 units by using HCl to pH 2.5, the results of bacterial growth showed better correspondence with the *in vivo* survival through the stomach. The strains with the best *in vivo* survival grew well at pH 3.5.

Colonization

It has been proposed that LAB as probiotics could colonize the piglet digestive tract and thereby continually exert their influence. For example, LAB may protect the host from infections by colonizing sites and thereby exclude invading pathogens. This latter concept is often referred to as competitive exclusion and has been extensively studied for complex probiotic type preparations for chickens since the first experiments reported by Nurmi and Rantala (1973). This was also studied in piglets for example by Corpet and Nicolas (1979). By colonization, one means that the LAB can remain and multiply within the digestive tract and are not eliminated from the system. This could be achieved either by attachment to epithelia and/or growth in the digesta. In the small intestine with its rapid flow of digesta, attachment to the epithelium is most likely a prerequisite for colonization, whereas in the stomach and large intestine the flow rate is lower and the LAB may survive solely by multiplication within the lumenal contents. In the stomach the squamous cells sloughed off from the *pars oesophagea* area with their attached LAB will also continuously inoculate the ingesta. The fact that the stomach seldom empties completely between meals and that the conditions will not become too acid also facilitates colonization in this site.

Colonization can be confirmed *in vivo* by detecting the strain in the tract beyond the time accepted as the lowest possible for transit of particles through the tract. Generally, detection of a strain for more than 7 days after dosage would be strong evidence of colonization. The detection can be made in faecal samples, although to obtain information as to what habitats are colonized within the digestive tract, samples from the various regions of the gut must be analysed for presence of the strain.

The potential for *in vivo* colonization is presently assessed *in vitro* by studying the capacity of the strains to attach to the epithelia of the digestive tract. When discussing adhesion to epithelial mucosa, it is important to distinguish between the various epithelial surfaces to which bacteria can attach. The *pars oesophagea* region consists of keratinized squamous non-secreting cells while the remainder of the epithelia in the digestive tract consists of columnar cells which are covered by a mucus layer. Barrow *et al.* (1980) found considerable differences in the capacity of LAB to adhere to squamous cells and also reported that strains with such ability did not associate with columnar cells. Subsequently it has been shown that stomach and colon are colonized by different *Lactobacillus* strains (Tannock *et al.*, 1990). It is most probable that different adhesion mechanisms mediate attachment to squamous and columnar secreting cells. Another site for colonization may be the mucus layer overlying the epithelia. This has recently been shown *in vitro* for pathogenic *E. coli* on piglet ileal mucosa (Conway *et al.*, 1990) and it is reasonable to propose that LAB may also colonize this niche.

Bacteria can adhere specifically or non-specifically to surfaces as defined by Rutter *et al.* (1984). The former mechanism involves adhesin–receptor interactions while the latter is mediated by physico-chemical factors such as, hydrogen bonding, surface charge and the degree of surface hydration. Specific adhesion occurs when an adhesin on the bacterial cell binds to a receptor on the epithelial cell, in what is often defined as a lock and key function. To date, no adhesin–receptor complexes have been characterized for adhesion of lactobacilli to porcine epithelia. It has been proposed that adhesion of lactobacilli to squamous cells was mediated by an extracellular polysaccharide (Barrow *et al.*, 1980). More recently, proteinaceous components have been proposed (Henriksson *et al.*, 1991) as demonstrated for adhesion of lactobacilli to rodent squamous cells (Conway and Kjelleberg, 1989). Subsequent studies suggest that the mechanism of adhesion may involve other components as well (Henriksson and Conway, manuscript. unpublished). Wadström *et al.* (1987) studied the adhesion of lactobacilli to cells from pig small intestinal epithelium *in vitro* and showed a correlation between hydrophobic bacterial cell surface and adhesion.

Non-specific adhesion, which is demonstrable *in vitro* in adhesion assays, may not have any significance in the colonization of epithelia *in vivo* but may possibly be important in the colonization of luminal contents. For example, non-specific adhesion may enhance substrate uptake and thus utilization for growth. Defined adhesin–receptor interactions have been extensively studied for pathogens and evidence is now appearing that some of these interactions may have no *in vivo* significance either. For example, the P-fimbriae on uropathogenic *E. coli* bind specifically to a gal-gal receptor but from recent studies it is apparent that this interaction is not important for *in vivo* colonization (C.Svanborg-Edén, pers. comm.). In this connection, it might also be emphasized that competitive colonization does not imply that the lactobacilli and the pathogenic bacteria will bind to the same receptors on the epithelial cells, but rather that the lactobacilli may sterically hinder the adhesion of the pathogens as demonstrated *in vitro* (Conway, 1985). It should also be pointed out that *in vivo* colonization of a site involves more aspects than simply adhesion to the epithelium. LAB growing at the site of colonization may utilize nutrients otherwise available for the pathogens. Furthermore, the growth of LAB may also result in production of metabolites inhibitory to the pathogens.

When testing the adhesion capacity of LAB *in vitro*, epithelial cell preparations or mucosal pieces washed free of the mucus layer have most often been used. After the bacteria and the epithelial cells are incubated together, adhesion is sometimes assessed by direct microscopy. This technique of not washing the epithelial cells prior to microscopy means that loosely bound bacteria with low affinity for the epithelial cells are not removed, thus some weakly adhesive strains may be considered adhesive. A standardized washing procedure is required after the incubation to remove loosely bound bacteria.

To check if the adhesion is specific or non-specific it can be tested if the bacteria have a higher affinity for epithelial cells than control surfaces such as protein-coated plastic as reported for *L. fermentum* adhesion to squamous cells (Henriksson *et al.*, 1991). If the affinity for the control surface is comparable or higher compared to the epithelial cells, non-specific adhesion might be involved. Unfortunately, these types of controls are often not reported.

Many *in vitro* investigations have been done to test the adhesive capacity of LAB and to show that this adhesion is host species-specific, thus confirming the rationale for selecting probiotic strains from the piglet microflora rather than from another animal host. These tests have not been strictly discriminating for species-specificity as some LAB strains from other host species can attach to piglet cells (Barrow *et al.*, 1980; Conway *et al.* 1987). In the latter investigation even some yoghurt strains adhered to human and pig ileal cells. Species-specificity of LAB in pigs was also demonstrated *in vivo* in gnotobiotic pigs by Tannock *et al.* (1982). Pedersen and Tannock (1989) tested eight strains of lactobacilli isolated from the digestive tract of pigs for their ability to adhere *in vitro* to cells collected from the stratified squamous epithelium of newborn piglets. *In vivo* colonization capacity of these strains was studied in piglets dosed with the individual strains. The results showed that the *in vitro* tests did not predict whether a *Lactobacillus* strain would associate with stratified epithelium *in vivo*. None of these strains was dominant in the digestive tract 7 days after the inoculation of the piglets with a single dose of the bacteria confirming the results obtained by Jonsson (1986). A heterofermentative *Lactobacillus* strain could be found by Pedersen and Tannock (1989) during the first 7 days in piglet faeces but not after 19 days. No permanent establishment on the *pars oesophagea* region was noted in spite of the *in vitro* adhesive properties of the strain to the *pars oesophagea* cells. So far, only on one occasion has long-term establishment been reported. The *L. murinus* used in this study was of porcine gastric origin and was extremely adhesive in an *in vitro* adhesion assay (Henriksson *et al.*, 1991). Piglets were dosed at either 2 days, 2 weeks or 4 weeks of age and harboured the strain continually for the duration of the 9-week study (Conway *et al.*, manuscript).

In summary, inconsistencies between *in vitro* adhesion and *in vivo* colonization can depend on factors such as the survival capacity and *in vivo* growth potential. In addition, it is also quite plausible that the design and interpretation of results of the *in vitro* adhesion assay should be closely examined. It is difficult to extrapolate from *in vitro* adhesion results stating how many bacteria bind per epithelial cell, because one cannot predict the threshold value below which it can be stated that the strain under investigation is not adhesive or rather will not be adhesive *in vivo*. It should also be noted that *in vitro* adhesive capacity *per se* does not necessarily confirm that the strain is equipped with colonization capacity. *In vitro* adhesive capacity can be combined with other parameters, such as survival capacity and growth in the gastrointestinal milieu, and the total picture of these *in vitro* analyses may yield a prediction of the colonization capacity which correlates better with the *in vivo* results.

Microbial interactions

Interations between the probiotic strain and other microbes can be discussed in terms of effects of the probiotic microbe on the indigenous microflora and its effects on potentially pathogenic microorganisms.

In vitro, microbial interactions are largely measured as effects on pathogenic bacteria by low and high molecular weight metabolites of the LAB (reviewed by Fernandes *et al.*, 1987; Lindgren and Dobrogosz, 1990). The effects on the indigenous microflora can also be assessed by monitoring changes in the metabolism and profile of indigenous population. This aspect has been studied *in vitro* by Aimutis *et al.* (1985). They isolated bacteria from the piglet small intestine and propagated them in different selective media and found that apart from the selective effect of the media the bacteria growing in the different substrates also showed antagonistic effects for *E. coli*. For example, *Bifidobacterium, Clostridium, Lactobacillus* and *Leptotrichia* spp. serially transferred in sow colostrum broth or piglet feed infusion broth inhibited the growth of *E. coli*.

In vivo, microbial interactions can be assessed in terms of the effects of administration of the probiotic strain on the levels of groups of microorganisms in any region of the digestive tract. The studies can be done as challenge studies or administration of probiotics to animals on farms with a known history of diseases. For the latter case, the results are more difficult to evaluate as there are many factors which cannot be controlled. For example, when enumerating the number of *E. coli* in the digestive tract, there are no convenient assays for pathogenic *E. coli* or for distinguishing them from the indigenous ones. Because the indigenous *E. coli* levels are not stable, if total *E. coli* levels are decreased as a result of administration of a probiotic microbes or its metabolites, one cannot state conclusively that the effect is on the pathogenic *E. coli*. Some *in vivo* studies have reported changes in levels of lactobacilli and coliforms after dosage of the probiotic strain, see section. 11.4.2. The effect of lactobacilli on the metabolism of the microflora has been studied in gnotobiotic mice colonized with human digestive tract microflora (Norin *et al.*, 1991). No changes in microflora associated biochemical characteristics were observed by these workers. Gnotobiotic piglets and gnotobiotic adult mice were associated with microflora from conventional piglets and adult pigs in a study by Ducluzeau *et al.* (1978b). They found that the ex-gnotobiotic mouse was a better model for studies of association of the flora from piglets than ex-gnotobiotic mice associated with the flora of adult pigs. Provided that the inoculation was made anaerobically, the gnotobiotic mouse was an adequate model to preserve the bacterial strains, but the equilibrium between the strains was markedly affected by the host. This equilibrium changed with time but upon transfer into a new piglet or mouse, the equilibrium was restored. It is therefore feasible to propose testing probiotic preparations on ex-gnotobiotic mice associated with piglet digestive tract microflora.

Probiotic strains producing well-characterised low and high molecular weight antagonistic metabolites may offer the only possibility of yielding *in vitro* and *in vivo* studies which can be directly compared. This will be possible if a probe can be prepared which is specific for the antagonistic metabolite. Such a probe would allow *in vivo* confirmation of the production of the active component. To our knowledge, no such probes are presently available for testing in pigs. Consequently, *in vitro* antagonistic activities can only be compared with indirect observations *in vivo* such as decreases in coliforms levels, decreased incidence of diarrhoea and death as well as improved weight gain. For example, an *L. fermentum* strain which produces low and high molecular weight metabolites antagonistic to growth and adhesion of *E. coli* K88 *in vitro*, was used to dose piglets. Improved weight gain and a decreased incidence of diarrhoea was noted in these piglets, which suggest that the strain expressed its antagonistic metabolites *in vivo* (Conway *et al.*, manuscripts). It should be emphasized that direct comparisons of *in vitro* data to *in vivo* results is presently only based on extrapolations. Another example is the study of Ducluzeau and Raibaud (1974) who found significant differences between *in vitro* and *in vivo* interactions; *E. coli* could suppress the *Shigella* population *in vitro* whereas they could coexist *in vivo*. Freter (1974) claimed that he could cultivate any two organisms in different ways to make either one show antagonism towards the other. This comment highlights the erroneous conclusions which can be drawn from *in vitro* studies if culturing conditions are not controlled appropriately.

The above comments on *in vivo* studies apply specifically to conventional animals. Gnotobiotic animals may be useful for eliminating some of the problems with these indirect measurements. This aspect is discussed further in the following section.

(b) Gnotobiotic versus conventional animals

Gnotobiotic animals have been very important for understanding of the interrelationships between host and microbes and between microbes *in vivo*. Results from such animals are not necessarily transferable to conventional animals, because of the absence of the contributions to the system from the indigenous microflora (cf. Coates and Gustafsson, 1984). Gnotobiotic animals may be valuable for testing probiotic strains, in particular, for testing the colonization capacity, survival and microbial interactions in the absence of the indigenous microflora. For the latter two points, the difficulties of detection as discussed below are eliminated. Gnotobiotic animals can also be valuable for evaluating the effects of the probiotic strain on host physiological factors, e.g the immune system. Luckey (1987) has raised the issue that such studies should pay close attention to the diet of control animals. He proposes that stimulation of the immune system may occur simply because of a deficiency in general antigens of the diet rather than a specific effect of the administered microorganism. The gnotobiotic animal can be a valuable first step for testing survival and colonization potential of a probiotic strain. That is, if the strain fails to establish in either the gnotobiotic animal or the ex-gnotobiotic animal colonized with a complex flora, one can conclude the strain lacks colonizing capacity. This type of a study has been performed using caesarian derived colostrum-deprived piglets by Underdahl *et al.* (1983) and Ushe and Nagy (1985) who studied *Ent. faecium* C68 and M74, respectively. *Ent. faecium* C68 established in the piglets while M74 did not. It should be added that non-indigenous LAB or even strains originated from dairy products have been reported to establish in the digestive tract of germ-free animals (Morishita *et al.*, 1971, Bianchi-Salvadori *et al.*, 1984). When studying colonization in the digestive tract of germ-free animals, it has to be kept in mind that the animals may be continuously reinoculated by contaminated equipment, faeces and left-over food. If the administered strain has a good survival capacity in the digestive tract, the levels found, especially in the digesta, might reflect the continuous addition of transient bacteria. The numbers found associated with the epithelia could be less influenced by such reinoculation or recycling of the probiotic strain.

Jonsson and Björck (submitted) tested a strain of *L. reuteri* in a series of consecutively harsher *in vivo* steps beginning with establishment in germ-free piglets and finishing with conventional pigs. The strain established in the digestive tract (mucosa and digesta) of the germ-free piglets but when challenged after a week with a faecal SPF-flora, the strain persisted but in decreased numbers. When the strain was administered at the same time as the faecal SPF-flora, it only established in the digesta but not on the epithelia. When fed once to conventional cannulated pigs, the strain decreased successively in numbers in the small intestinal digesta and after 3 days the numbers had declined to below the detection level. This series showed that the harsher the pressure on the ex-gnotobiotic animal, the lower the establishment of this strain. It may be concluded from this study that this strain will probably not colonize conventional pigs. This type of experiment, in which colonization is tested in a range of progressively more rigorous conditions, could be useful as a standard test system.

To obtain comparable results when using conventional animals, considerably larger numbers of animals need to be investigated because of variable factors such as genetic background and environmental stresses. For experimental purposes, cannulated pigs could be used (e.g. Jonsson, 1985), which enable more controlled studies that can be repeated in the same individual animal, hence minimizing the influence of the above-mentioned factors.

The difficulties of directly demonstrating *in vivo* antagonistic activities of probiotic strains on specific pathogens are eliminated when using gnotobiotic animals. For such studies, the animals can be monoassociated with the probiotic and then challenged with the pathogen, and *vice versa*. Ushe and Nagy (1985) showed decreased ileal colonization by enterotoxigenic *E. coli* in caesarian derived colostrum-deprived piglets dosed with *Ent. faecium* M74. As another alternative to gnotobiotic animals, piglets have been treated with antibiotic at birth and then artificially reared (Muralidhara *et al.*, 1977). These workers showed the successful competition of an *L. lactic* strain of human origin over pathogenic *E. coli*.

A promising alternative to gnotobiotic or conventional animals has been developed in mice by Tannock and co-workers. Specially derived mice which have a gastrointestinal microflora devoid of lactobacilli have been established and used to confirm species-specificity of lactobacilli (Tannock and Archibald, 1984). This animal model is closer to conventional animals and holds promise as a tool for evaluating directly the role of lactobacilli in the mouse.

While these types of studies provide controlled conditions, it must be emphasized that from such results it cannot be assumed that we can predict exactly what would occur in conventional animals. Ultimately, probiotics which appear potentially valuable must be evaluated in conventional animals in the open environment to make sure of their commercial potential.

When studying survival of administered LAB in the digestive tract, the administered strain has to be redetected. In gnotobiotic animals and ex-gnotobiotic animals with a known microflora this is easier than within the rich LAB microflora of conventional pigs. Most investigators have not attempted to identify the administered bacteria within the microflora of the digestive tract but have only analysed the total count of the species used. Different methods have been used for detection of lactobacilli of intestinal origin and these include biochemical and serological tests (Muralidhara *et al.*, 1977; Jonsson *et al.*, 1985; Conway *et al.*, manuscript), genetic probes (Betzl *et al.*, 1990, Tannock *et al.*, 1990) and antibiotic resistance (Jonsson and Björck, submitted). Yoghurt lactobacilli have been detected using selective inhibitory media (Ratcliffe *et al.*, 1986). While serology is extremely valuable in confirming detection of an administered strain, caution should be taken to ensure the antibody is treated against cells growing under conditions comparable to those used for preparation of the antibodies as described by Conway and co-workers (manuscript). Bacteria can express different surface antigens depending on growth conditions (P.S. Cohen, pers. comm.). Genetic markers may make it possible to redetect strains conveniently and with certainty in conventional animals in the future. It may be difficult, however, to find naturally occurring traits in the administered strain that are not common to a relatively large part of the present indigenous microflora. Introduction of such markers into a strain may also decrease its competitive ability within the digestive microflora (Wells *et al.*, 1979).

11.6 GENERAL DISCUSSION

Piglets are ideal candidates for probiotic administration, especially preparations containing lactic acid bacteria, because of the large population of these microorganisms in the digestive tract of the healthy animal. Unfortunately, beneficial effects of probiotic administration on growth and health of pigs are not always achieved. Many factors could influence the effectiveness of such preparations, for example the genetic, physiological or health status of the animal, as well as the fact that the diet or the environment and its microbial load could vary between herds. In addition, the probiotic preparation could vary with regard to the type(s) of microbe used, its (their) viability and physiological state as well as the form, time and levels of dosage.

Although pronounced beneficial effects on growth and health have not always been demonstrable, there are often trends towards improvement especially in neonatal and pre- and post-weaning piglets. For example, administering lactobacilli has eliminated long-term diarrhoea in some herds and administration of streptococci in some cases has prevented diarrhoea in piglets. These facts support the belief that the digestive microflora of young pigs could be favourably influenced by administering LAB. For these animals, the digestive ecosystem is less stable than in the adults and is, hence, more easily invaded by opportunistic pathogens.

The probiotic preparation and the time for administering would have great influence on the induced effects. The preparations can be either based on preformed antagonistic substances or on viable microorganisms. For the former type, the activity of the preparations is related to the type(s) and concentration of the compound(s). This concept may be the easier to monitor and to control. For the latter type, many more factors have to be considered, for example, the viability and genetic stability of the organisms in the preparation as well as their effect on the digestive ecosystem. Presently, it seems as if it is very difficult in most situations to implant permanent probiotic organisms in the digestive tract. Although one report points in this direction, most studies indicate that the indigenous microflora are very efficient in preventing new organisms from establishing permanently.

To obtain more consistent effects of both types of probiotics, more basic knowledge of the digestive ecosystem is needed. For example, the interactions between animal, diet, microbes, digestion and the immune response need to be better understood (see Fig 11.3). We postulate that probiotics would have greater influences when pigs are exposed to negative stress of some kind. This stress may be imposed from the outside of, or from within, the animal, and directly predispose the animal to infections or indirectly influence the health. Much research has been put into optimizing the rearing systems, however, there may still be further improvements to be obtained, especially when the genetic material of the stock is changing due to the breeding programs. In addition, the expanding knowledge of animal behaviour and its influence on the physiology of the animal will be valuable in finding ways to improve conditions for raising of the pigs and their well-being and performance.

To understand better the digestive ecosystem and hence the modes of action of probiotics, improved methods for studying all the complex interactions *in situ* are required. One possibility could be to concentrate on the microenvironments in the digestive tract and thus understand the conditions for the microbes in these sites. Increased knowledge about hormonal reactions and the immune responses are probably also relevant for this kind of study. Benefit can probably also be gained by connecting knowledge from the spheres of digestive microbiology with that of physiology. One example of this can be seen from the early studies of germ-free rodents which have enlarged caeca. The reason for this enlargement was not obvious until it was apparent that the digestive tract microflora have mucinolytic activity and that there is a retrograde transportation of digesta in the hindgut of these animals. A better understanding of the complex interactions within the digestive tract will allow predictions as to which situations could be improved by probiotic treatments.

In summary, the current awareness of the risks involved with excessive antibiotic usage and the recognition that in the past, many of the probiotic preparations were largely ineffective, herald a new generation of probiotic strains. The signature of these new probiotics is that the functional characteristics are defined. In addition, the target and associated causative agents must be identified in order to establish which conditions can be improved by which probiotic preparations. Conversely, it is also important to define conditions of health which

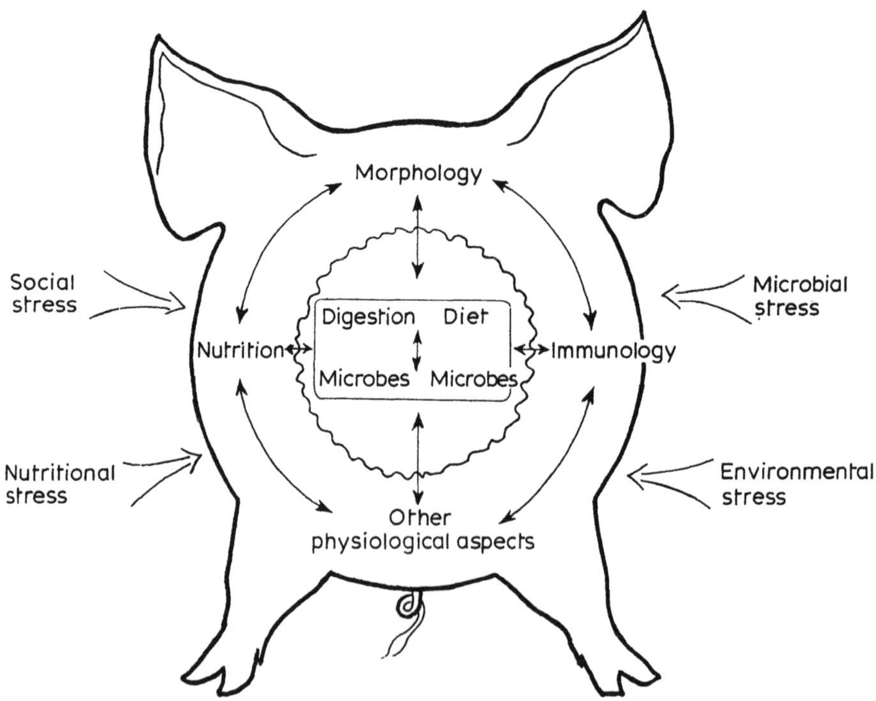

Figure 11.3 To further understand how probiotics exert their effects, studies on the total ecosystem of the pig in its specific environment will be needed.

are unlikely to be improved by known probiotic activities. Situations most certainly exist whereby probiotic strains function cooperatively with other developing concepts in the treatment and care of piglets, thus providing the opportunity to have multifaceted prophylactic piglet care.

From studies to date, it is feasible to postulate that probiotic preparations could contain microbes with the capacity to improve piglet health by their direct inhibitory effects on enteropathogenic bacteria. The *in vitro* and *in vivo* demonstration and characterization of inhibitory components will ensure functional preparations. Identification of adhesins which mediate *in vivo* colonization will allow better predictions if probiotic preparations have the capacity to colonize the digestive tract. Probiotic strains with demonstrable direct effects may also function indirectly by stabilizing the digestive microflora at times when the ecosystem is stressed. This latter vital role is essentially restricted until the complex interactions within the digestive tract are better understood.

REFERENCES

Abrams, G.D., Bauer, H. and Sprinz, H. (1963) Influence of the normal flora on mucosal morphology and cellular renewal in the ileum. *Lab. Invest.*, **12**, 355–64.

Aimutis, W.R., Kornegay, E.T. and Eigel, W.N. (1985) *In vitro* antagonism of *Escherichia coli* by piglet small intestinal bacteria isolated by selective enrichment in media containing sow colostrum, piglet feed or brain heart infusion. *J. Gen. Microbiol.*, **31**, 135–46.

Algers, B. (1980) Djurhälsa och beteende vid burhållning av avvanda smågrisar (Animal health and behaviour in flatdeck rearing of weaned piglets). Dept animal hygiene with farrier's school, Swed. Univ. Agric. Sci., Skara, Sweden, Report, No. 4, pp. 1–82.

Axelsson, L. and Lindgren, S (1987). Characterization and DNA homology of *Lactobacillus* strains isolated from pig intestine. *J Appl. Bacteriol.*, 62, 433–40.

Auffray, P. (1965) Cited in Rerat *et al.* (1976).

Bäckström, L. (1973) Environment and health in piglet production. A field study of incidences and correlations. *Acta Vet. Scand. Suppl.*, **41**, 1–240.

Barnett, J.L., Hemsworth, P.H., Winfield, C.G. and Fahy, V.A. (1987) The effects of pregnancy and litter number on behaviour and physiological responses related to welfare status of individual and group housed pigs. *Appl. Anim. Behaviour Sci.*, **17**, 229–43.

Barrow, P.A., Fuller, R. and Newport, M.J. (1977) Changes in the microflora and physiology of the anterior intestinal tract of pigs weaned at 2 days, with special reference to the pathogenesis of diarrhea. *Infect. Immun.*, **18**, 586–95.

Barrow, P.A., Brooker, B.E., Fuller, R. and Newport, M.J. (1980) The attachment of bacteria to the gastric epithelium of the pig and its importance in the microecology of the intestine. *J. Appl. Bacteriol.*, **48**, 147–54.

Betzl, D., Ludwig, W. and Schleifer, K.N. (1990) Identification of lactococci and enterococci by colony hybridization with 23S rRNA-targeted oligonucleotide probes. *Appl. Environ. Microbiol.*, **56**, 2927–9.

Bianchi-Salvadori, B., Camaschella, P. and Bazzigaluppi, E. (1984) Distribution and adherence of *Lactobacillus bulgaricus* in the gastroenteric tract of germ-free animals. *Milchwissenschaft*, **39**, 387–91.

Björk, A.K.K., Christensson, E., Olsson, N.-G. and Martinsson, K.B. (1984) The clinical effect of amperozide in pig production. I.Effects of amperozide on aggression and performance in weaners. *Proc. Int. Vet. Soc. 8th Congress, Ghent, Belgium, 27–31 August, 1984*, XV, p. 335.

Björk, A.K.K., Olsson, N.-G.E., Martinsson, K.B. and Göransson, L.A.T. (1987) A note on the role of behaviour in pig production and the effect of amperozide in growth performance. *Anim. Prod.*, **45**, 523–6.

Blomberg, L. and Conway, P.L. (1989) An *in vitro* study of ileal colonisation resistance to *Escherichia coli* strain Bd 1107/7508 (K88) in relation to indigenous squamous gastric colonisation in piglets of varying ages. *Microbial Ecol. Health Dis.*, **2**, 285–91.

Borriello, S.P. and Barclay, F.E. (1985) Protection of hamsters against *Clostridium difficile* ileocaecitis by prior colonisation with non-pathogenic strains. *J. Med. Microbiol.*, **19**, 339–50.

Boucourt, R. and Ly, J. (1975) Microflora and fermentation in the gastrointestinal tract of the young pig. 2. Bacterial population. *Cuban J. Agric.*, **9**, 163.

Bourlioux, P. and Pochart, P. (1988) Nutritional and health properties of yoghurt. *Wld Rev. Nutr. Diet.*, **56**, 217–58.

Brandenburg, A.C and Wilson, W.R. (1973) Immunity to *Escherichia coli* in pigs; IgG immunoglobulin in passive immunity to *E. coli* enteritis. *Immunology*, **24**, 119–27.

Braude, R., Fulford, R.J. and Low, A.G. (1976), Studies on digestion and absorption in the intestines of growing pigs. Measurements of the flow of digesta and pH. *Brit. J. Nutr.*, **36**, 497–510.

Brennan, M., Wanismail, B., Johnson, M.C. and Ray, B. (1986) Cellular damage in dried *Lactobacillus acidophilus*. *J. Food Protect.*, **49**, 47–53.

Brockett, M. and Tannock, G.W. (1981) Dietary components influence tissue-associated lactobacilli in the mouse stomach. *Can. J. Microbiol.*, **27**, 452–5.

Brownlee, A. and Moss, W. (1961) The influence of diet on lactobacilli in the stomach of the rat. *J. Pathol. Bacteriol.*, **82**, 513–16.

Burnett, G.S. and Neil, E.L. (1977) A note on the effect of probioticum feed additive on the live-weight gain, feed conversion and carcass quality of bacon pigs. *Anim. Prod.*, **25**, 95–8.

Cain, C.J. (1989) Observations on indigenous and non-indigenous lactic acid bacteria as potential probiotic organisms in the pig. M.Agr.Sc. Thesis, La Trobe University, Australia.

Campbell, R.G. (1989) The nutritional management of weaner pigs, in *Manipulating Pig Production*, II (eds J.L. Barnett. and D.P. Hennesy.), Australasian Pig Science Association, Werribee, Victoria, Australia, pp. 170–5.

Carghill, C.F. (1982) Control of *E. coli* infections in pigs. *Austral. Adv. Vet. Sci.*, pp. 206–7.

Chapman, J.D. (1988) Probiotics, acidifiers and yeast culture: a place for natural additives in pig and poultry production, in *Biotechnology in the Feed Industry, Proc. Alltech's 4th Ann. Symp.* (ed. T.P. Lyons), pp. 219–33.

Clemens, E.T., Stevens, C.E. and Southworth, M. (1975) Sites of organic acid production and pattern of digesta movement in the gastrointestinal tract of swine. *J. Nutr.*, **105**, 759–68.

Coates, M.E. (1973) Gnotobiotic animals in nutrition research. *Proc. Nutr. Soc.*, **32**, 53–8.

Coates, M.E. and Gustafsson, B.E. (eds) (1984) *The Germfree Animal in Biomedical Research*, Laboratory Animals, London.

Collins, M.D., Farrow, J.A.E., Katic, V. and Kandler, O. (1984) Taxonomic studies on streptococci of serological groups E, P, U and V: Description of *Streptococcus porcinus* sp. *nov. System. Appl. Microbiol.*, **5**, 402–13.

Conway, P.L. (1985) *Lactobacillus* spp. of the intestinal tract and mechanisms of attachment. *SBF-Nytt*, **31**, 32–6.

Conway, P.L. (1989) Lactobacilli: Fact and fiction, in *The Regulatory and Protective Role of the Normal Microflora*, (eds R. Grubb, *et al.*), Macmillan Press, Basingstoke, pp. 263–81.

Conway, P.L. and Kjelleberg, S. (1989) Protein-mediated adhesion of *Lactobacillus fermentum* strain 737 to mouse stomach squamous epithelium. *J. Gen. Microbiol.*, **135**, 1175–86.

Conway, P.L., Maki, J., Mitchell, R. and Kjelleberg, S. (1986) Starvation of marine flounder, squid and laboratory mice and its effect on the intestinal microbiota. *FEMS Microbiol. Ecol.*, **38**, 187–95.

Conway, P.L., Gorbach, S.L. and Goldin, B.R. (1987) Survival of lactic acid bacteria in the human stomach and adhesion to intestinal cells. *J. Dairy Sci.*, **70**, 1–12.

Conway, P.L., Welin, A. and Cohen, P.S. (1990) Presence of K88 specific receptors in porcine ileal mucus is age dependent. *Infect. Immun.*, **58**, 3178–82.

Conway, P.L., Blomberg, L., Welin, A. and Cohen, P.S. (1991) The role of piglet intestinal mucus in the pathogenicity of *Escherichia coli* K88, in *Molecular Pathogenesis of Gastrointestinal Infections*, (eds T. Wadström, et al.), Plenum Press, New York. pp 335–8

Corpet, D. and Nicolas, J.-L. (1979) Antagonistic effect of intestinal bacteria from the microflora of holoxenic (conventional) piglets, against *Clostridium perfringens* in the digestive tract of gnotoxenic mice and gnotoxenic piglets, in *Clinical and Experimental Gnotobiotics, Zbl. Bakteriol. Suppl* 7, Gustav Fischer, Stuttgart, pp. 167–74.

Cowman, G.L., Davis, L.W., Duncan, M.S. and Trammell, J.H. (1978) Dose-response relationship in post-weaned pigs fed a non-viable *Lactobacillus* fermentation product. *J. Anim. Sci.*, **47**, (Suppl. 1), 298

Cranwell, P.D. (1968) Microbial fermentation in the alimentary tract of the pig. *Nutr. Abstr. Rev.*, **38**, 721–30

Cranwell, P.D. (1990) Development of the gastrointestinal microflora in the milk-fed pig and milk-fed human infant. La Trobe University, School of Agriculture, Melbourne, Bundoora 3083, Australia, Occasional Paper 21, 1–40.

Cranwell, P.D. and Moughan, P.J. (1989) Biological limitations imposed by the digestive system to the growth performance of weaned pigs, in *Manipulating Pig Production* II, (eds J.L. Barnett, and D.P. Hennesy), Australasian Pig Science Association, Werribee, Victoria, Australia, pp. 140–59.

Cranwell, P.D., Noakes, D.E. and Hill, K.J. (1976) Gastric secretion and fermentation in the suckling pig. *Brit. J. Nutr.*, **36**, 71–86.

Danek, P. (1986) Zmeny fekalni mikroflory selat behem aplikace *Streptococcus faecium* M-74 (Lactiferm) v krmne smesi. *Zivocisna Vyroba*, **31**, 1019–27.

Davidson, J.N. and Hirsch, D.C. (1976) Bacterial competition as a means of preventing neonatal diarrhea in pigs. *Infect. Immun.*, **13**, 1773–4.

De Graaf, F.K. and Mooi, F.R. (1986) The fimbrial adhesins of *Escherichia coli*. *Adv. Microb. Physiol.*, **28**, 65–143.

Deprez, P., van den Branden, J., de Geest, J. and Muylle, E. (1989) De invloed van *Streptococcus faecium* toediening op de excretie van *Escherichia coli* en het coorkomen van slingerziekte bij gespeende biggen. *Vlaams Diergeneeskde Tijdschr.*, **58**, 113–17.

De Valdez,G.F., de Giori, G.S., Ruiz Holgardo, A.P. and Oliver, G. (1985) Effect of drying medium on residual moisture content and viability of freeze-dried lactic acid bacteria. *Appl. Environ. Microbiol.*, **49**, 413–15.

Drew, M.D. & Owen, B.D. (1988) The provision of passive immunity to colostrum-deprived piglets by bovine or porcine serum immunoglobulins. *Can. J. Anim. Sci.*, **68**, 1277.

Ducluzeau, R. (1985) Implantation and development of the gut flora in the newborn piglet. *Pig News Inform.*, **6**, 415–18.

Ducluzeau, R. and Raibaud, P. (1974) Interaction between *Escherichia coli* and *Shigella flexneri* in the digestive tract of 'gnotobiotic' mice. *Infect. Immun.*, **9**, 730–3.

Ducluzeau, R., Dubos, F., Raibaud, P. and Abrams, G.D. (1978a) Production of an antibiotic substance by *Bacillus licheniformis* within the digestive tract of gnotobiotic mice. *Antimicrob. Agents Chemother.*, **13**, 97–103

Ducluzeau, R., Rapine, P., Courvalin, C. and Raibaud, P. (1978b) Transfert de la flore microbienne fécale de porcelets et de porcs adultes holoxéniques à des souris adultes et des porcelets axéniques: Effet de l'animal hôte et du régime alimentaire sur le faciès microbien du tube digestif des divers animaux. *Ann. Microbiol (Inst. Pasteur).*, **129B**, 597–612.

Ducluzeau, R., Ladire, M. and Laplace, J.P. (1986) Évolution de la flore microbienne caecale après isolement chirurgical du gros intestin chez le porc. *Ann. Microbiol (Inst. Pasteur).*, **137B**, 123–8.

Duval-Iflah, Y., Chappuis, J.P., Ducluzeau, R. and Raibaud, P. (1983) Intraspecific interactions between *Escherichia coli* strains in human newborns and in gnotobiotic mice and piglets. *Prog. Food Nutr. Sci.*, **7**, 107–16

English, P.R. (1984) Establishing the early weaned pig. *Pig Vet. Soc. Proc.*, **7**, 29–37.

Ervolder, T.M., Gudkov, A.V., Gudkov, S.A. *et al.* (1984) Use of bifidobacteria in control of gastrointestinal disease in piglets and broiler chicks. *Molochnaya Promyshlennost*, **8**, 18–20. (*in Dairy Sci. Abstr.*, 1985, **47**, 714).

Fahy, V.A., Connaughton, D., Driesen, S.J. and Spicer, E.M. (1987a) Preweaning colibacillosis, in *Manipulating Pig Production*. (eds APSA Comittee), Australian Pig Science Association, Werribee, Victoria, Australia, pp. 176–88.

Fahy, V.A., Connaughton, D., Driesen, S.J., and Spicer, E.M. (1987b) Postweaning colibacillosis, in *Manipulating Pig Production*, (eds APSA Committee), Australian Pig Science Association, Werribee, Victoria, Australia, pp. 189–201.

Fernandes, C.F., Shahani, K.M. and Amer, M.A. (1987) Therapeutic role of dietary lactobacilli and lactobacillus fermented dairy products. *FEMS Microbiol. Rev.*, **46**, 343–56.

Fewins, B.G., Newland, L.G.M. and Briggs, C.A.E. (1957) The normal intestinal flora of the pig. III. Qualitative studies of lactobacilli and streptococci. *J. Appl. Bacteriol.*, **20**, 234–42.

Fossum, K. and Liven, E. (1974) The distribution of enzymes and bacteria in the small intestines of slaughter pigs. *Acta Path. Microbiol. Scand.*, **382**, 644–52.

Freter, R. (1974) Interactions between mechanisms controlling the intestinal microflora. *Amer. J. Clin. Nutr.*, **27**, 1409–16.

Freter, R. (1983) Mechanisms that control the microflora in the large intestine, in *Human Flora in Health and Disease*. (ed. D. Hentges), Academic Press, New York, pp. 33–54.

Friend, D.W., Cunningham, H.M. and Nicholson, J.W.G. (1963) The production of organic acids in the pig. II. The effect of diet on the levels of volatile fatty acids and lactic acid in sections of the alimentary tract. *Can. J. Anim. Sci.*, **43**, 156–68.

Fuller, R. (1986) Probiotics. *J. Appl. Bacteriol. Suppl.*, 1S-7S.

Fuller, R. (1989) Probiotics in man and animals. *J. Appl. Bacteriol.*, **66**, 365–78.

Fuller, R., Barrow, P.A. and Brooker, B.E. (1978) Bacteria associated with the gastric epithelium of neonatal pigs. *Appl. Environ. Microbiol.*, **35**, 582–91.

Gilliland, S.E. and Rich, C.N. (1990) Stability during frozen and subsequent refrigerated storage of *Lactobacillus acidophilus* used as a dietary adjunct to produce a hypocholesterolemic effect in humans. *J. Dairy Sci.*, **73**, 1187–92.

Gilliland, S.E. and Walker, D.K. (1990) Factors to consider when selecting a culture of *Lactobacillus acidophilus* grown at different pH. *J. Dairy Sci.*, **73**, 905–11.

Gordon, D., Macrae, J. and Wheater, D. (1957) A *Lactobacillus* preparation for use with antibiotics. *Lancet*, May 4, 899–901.

Gouet, P., Fonty, G., Jouany, J.P. and Nebout, J.M. (1984) Cellulolytic bacteria establishment and rumen digestion in lambs isolated after birth. *Can. J. Anim. Sci. Suppl.* **64**, 163–4.

Graham, H., Hesselman, K., Jonsson, E. and Åman, P. (1986) Influence of β-D-glucanase supplementation on digestibility of a barley-based diet in the pig gastrointestinal tract. *Nutr. Rep. Intern.*, **34**, 1089–96.

Greenwood, P.E. and Tzipori, S. (1987) Protection of suckling piglets from diarrhoea caused by enterotoxigenic *Escherichia coli* by vaccination of the pregnant sow with recombinant DNA derived pilus antigens, in *Manipulating Pig Production*, (eds APSA Comittee), Australian Pig Science Association, Werribee, Victoria, Australia, p. 230.

Gualtieri, M. and Betti, S. (1984) Effects of the administration of *Streptococcus faecium* to suckling piglets. *Riv. Suinicoltura*, **25**, 39–43.

Gudding, R. and Larssen, R.B. (1985) Some effects of *Streptococcus faecium* M 74 in piglets. *Nord. Vet. Med.*, **37**, 48–9.

Hale, O.M. and Newton, G.L. (1979) Effects of a nonviable *Lactobacillus* species fermentation product on performance of pigs. *J. Anim. Sci.*, **48**, 770–5.

Han, I.K., Kim, J.D., Lee, J.H., *et al.* (1984) Studies on the growth promoting effects of probiotics. III. The effects of *Clostridium butyricum* ID on the performance and the changes in the microbial flora of the feces of growing pigs. *Korean J. Anim. Sci.*, **26**, 166–71.

Hardie, J.M. (1986) Genus Streptococcus, in *Bergey's Manual of Systematic Bacteriology*, vol. 2. (ed. P.H.A. Sneath), Williams & Wilkins, London, pp. 1043–7.

Hartmann, P.E., McCauley, I., Gooneratne, A.D. and Whitely, J.L. (1984) Inadequacies of sow lactation: survival of the fittest, in *Physiological Strategies in Lactation* (eds M. Peaker *et al.*), Academic Press, London, pp. 301–26.

Henriksson, A., Szewzyk, R. and Conway, P.L. (1991) Characteristics of the adhesive determinants of *Lactobacillus fermentum* 104, *Appl. Environ. Microbiol.* **57**, 499–502.

Hoefling, D. (1989) Tracking the culprits behind diarrhea in neonatal pigs. *Vet. Med.*, April, 427.

Horvath, D.J., Seeley, H.W., Warner, R.G. and Loosli, J.K. (1958) Microflora of intestinal contents and feces of pigs fed different diets including pigs showing parakeratosis. *J. Anim. Sci.*, **17**, 714–22.

Houpt, K.A., Zahorik, D.M., Anika, S.M. and Houpt, T.R. (1979) Taste aversion learning in suckling and weanling pigs. *Vet. Sci. Comm.*, **3**, 165–9.

Hungate, R.E. (1984) Microbes of nutritional importance in the alimentary tract. *Proc. Nutr. Soc.*, **43**, 1–11.

Ingram, S.H., Lennon, A.M., Albin, R.C. *et al.* (1973) *Lactobacillus acidophilus* for pigs at weaning. *J. Anim. Sci.*, **36**, 207

Jensen, H. (1974) Biological effects of feeding pigs with *Lactobacillus acidophilus*. *Proc. XIX Intern. Dairy Congr., 1974, India.* 1E, p. 579.

Jensen, P. and Recén, B. (1989) When to wean – observations from free-ranging domestic pigs. *Appl. Anim. Behaviour Sci.*, **23**, 49–60.

Jonsson, E. (1985) Lactobacilli as probiotics to pigs and calves. Dissertation, Dept. Animal Nutrition Management, Swed. Univ. Agric. Sci., No. 148, pp. 1–65.

Jonsson, E. (1986) Persistence of a *Lactobacillus* strain in the gut of sucking piglets and its influence on performance and health. *Swed. J. Agric. Res.*, **16**, 43–7.

Jonsson, E. and Henningsson, S. (1991) Establishment in the piglet gut of lactobacilli capable of degrading mixed-linked β-D-glucans. *J. Appl. Bacteriol.* **70**, 512–6.

Jonsson, E., Björck, L. and Claesson, C.O. (1985) Survival of orally administered *Lactobacillus* strain in the gut of cannulated pigs. *Livest. Prod. Sci.*, **12**, 279–85.

Jørgensen, J.H. (1988) Probiotika till svin. En litteraturoversigt og en afprøvning at 2 probiotikas effekt under forskellige produktionsformer. *Dansk Vet. Tidskr.*, **71**, 1211–21.

Jucker, H., Pfirter, H.P., Wenk, C. and Scgürch, A. (1973) Untersuchungen über die Wirkung antimikrobiell wirksamer Zulagen zum Futter von Ferkeln und Mastschweinen. *Wien Tierärtzl. Monatschr.*, **60**, 100–3.

Kandler, O. and Weiss, N. (1986) Regular, non-sporing gram-positive rods in *Bergey's Manual of Systematic Bacteriology*, vol. 2. (ed. P.H.A. Sneath.), Williams & Wilkins, London, pp. 1208–34.

Kenworthy, R. (1973) Intestinal microflora of the pig. *Adv. Appl. Microbiol.*, **16**, 31–54.

Kenworthy, R. and Crabb, W.E. (1963) The intestinal flora of young pigs, with reference to early weaning, *Escherichia coli* and scours. *J. Comp. Pathol.*, **73**, 215–28.

Kidder, D.E. and Manners, M.J. (1978) *Digestion in the Pig*, Scientechnica, Bristol.

Kimura, N., Yoshikane, M., Kabayashi, A. and Mitsuoka, T. (1983) An application of dried bifidobacteria preparation to scouring animals. *Bifidobacteria Microflora*, **2**, 41–55.

Kinsey, C.M. (1980) Use of microbial additives in feed. A literature review. *Proc. 40th Semiannual Meeting, San Antonio, Texas, USA, Nov. 12–14, 1980*, pp. 25–30.

Kjelleberg, S., Nyström, T., Albertson, N. and Flärdh, K. (eds) (1990). Symposium 'Nutrient limitation: Global responses and prokaryotic development'. *FEMS Microbiol. Ecol.*, **74**, 91–241.

Klaenhammer, T.R. (1982) Microbial considerations in selection and proparation of *Lactobacillus* strains for use as dietary adjuncts. *J. Dairy Sci.*, **65**, 1339–49.

Klaenhammer, T.R. and Kleeman, E.G. (1981) Growth characteristics, bile sensitivity, and freeze damage in colonial variants of *Lactobacillus acidophilus. Appl. Environ. Microbiol.*, **41**, 1461–7.

Kluber, E.T., Pollman, D.S. and Blecha, F. (1985) Effect of feeding *Streptococcus faecium* to artificially reared pigs on growth, haematology, and cell-mediated immunity. *Nutr. Rep. Intern.*, **32**, 57–66.

Kornegay, E.T. (1985/86) Effect of dosing or feeding *Lactobacillus acidophilus* on blood cholesterol levels and growth rate of weanling and growing pigs. Virginia Polytechnic Institute and State University, Animal Science Research Report, No. 5, Blacksburg, Virginia, USA, pp. 14–19.

Kornegay, E.T. and Thomas, H.R. (1973) Bacterial and yeast preparations for starter and grower rations. Virginia Polytechnic Institute and State University, Research Division Report, No. 151, Blacksburg, Virginia, USA.

Kovacs, F., Nagy, B. and Sinkovics, G. (1972) The gut bacterial flora of healthy early weaned piglets, with special regard to factors influencing its composition. *Acta Vet. Acad. Sci. Hung.*, **22**, 327–38.

Krarup, L.T. (1987) Forebyggelse af diarré hos spædgrise. *Dansk Vet. Tidskr.*, **70**, 324–8.

Kurman, J.A. (1983) The development and significance of new cultures with bifidobacteria as an example. *North Eur. Dairy J.*, **3**, 65–74.

Kvasnitskii, A.V. (1951), *Voprosy fiziologii pishchevareniya u svinei*. Sel'khozgir, Moscow. (Cited from Kidder and Manners, 1978.)

Landsutvalget for svin (1988) Probilact H til smågrise. Den rullende Afprøfning af stald- og produktionssystemer, Copenhagen, Meddelse, No. 144, pp. 1–6.

Landsutvalget for svin (1989) Bioplus 2B till smågrise. Den rullende Afprøfning af stald- og produktionssystemer, Copenhagen, Meddelse, p. 162.

Lawrence, T.L.J. (1970) Some effects of including differently processed barley in the diet of the growing pig. 2. *In vivo* gastric pH changes. *Anim. Prod.*, **12**, 151–63.

Lawrence, T.L.J. (1983) Dietary manipulation of the environment within the gastrointestinal tract of the growing pig and its possible influence on disease control: some thoughts. *Pig Vet. Soc. Proc., Cambridge*, **10**, 40–9.

Lencner, A.A., Lencner, C.P., Mikelsaar, M.E. *et al.* (1984) Die quantitative Zusammensetzung der Lactoflora des Verdauungstrakts vor und nach kosmischen Flügen unterschiedlicher Dauer. *Die Nahrung*, **28**, 607–13.

Lessard, M. and Brisson, G.J. (1987) Effect of a *Lactobacillus* fermentation product on growth, immune response and fecal enzyme activity in weaned pigs. *Can. J. Anim. Sci.*, **67**, 509–16.

Lindgren, S. and Dobrogosz, W.J. (1990) Antagonistic activities of lactic acid bacteria in food and feed fermentations. *FEMS Microbiol. Rev.*, **87**, 149–64.

Lipkin, M. (1987) Proliferation and differentiation of normal and diseased gastrointestinal cells, in *Physiology of the Gastrointestinal Tract*, 2nd edn (ed. L.R. Johnson), Raven Press, New York, pp. 255–84.

Lönnroth, I., Martinsson, K. and Lange, S. (1988) Evidence of protection against diarrhoea in suckling piglets by a hormone-like protein in sow milk. *J. Vet. Med.*, **356**, 628–35.

Luckey, T.D. (1987) Misconceptions in gnotobiology. *Microecol. Ther.*, **17**, 19–20.

Lyons, T.P. (1988) Probiotics: An alternative to antibiotics. *The Bovine Practitioner*, **23**, 64–9.

Maeng, W.J., Kim, C.W. and Shin, H.T. (1989) Effects of feeding lactic acid bacteria concentrate (LBC, *Streptococcus faecium* Cernelle 68) on the growth rate and prevention of scouring in piglets. *Korea J. Anim. Sci.*, **31**, 318–23.

Manners, M.J. (1976) The development of digestive function in the pig. *Proc. Nutr. Soc.*, **35**, 49–55.

Mäyrä-Mäkinen, A., Manninen, M. and Gyllenberg, H. (1983) The adherence of lactic acid bacteria to the columnar epithelial cells of pigs and calves. *J. Appl. Bacteriol.*, **55**, 241–5.

Maxwell, C.V., Reimann, E.M., Hoekstra, W.G. *et al.* (1970) Effect of dietary particle size on lesion development and on the contents of various regions of the swine stomach. *J. Anim. Sci.*, **30**, 911–22.

McAllister, J.S., Kurtz, H.J. and Short, E.C. (1979) Changes in the intestinal flora of young pigs with postweaning diarrhea or edema disease. *J. Anim. Sci.*, **49**, 868–79.

McGillivery, D.J., Cranwell, P.D., Cain, C.J. and Chandler, K.D. (1984) The microbiology of the *pars oesophagea* of sucking and weaned pigs. *Int. Pig Vet. Soc. Proc.*, **8**, 235.

Moen, J.H. (1982) *Streptococcus faecium* mot spedgrisdiaré. *Norsk Vet. Tidskr.*, **94**, 629–31.

Møllgaard, H. (1946) On Phytic acid, its importance in metabolism and its enzymic cleavage in bread supplemented with calcium *Biochem. J.*, **40**, 589–603.

Møllgaard, H. (1947) Resorptionen af kalcium og fosforsyre. *Beretning fra forsøgslab.*, **228**, 1–55.

Morishita, Y. and Ogata, M. (1970) Studies on the alimentary flora of pig. V. Influence of starvation on the microbial flora. *Japan. J. Vet. Sci.*, **32**, 19–24.

Morishita, Y., Mitsuoka, T., Kaneuchi, C. *et al.* (1971) Specific establishment of lactobacilli in the digestive tract of germ-free chickens. *Japan. J. Microbiol.*, **15**, 531–8.

Mumpton, F.A. and Fishman, P.H. (1977) The application of natural zeolites in animal science and aquaculture. *J. Anim. Sci.*, **45**, 1118–1203

Mundt, J.O. (1986) Enterococci, in *Bergey's Manual of Systematic Bacteriology*, vol. 2, (ed P.H.A. Sneath), Williams & Wilkins, London, pp. 1063–5.

Muralidhara, K.S., Sheggeby, G.G., Elliker, P.R. *et al.* (1977) Effect of feeding lactobacilli on the coliform and *Lactobacillus* flora of intestinal tissue and feces from piglets. *J. Food Protect.*, **40**, 288–95.

Nagy, L.K., Bhugal, B.S. and McKenzie, T. (1979) Duration of anti-adhesive and bactericidal activities of milk from vaccinated sows on *Escherichia coli* O149 in the digestive tract of piglets during the nursing period. *Res. Vet. Sci.*, **27**, 289–96.

Nakamura, M., Yoshimura, H., Koeda, T. and Sato, S. (1983) Prevention of drug-resistant *Escherichia coli* colonization in chickens by treatment with a faecal fluid. *J. Appl. Bacteriol.*, **55**, 225–31.

Neupert,B. (1988) Milchsäurebakterien CYCLACTIN* in Verbindung mit AVOTAN* erhöhen den Deckungsbeitrag in der Schweinemast. *Kraftfutter*, **4**, 126.

Newby,T.J., Miller,B., Stokes,C.R. *et al.* (1984) Local hypersensitivity response to dietary antigens in early weaned pigs, in *Recent Advances in Animal Nutrition* (eds W. Haresign. and D.J.A. Cole), Butterworth, London, pp. 49–59.

Nickel, R., Schummer, A. and Seiferle, E. (1967) *Lehrbuch der Anatomie der Haustiere. Band 2. Eingeweide*, 2nd edn., Paul Parey, Berlin, p. 138.

Noakes, D.E. (1971) Gastric function in the young pig. Ph.D. Thesis, University of London. (Cited from Cranwell and Moughan, 1989.)

Norin, K.E., Persson, A-K., Saxerholt, N. and Midtvedt, T. (1991) Establishment of *Lactobacillus* and *Bifidobacterium* species in germ-free mice and their influence on some microflora associated characteristics. *Appl. Environ. Microbiol.* **57**, 1850–2.

Nurmi, I.E. and Rantala, M. (1973) New aspects of *Salmonella* infection in broiler production. *Nature*, **241**, 210–11.

Ogata, M. and Morishita, Y. (1969) Studies on the alimentary flora of pigs. IV. The alimentary flora of pigs infected with hog cholera. *Japan. J. Vet. Sci.*, **31**, 71–82.

Ogata, M., Morishita, Y., Tanaka, Y. and Shinjo, T. (1968) Studies on the alimentary flora of pig. III. Influence of fermented chlorella-diet on the fecal flora. *Japan. J. Vet. Sci.*, **30**, 281–7.

Ogle, R.B. and Inborr, J. (1987) Alternatives to low dose antibiotics in piglet feeds in Sweden. *38th Meeting of the European Association for Animal Production, Portugal, Commission on Pig Production*, Session 4, pp. 1–6.

Olsson, T. (1966) Tarmrubbningar på svin – profylax och terapi med *Lactobacillus acidophilus*. *Svensk Vet. Tidn.*, **18**, 355–63.

Ozawa, K., Yokota, H., Kimura, M. and Mitsuoka, T. (1981) Effects of administration of *Bacillus subtilis* strain BN on intestinal flora of weanling piglets. *Japan. J. Vet. Sci.*, **43**, 771–5.

Ozawa, K., Yabu-Ochi, K., Yamanaka, K., *et al.*, (1983) Effect of *Streptococcus faecalis* BIO-4R on intestinal flora of weanling piglets and calves. *Appl. Environ. Microbiol.*, **45**, 1513–18.

Parker, R.B. (1974) The other half of the antibiotics story. *Anim. Nutr. Health*, **29**, 4–8.

Partridge, I.G. (1989) Alternative feeding strategies for weaner pigs, in *Manipulating Pig Production* II, (eds J.L. Barnett and D.P. Hennesy), Australasian Pig Science Association, Werribee, Victoria, Australia, pp. 140–59.

Pedersen, K. and Tannock, G.W. (1989) Colonization of the porcine gastrointestinal tract by lactobacilli. *Appl. Environ. Microbiol.*, **55**, 279–83.

Perdigón, G., Nader de Macias, M.E., Alvarez, S. *et al.* (1987) Enhancement of immune response in mice fed with *Streptococcus thermophilus* and *Lactobacillus acidophilus*. *J. Dairy. Sci.*, **70**, 919–26.

Pettersson, H. (1990) *Genotype x nutrition interactions in the performance testing of pigs*. Dissertation, Dept Animal Breeding and Genetics, Swed. Univ. Agric. Sci., No. 93, p. 157.

Pollman, D.S. (1986) Probiotics in pig diets, in *Recent Advances in Animal Nutrition* (eds W. Haresign and D.J.A. Cole), Butterworth, London, pp. 193–205.

Pollman, D.S. and Bandyk, C.A. (1984) Stability of *Lactobacillus* products. *Anim. Feed Sci. Technol.*, **11**, 261–7.

Pollman, D.S., Danielson, D.M. and Peo, E.R. Jr (1980a) Effect of microbial feed additives on performance of starter and growing-finishing pigs. *J. Anim. Sci.*, **51**, 577–81.

Pollman, D.S., Danielson, D.M., Wren, W.B. *et al.* (1980b) Influence of *Lactobacillus acidophilus* inoculum on gnotobiotic and conventional pigs. *J. Anim. Sci.*, **51**, 629–37.

Pollman, D.S., Danielson, D.M. and Peo, E.R. Jr (1980c) Effect of *Lactobacillus acidophilus* on starter pigs fed a diet supplemented with lactose. *J. Anim. Sci.*, 51, 638–44.

Pollman, D.S., Kennedy, G.A., Koch, B.A. and Allee, G.L. (1984) Influence of nonviable *Lactobacillus* fermentation product in artificially reared pigs. *Nutr. Rep. Int.*, **29**, 977–82.

Porter, P. (1969) Transfer of immunoglobulins IgG, IgA and IgM to lacteal secretions in the paturient sow and their absorption by the neonatal piglet. *Biochem. Biophys. Acta*, **18**, 381–92.

Porter, P. and Linggood, M. (1983) Development of oral vaccine for prevention of scouring caused by enteropathogenic *Escherichia coli*. *J. Infect.*, **6**, 111–21.

Premi, L. and Bottazzi, V. (1974) Use of lactobacilli in the control of intestinal disturbances of pigs. *Proc. XIX Int. Dairy Congr., 1974 India*. 1E, p. 89.

Raibaud, P., Caulet, M., Galpin, J.V. and Mocquot, G. (1961) Studies on the bacterial flora of the alimentary tract of pigs. II. Streptococci: selective enumeration and differentiation of the dominant group. *J. Appl. Bacteriol.*, **24**, 285–306.

Ratcliffe, B. (1985) The influence of the gut microflora on the digestive processes. *Beretning fra Statens Husdyrbrugsforsøg*, **580**, 245–67.

Ratcliffe, B., Cole, C.B., Fuller, R. and Newport, M.J. (1986) The effect of yoghurt and milk fermented with a porcine intestinal strain of *Lactobacillus reuteri* on the performance and gastrointestinal flora of pigs weaned at two days of age. *Food Microbiol.*, **3**, 203–11.

Redmond, H.E. and Moore, R.W. (1965) Biologic effect of introducing *Lactobacillus acidophilus* into a large swine herd experiencing enteritis. *Southw. Vet.*, **18**, 287–8.

Reiter, B. (1985) Protective proteins in milk, biological significance and exploitation. *Int. Dairy Fed. Bull.*, **191**, 1–35.

Rerat, A., Corring, T. and Laplace, J.P. (1976) Protein digestion and absorption, in *Protein Metabolism and Nutrition* (eds D.J.A., Cole *et al.*), Butterworth, London, pp. 97–138.

Robinson, I.M., Allison, M.J. and Bucklin, J.A. (1981), Characterization of the cecal bacteria of normal pigs. *Appl. Environ. Microbiol.*, **41**, 950–5.

Robinson, I.M., Whipp S.C., Bucklin, J.A. and Allison, M.J. (1984) Characterization of predominant bacteria from the colons of normal and dysenteric pigs. *Appl. Environ. Microbiol.*, **48**, 964–9.

Robinson, I.M., Stromley, J.M., Varel, V.H. and Cato, E.P. (1988) *Streptococcus intestinalis*, a new species from the colons and feces of pigs. *Int. J. Syst. Bacteriol.*, **38**, 245–8.

Roth,F.X. and Kirchgessner,M. (1986) Zur nutritiver Wirksamkeit von *Streptococcus faecium* (Stamm M74) in der Ferkelaufzucht. *Landwirtsch. Forsch.*, **39**, 198–205.

Roth, F.X. and Kirchgessner, M. (1988) Nutritive Wirksamkeit von Toyocerin. 1. Mitteilung: Ferkelaufzucht. *Landwirtsch. Forsch.*, **41**, 58–62.

Russel, E.G. (1979) Types and distribution of anaerobic bacteria in the large intestine of pigs. *Appl. Environ. Microbiol.*, **37**, 187–93.

Rutter, P.R., Dazzo, F.B., Freter, R., Gingell, D. *et al.* (1984) Mechanisms of adhesion, in *Microbial Adhesion and Aggregation*, (ed. K.C. Marshall), Springer-Verlag, Berlin, pp. 5–19.

Salanitro, J.P., Blake, I.G. and Muirhead, P.A. (1977) Isolation and identification of fecal bacteria from adult swine. *Appl. Environ. Microbiol.*, **33**, 79–84.

Savage, D.C. (1977) Microbial ecology of the gastrointestinal tract. *Ann. Rev. Microbiol.*, **31**, 107–33.

Savage, D.C., Siegel, J.E., Snellen, J.E. and Whitt, D.D. (1981) Transit time of epithelial cells in the small intestines of germfree mice and ex-germfree mice associated with indigenous microorganisms. *Appl. Environ. Microbiol.*, **42**, 996–1001.

Schulze, F. (1977) Quantitative Magen-Darm-Flora-Analysen beim Ferkel vor und nach dem Absetzen unter Berücksichtigung der Pathogenese der Kolienterotoxämie. *Arch. Exp. Vet. Med., Leipzig*, **31**, 299–316.

Schulze, F. (1978) Zur quantitativen Zusammensetzung der Magen-Darm-Flora bei Ferkeln in den ersten zwei Lebenswochen. 1. Mitteilung: Zur Entwicklung der Magen-Darm-Flora bei natürlich aufgezogenen Ferkeln. *Arch. Exp. Vet. Med., Leipzig*, **32**, 155–70.

Schulze, F. (1987) Die Gastrointestinalflora des Schweines und ihre Regulationsmechanismen. *Fortschrittsber. Landw. Nahrungsgüterwirtsch.*, **25**, 1–72.

Schulze, F. and Bathke, W. (1977) Zur quantitativen Zusammensetzung der Magen-Darm-Flora beim Läuferschwein. *Arch. Exp. Vet. Med., Leipzig*, **31**, 161–85.

Schulze, F. and Jacob, H.-E. (1981) Das Redoxpotential im Magen-Darm-Kanal von Schweinen unterschiedlichen Alters. *Arch. Exp. Vet. Med., Leipzig*, **35**, 349–57.

Schulze, F. and Müller, G. (1980) Lysozym in der Sauenmilch und seine Bedeutung für die bakterielle Besiedlung des Magen-Darm-Kanals beim Saugferkel. *Arch. Exp. Vet. Med., Leipzig*, **34**, 317–24.

Shahani, K., Fernandes, H. and Amer, V. (1989) Immunologic and therapeutic modulation of gastrointestinal microecology by lactobacilli. *Microecol. Ther.*, **18**, 103–4.

Sharpe, M.E. (1981) The genus *Lactobacillus*, in *The Procaryotes, a Handbook on Habitats, Isolation and Identification of Bacteria* (eds M.P. Starr *et al.*), vol. 2, Springer-Verlag, Berlin, pp. 1653–79.

Sharpe, M.E., Hill, L.R. and Lapage, S.P. (1973) Pathogenic lactobacilli. *J. Med. Microbiol.*, **6**, 281–6.

Sinkovics, G. and Juhasz, B. (1974) Development of the intestinal flora in suckling pigs. *Acta Vet. Acad. Scient. Hungar.*, **24**, 375–81.

Sissons, J.W. (1988) The potential of probiotic organisms to prevent diarrhoea and promote digestion. *Proc. 4th Int. Sem. on Digestive Physiology in the Pig, Jablonna, Poland, 1988*, pp. 357–70.

Sissons, J.W. (1989) Potential of probiotic organisms to prevent diarrhoea and promote digestion in farm animals – a review. *J.Sci. Food Agric.*, **49**, 1–13.

Smith, H.W. (1961) The development of the bacterial flora of the faeces of animals and man: The changes that occur during ageing. *J. Appl. Bacteriol.*, **24**, 235–41.

Smith, H.W. (1965) The development of the flora of the alimentary tract in young animals. *J. Pathol. Bacteriol.*, **90**, 495–513.

Smith, H.W. and Crabb, W.E. (1961) The faecal bacterial flora of animals and man: its development in the young. *J. Pathol. Bacteriol.*, **82**, 53–66.

Smith, H.W. and Huggins, M.B. (1983) Effectiveness of phages in treating experimental *Escherichia coli* diarrhoea in calves, piglets and lambs. *J. Gen. Microbiol.*, **129**, 2659–75.

Smith, H.W. and Jones, J.E.T. (1963) Observations on the alimentary tract and its bacterial flora in healthy and diseased pigs. *J. Pathol. Bacteriol.*, **86**, 387–412.

Smith, M.W. and Jarvis, L.G. (1978) Growth and cell replacement in the new-born pig intestine. *Proc. R. Soc. London, B.*, **203**, 69–89.

Söderlind, O., Olson, E., Smyth, C.J. and Möllby, R. (1982) Effect of parenteral vaccination of dams and intestinal *Escherichia coli* in piglets with diarrhea. *Infect. Immun.*, **36**, 900–6.

Söderlind, O., Thafvelin, B. and Möllby, R. (1988) Virulence factors in *Escherichia coli* strains isolated from Swedish piglets with diarrhea. *J. Clin. Microbiol.*, **26**, 879–84.

Søgaard, H. (1987) Det teoretiske grundlag for probiotikakonceptet. *Dansk Vet. Tidskr.*, **70**, 310–20.

Spicer, E.M., Driesen, S.J., Fahy, V.A., *et al.* (1986) Causes of preweaning mortality on a large intensive piggery. *Austral. Vet. J.*, **63**, 71–5.

Spriet, S.M., Decuypere, J.A. and Henderickx, H.K. (1987) Effect of *Bacillus toyoi* (Toyocerin) on the gastrointestinal microflora, concentration of some bacterial metabolites, digestibility of the nutrients and the small intestinal mean retention time in pigs. *Med. Fac. Landbouww. Rijksuniv. Gent*, **52**, 1673–83.

Stevens, A.J. and Blackburn, P.W. (1967) An indirect approach to piglet scours *Vet. Record*, **80**, 637–8.

Tannock, G.W. (1984) Control of gastrointestinal pathogens by normal flora, in *Current Perspectives in Microbial Ecology* (eds M.J. Klug and C.A. Reddy), Proc. 3rd Int. Symp. Microbial Ecology, Michigan State Univ., 7–12 Aug 1983. American Society for Microbiologists, Washington, DC, pp. 374–82.

Tannock, G.W. (1990) The microecology of lactobacilli inhabiting the gastrointestinal tract, in *Advances in Microbial Ecology*, vol. 11 (ed. K.C. Marshall), Plenum Press, New York, pp. 147–71.

Tannock, G.W. and Archibald, R.D. (1984) Derivation and use of mice which do not harbour lactobacilli in the gastrointestinal tract. *Can. J. Microbiol.*, **30**, 849–53.

Tannock, G.W. and Savage, D.C. (1974) Influences of dietary and environmental stress on microbial populations in the murine gastrointestinal tract. *Infect. Immun.*, **9**, 591–8.

Tannock, G.W. and Smith, J.M.B. (1970) The microflora of the pig stomach and its possible relationship to ulceration of the *pars oesophagea*. *J. Comp. Pathol.*, **80**, 359–67.

Tannock, G.W., Szylit, O., Duval, Y. and Raibaud, P. (1982) Colonization of tissue surfaces in the gastrointestinal tract of gnotobiotic animals by *Lactobacillus* strains. *Can. J. Microbiol.*, **28**, 1196–8.

Tannock, G.W., Fuller, R. and Pedersen, K. (1990) *Lactobacillus* succession in the piglet digestive tract demonstrated by plasmid profiling. *Appl. Environ. Microbiol.*, **56**, 1310–16.

Thacker, P.A. (1988) Novel approaches to growth promotion in the pig, in *Recent Advances in Animal Nutrition* (eds W. Haresign and D.J.A. Cole), Butterworth, London, pp. 73–84.

Tournut, J. (1989) Applications of probiotics to animal husbandry. *Rev. Sci. Tech. Off. Int. Epiz.*, **8**, 551–66.

Trevallyn-Jones, J. (1987) Field trials with a recombinant DNA scours vaccine, in *Manipulating Pig Production* (eds APSA Comittee), Australian Pig Science Association, Werribee, Victoria, Australia, p. 231.

Tuschy, D. (1986) Verwendung von 'Probiotika' als Leistungsförderer in der Tierernährung. *Übers. Tierernährg*, **14**, 157–78.

Tzipori, S. (1985) The relative importance of enteric pathogens affecting neonates of domestic animals. *Adv. Vet. Sci. Comp. Med.*, **29**, 103–206.

Underdahl, N.R. Torres-Medina, A. and Doster, A.R. (1983) Effect of *Streptococcus faecium* C-68 in control of *Escherichia coli*-induced diarrhea in gnotobiotic pigs. *Amer. J. Vet. Res.*, **43**, 2227–32.

Ushe, T.C. and Nagy, B. (1985) Inhibition of small intestinal colonization of enterotoxigenic *Escherichia coli* by *Streptococcus faecium* M74 in pigs. *Zbl. Bakteriol. Hyg., I Abt. Orig. B*, **181**, 374–82.

Vanbelle, M., Teller, E. and Focant, M. (1989) Probiotics in animal nutrition; a review. *Lab. Biochim. Nutr., Belgium*, **55**, 1–59.

Visek, W.J. (1978) The mode of growth promotion by antibiotics. *J. Anim. Sci.*, **46**, 1447–69.

Vodovar, N., Flanzy, J. and François, A.C. (1964) Intestin grele du porc. I. – Dimensions en fonction de l'age et du poids, étude de la jonction du canal cholédoque et du canal pancréatique a celui-ci. *Ann. Biol. Anim. Biochem. Biophys.*, **4**, 27–34.

Wadström, T., Andersson, K., Sydow, M. *et al.* (1987) Surface properties of lactobacilli isolated from the small intestine of pigs. *J. Appl. Bacteriol.*, **62**, 513–20.

Walton, J.R. (1980) Modes of action of growth promoting agents. *Fortschr. Vet. Med.*, **33**, 77–82.

Wells, C.L., Johnson, W.J., Kan, C.M. and Balish, E. (1979) Inability of a genetically engineered strain of *Escherichia coli* (c 1776) to colonize germfree animals, in *Clinical and Experimental Gnotobiotics* (eds Fliedner *et al.*), *Zbl. Bakteriol. Suppl.*, **7**, 197–200.

White, F. Wenham, G., Sharman, G.A.M. *et al.* (1969) Stomach function in relation to a scour syndrome in the piglet. *Brit. J. Nutr.*, **23**, 847–58.

Wilson, W.R. and Svendsen, J. (1971) Immunity to *Escherichia coli* in pigs. The role of milk in protective immunity to *E. coli* enteritis. *Can. J. Comp. Med.*, **35**, 230–43.

Wolter, R. and Henry, N. (1982) Les 'probiotiques' en alimentation animale. *Rec. Méd. Vét.*, **158**, 283–90.

Zani, G., Biavati, B., Crociani, F. and Matteuzzi, D. (1974) Bifidobacteria from the faeces of piglets. *J. Appl. Bacteriol.*, **37**, 537–47.

Chapter Twelve

Probiotics for ruminants

R. JOHN WALLACE AND C. JAMES NEWBOLD

12.1. INTRODUCTION

The digestive anatomy and physiology of ruminants is markedly different to that of non-ruminant animals, including pigs and man. The ruminant has two additional digestive organs at the anterior end of the tract. The first of these, the rumen, contains in the dairy cow, for example, a liquid volume of 60–100 litres. This organ is essentially a fermentation chamber, containing about 10^{10} bacteria and 10^5–10^6 ciliate protozoa ml^{-1}, together with an unknown, probably small, number of anaerobic fungi. The partly fermented food and the microorganisms then pass through the omasum, which is much smaller than the rumen (Figure 12.1). Its function is primarily to absorb both water and soluble nutrients.

Digestion of food in the rumen occurs by a combination of microbial fermentation and physical breakdown during regurgitation of the food by rumination. No host animal enzymes are thought to be involved. Ruminants have evolved to utilize the volatile fatty acids (VFA) resulting from the fermentation as their main substrates for growth and lactation. These VFA comprise mainly acetic, propionic and butyric acids, with smaller amounts of higher and branched acids. The advantage of this foregut fermentation compared with a hindgut fermentation is that the microbial cells formed as a result of the fermentation are available to the host as they pass down the tract to the ruminant's gastric stomach, the abomasum. Indeed, microbial protein is the most important source of amino acids for absorption.

During early life, when milk is the main ingredient, food tends to bypass the rumen. The distance between the end of the oesophagus and the reticulo-omasal orifice, through which food leaves the rumen, is small (Figure 12.1). Young animals have a reflex which closes the vestigial oesophagus between these two orifices, the so-called oesophageal groove, so that food passes directly from the oesophagus to the omasum and then to the abomasum. Some food does enter the rumen, however, and the rumen becomes inoculated with bacteria, protozoa and fungi in preparation for weaning, when the adult, fibre-digesting population takes over.

The potential benefits of probiotics to ruminants are therefore perhaps even greater than with monogastric animals. The prevention of scouring in calves is essentially the same problem as in other species. In ruminants, however, benefits may also be obtained by enhancing the rate at which the rumen flora and fauna develop or, once the fermentation has fully established in the adult animals, by stimulating fermentation in adult ruminants.

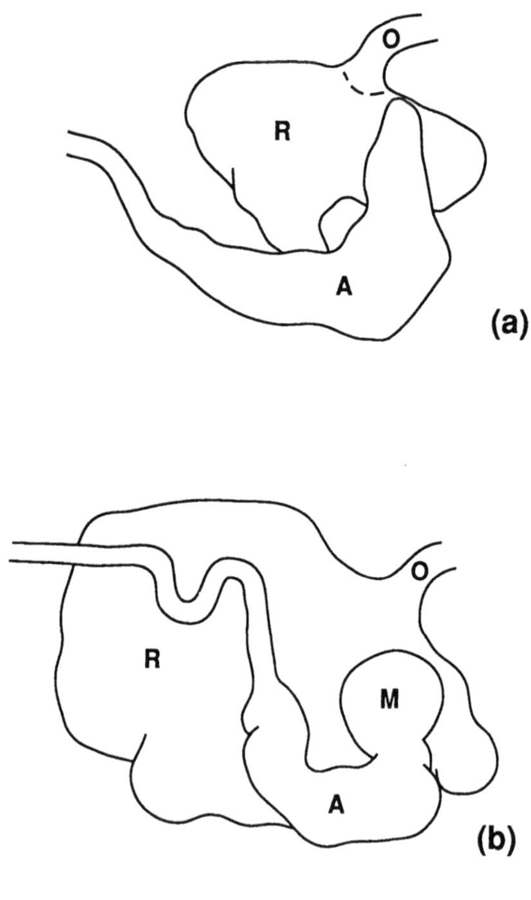

Figure 12.1 Schematic differences between (a) the young ruminant and (b) the adult ruminant. O – oesophagus, R – rumen, M – omasum, A – abomasum. The dashed line in (a) represents the oesophageal groove, which closes during suckling. From Orskov (1982).

12.2. PROBIOTICS FOR YOUNG RUMINANTS

12.2.1 Development of digestive function

The development of a functional microbial population in the gut of the newborn ruminant facilitates not only the digestion of fibre by the host but also helps protect the gut from infection by pathogenic organisms. Like other young mammals (Savage, 1977), ruminants are born with a sterile gastrointestinal tract (Cushnie *et al.*, 1981). However, bacterial colonization is rapid, with *Escherichia coli* detectable in all areas of the digestive tract of lambs and calves 8 h after birth, and lactobacilli and streptococci detectable from 24 h onwards (Smith, 1965). In healthy animals, lactobacilli quickly colonize the gut, displacing coliforms and reaching populations of 10^7–10^9 g^{-1} throughout the intestines by 1 week of age (Smith, 1965; Karney *et al.*, 1986).

The ingestion of solid food by the animal marks the second stage in the establishment of an intestinal flora in young ruminants, with the development of a rumen fermentation. Microbial colonization of the rumen occurs rapidly after birth, with a large population of strictly anaerobic bacteria present after 48 h (Ziolecki and Briggs, 1961; Fonty *et al.*, 1987). During the early days of life, the rumen is not fully functional. Most of the milk bypasses the rumen *via* the oesophageal groove (Hungate, 1966). As the animal begins to consume solid feed, the microbial population in the rumen increases and begins to resemble that of the adult ruminant (Fonty *et al.*, 1987; Dehority and Orpin, 1988). The end-products of microbial fermentation encourage the development and enlargement of the rumen (Warner *et al.*, 1953, 1955), such that around the time of weaning the rumen is fully developed both as a digestive and absorptive organ (Thivend *et al.*, 1979). Rapid development of the rumen is important to a successful transition from liquid to solid feed which is of great importance in the profitability of modern stock rearing operations, both in terms of reduced labour and feed costs and because digestive disorders are less frequent in weaned as opposed to liquid-fed calves (van Horn *et al.*, 1976; James *et al.*, 1984; Roy, 1980).

12.2.2 Prevention of diarrhoea

Diarrhoea caused by enterotoxigenic bacteria colonizing the gut represents a serious economic constraint on the rearing of young animals. Massip and Pondant (1975) reported that up to 6.5% of Belgian calves died due to intestinal disorders during the first month of life, while less severe cases of infection might reduce intestinal nutrient absorption and animal performance (Youanes and Herdt, 1987). Increased coliform counts have been reported in the intestine of calves suffering from diarrhoea (Smith, 1971; Youanes and Herdt, 1987). Guard (1986) states that *E. coli* tends to cause diarrhoea mainly in young animals (< 1 week of age) while increased coliform counts were also noticeable around the time of weaning (Karney *et al.*, 1986). To induce diarrhoea by enterotoxin production, *E. coli* must first colonize the gut (Guard, 1986). It has been suggested that probiotics might be used either to displace enterotoxigenic *E. coli* from the gut wall or to promote a healthy bacterial population which excludes coliforms from the gut (Fuller, 1989).

Lactobacillus acidophilus has been reported to reduce coliform numbers in the intestine of calves (Ellinger *et al.*, 1978; Bruce *et al.*, 1979). Gilliland *et al.* (1980) noted that *L. acidophilus* strains originally isolated from calves were more effective in this respect than those isolated from pigs. *L. acidophilus* alone or in combination with other lactobacilli has been reported to reduce scouring and increase liveweight gain in calves in some (Bechman *et al.*, 1977; Beeman, 1985; Bonaldi *et al.*, 1986) but not all trials (Jonsson, 1985; Jonsson and Olsson, 1985). *Lactobacillus* mixtures have also been effective in reducing mortality in wean-stressed lambs (Pond and Goode, 1985). Aldrovandi *et al.* (1984) and Wolter *et al.* (1987) both noted that live lactobacilli are more effective than dead cells. Other bacteria have also been used. *Enterococcus faecium* has been reported to reduce scouring and improve weight gain between birth and weaning (Hefel, 1980; Maeng *et al.*, 1987; Svozil *et al.*, 1987; Tournut, 1989). Tournut (1989) reported that a mixture of *L. acidophilus* and *Ent. faecium* reduced the incidence of diarrhoea by almost 70% and mortality by 99% when fed to calves between birth and 5 days of age. *L. acidophilus* and *Bacillus toyoi* have both been reported to reduce scouring in young calves (Hatch *et al.*, 1973; Tournut, 1989).

Although the above results demonstrate that lactobacilli and enterococci can reduce diarrhoea in young ruminants, their mode of action remains elusive. As noted above, lactobacilli have been shown to prevent coliform colonization of the gut in calves (Ellinger *et al.*, 1978; Bruce *et al.*, 1979), while streptococci have been shown to prevent coliform proliferation in the intestines of non-ruminant species (Underdahl *et al.*, 1982; Wadstrom, 1984). Several explanations have been put forward to explain the effect of lactobacilli and streptococci on *E. coli* in the gut. Adhesion to the gut wall may prevent colonization by coliforms (Muralidhara *et al.*, 1977; Barrow *et al.*, 1980). Alternatively, these bacteria may in some way neutralize enterotoxin. Lactobacilli have been shown to produce an as yet unidentified metabolite capable of neutralizing *E. coli* enterotoxin in pigs (Mitchell and Kenworthy, 1976). They may produce organic acids and thereby reduce gastric pH.

Acid conditions inhibit the *in vitro* growth of *E. coli*. As many strains of lactobacilli and streptococci produce large quantities of lactic acid under *in vitro* conditions (Holdeman *et al.*, 1977), it has been suggested that they might reduce intestinal pH and thus reduce overgrowth of *E. coli* (Fox, 1988). Probiotic strains may possess bactericidal activity. Lactobacilli have been reported to produce hydrogen peroxide, which is bactericidal *in vitro* (Reiter *et al.*, 1980). *L. lactis* stimulated the lactoperoxidase thiocyanate system in the intestines of calves (Reiter, 1978), which reduced the ability of *E. coli* to survive in the gut (Reiter *et al.*, 1980). *E. coli* colonized the gut if reducing agents were used to reverse the effect of lactoperoxidase thiocyanate (Reiter *et al.*, 1980). Newman *et al.* (1990) recently identified a heat-stable, >5000 Da factor produced by *Ent. faecium* which was capable of inhibiting the growth of *E. coli*, *Ent. faecalis* and other related bacteria. However, the importance of such substances *in vivo* remains unclear (Fuller, 1989).

12.2.3 Use of probiotics in development of rumen fermentation

In addition to the prevention of diarrhoea, bacterial and fungal probiotics have also been used to enhance the development and maintenance of a stable rumen fermentation. Both lactobacilli and *Ent. faecium* have been shown to improve feed intake and liveweight gain in young cattle entering feedlots (Wren, 1987; Lee and Botts, 1988). In contrast, Kay and Poole (1988) found that *L. acidophilus* actually reduced the intake of weaning diet by calves. Umberger *et al.* (1989) reported that *Lactobacillus* stimulated feed intake and liveweight gains in lambs entering a feedlot. Ozawa *et al.* (1983) found that *Ent. faecalis* stabilized the intestinal flora and stimulated liveweight gain in calves following antibiotic treatment.

Probiotic inocula containing rumen organisms may also be useful. Theodorou *et al.* (1990) reported that a probiotic based on an anaerobic rumen fungus (*Neocallimastix* sp.) increased intake and liveweight gain in calves following weaning, while Ziolecka *et al.* (1984) and Ziolecki *et al.* (1984) reported that a stabilized rumen extract enhanced liveweight gain and stimulated rumen development in calves during weaning.

Products based on yeasts or aerobic fungi can be used in young as well as adult ruminants (see below). *Saccharomyces cerevisiae* increased food intake and liveweight gain after weaning in both calves (Fallon and Harte, 1987; Hughes, 1988) and lambs (Wells and Mason, 1976; Jordan and Johnston, 1990). Phillips and von Tungeln (1985) found that *Sac. cerevisiae* improved feed intake and liveweight gain in calves following transport stress. Allison and McCraw (1989) reported that *Aspergillus oryzae* fermentation extract stimulated weight gain over the first 28 days in calves entering a feedlot. Beharka *et al.* (1990) found that

A. oryzae extract stimulated dry matter intake in calves and allowed them to be weaned earlier. Rumen development was stimulated by *A. oryzae*, with higher counts of total, amylolytic, pectinolytic, cellulolytic and hemicellulolytic bacteria from week 2 of life onwards. A fungal extract supplemented with *Ent. bovis* stimulated bacterial numbers in the rumen of calves over the first 30 days after birth (Kmet *et al.*, 1988). A similar preparation stimulated rumen fermentation in newly weaned lambs (Bara and Kmet, 1987).

12.3. FUNGAL FEED ADDITIVES FOR ADULT RUMINANTS

12.3.1 Past and present use

Yeast and yeast-containing by-products have been used in ruminant diets for many years as sources of protein and energy. (Eckles and Williams, 1925; Carter and Phillips, 1944). However, since the late 1980s there has been an enormous increase in interest in products based on yeast and/or filamentous fungi that are analogous to probiotics and which enhance gut function. Products based on yeast or fungi are fed to adult ruminants to achieve a production response that is unrelated to the prevention of diarrhoea. The products nevertheless improve the nutrition of growing or adult ruminants much more than would be expected from their gross nutrient composition. The products in current use contain either the yeast *Sac. cerevisiae* or the aerobic fungus *A. oryzae*, or sometimes both together, hence they are sometimes described loosely as *fungal feed additives* or *fungal probiotics*.

The reason for the increased interest is primarily commercial. Partly, however, scepticism about the products' efficacy has been allayed by recent production trials, which have identified to some extent when the products are effective and when they are not, and by research into their mode of action.

Sac. cerevisiae and *A. oryzae* have been used in making food and beverages for millennia, and they are therefore perceived as, and almost certainly are, completely safe. Furthermore, only small quantities of fungal feed additives need to be added to the ruminant's diet. Thus the worst possible consequence that can be anticipated from the inclusion of fungal feed additives in a feedstuff is an absence of production response.

The effectiveness of fungal feed additives seems to stem from their influence on rumen fermentation, so they fall into the category of rumen modifiers among the so-called growth promoters. The other main category of rumen modifiers in current use, the feedlot ionophores and antibiotics, have made a tremendous impact in terms of cost-effective improvements in feed efficiency for ruminants. In trying to determine their mode of action, much has been learned about the rumen microbial ecosystem and its relation to the animal's nutritional requirements. Ionophores and antibiotics are not perceived to have the same naturalness as the fungal products, however, despite the exhaustive safety standards that they must meet before being licensed for use in the industry. Furthermore, ionophores may be hazardous if used in amounts that exceed the recommended dosage (Collins and McCrea, 1978), or if the ruminant feed is consumed by other domesticated non-ruminant species, as has been reported for horses (Whitlock *et al.*, 1979) and turkeys (Stuart, 1978). In contrast, fungal feed additives may actually benefit other animal species (Terrell *et al.*, 1984; Hollister *et al.*, 1990; Lyons, 1987, 1990; Pagan, 1990a).

Fungal feed additives can be used in both meat and milk production, unlike ionophores which may give rise to residues in milk and are therefore used primarily in meat production. Figures for the financial value of fungal feed additives have not been published, but their potential value can be surmised by comparison with the ionophores. It was estimated that ionophores fed to beef cattle had an annual value of $70 million in the US (Russell and Strobel, 1989). Only about one-third of the total 100 million cattle in the US are beef cattle (US Department of Agriculture, 1988). Thus if the efficacy of fungal feed additives is comparable to ionophores, which some studies suggest may be the case, the total potential market is likely to exceed $200 million.

12.3.2 The products

Increasing numbers of products are becoming available internationally. There is no reason to suppose that these are not effective, especially if production data are available. Equally, as is described below, not all yeast or *A. oryzae* preparations produce the same effects on fermentation as others, and therefore not all yeasts or fungi would be expected to have similar nutritional effects.

Yeast products are supplied as mixtures of live and dead yeast cells together with an element of the medium in which the yeast was grown, or distillers' dried solubles. Because the medium component is claimed to be important in the products' activity, the accepted terminology for the supplement is 'yeast culture' (YC) rather than simply yeast. The definition of yeast culture is: '. . . a dry product composed of yeast and the media on which it was grown, dried in such a manner as to preserve the fermenting capacity of the yeast. The media must be stated on the label' (AAFCO, 1986).

A. oryzae fermentation extract (AO), on the other hand, consists of fungal spores and mycelium dried on to a base of wheat bran. The viability of the preparations appears to be quite different. Yeast culture has a viability of 10^9–10^{10} live cells g^{-1} (Dawson *et al.*, 1990) or 2×10^7 live cells g^{-1} (C.W. Stone, pers. comm.) depending on the product, whereas *A. oryzae* fermentation extract contained 1.6×10^3 viable cells g^{-1} (Newbold *et al.*, 1991). Fungal feed additives can be used either by sprinkling on the feed or by incorporation into a compound diet. Experiments have also been done where *A. oryzae* was administered as an inoculant to silage (Harris *et al.*, 1983).

12.3.3 Responses in milk and meat production

Fungal probiotics arguably suffer from as much scepticism among farmers and nutritionists as the bacterial probiotics that are used for non-ruminant animals. Their doubts are probably founded partly on trials that did not result in quantifiable benefits, the reasons for which we are now beginning to understand and are described below, and partly to overzealous marketing. It is equally important to know when trials gave no detectable effect as it is to determine when benefits of feeding fungal probiotics occurred. Inevitably, as with any feeding trial, there is less incentive to publish results of trials which showed no effect. We have no idea of the degree of bias that is introduced into our survey by this tendency, but we suspect that it does exist.

The general pattern with ruminants receiving fungal feed additives is that production, whether of meat or milk, is improved. Williams and Newbold (1990) reviewed this area, and they noted that eight trials with AO produced an average 4.3% improvement in milk yield. A similar analysis of nine YC trials resulted in an average improvement of 5.1%. The responses from which these averages were calculated ranged from 91.0–112.0% for AO and 96.3–116.7% for YC, and the averages may therefore conceal an even better response under optimum dietary or nutritional circumstances.

Less information is available for growing ruminants than lactating animals. Improved liveweight gain has, like milk production, been observed in some studies but not in others. Adams *et al.* (1981) found that steers had an improved daily weight gain of 1.39 kg with YC compared with 1.34 kg in controls. As with many responses of this magnitude, the increase did not reach statistical significance. The same is true of many trials with ionophores, unless the number of animals is very large (Goodrich *et al.* 1984). Edwards *et al.* (1990) found no significant improvement with YC in liveweight gain in bulls from 135 kg to slaughter, although once more the trend was favourable. The opposite trend was observed by Deaville and Galbraith (1990) with Angora goats. Beef cows and calves fed a poor-quality pasture improved weight gain from 0.57 to 0.80 kg day^{-1} with AO (Wiedmeier, 1989).

A crucial feature of the effectiveness of YC and AO seems to be the diet and the nutritional demands of the animal. Williams *et al.* (1991) demonstrated how sensitive the effects of YC can be to a relatively small change in dietary composition. A milk yield response of 4.1 kg day^{-1} to YC added to a 40 : 60 hay : concentrate diet fell to zero when the ratio was 50 : 50. Gomez-Alarcon (1988) found similar interactions with forage : concentrate ratio for AO in cows, and Huber *et al.* (1985) observed that an AO supplement increased milk production of cows fed normal but not high forage diets. One might therefore imagine that a dietary response curve could have a form of the type outlined in Figure 12.2. YC or AO might move the curve for whatever property might be important – the present graph might represent the effects of concentrate inclusion on fibre digestion (Mould *et al.*, 1983) – and a window of opportunity would be created. At dietary compositions outside this range, no benefit would be expected from adding fungal feed additives to the diet. Presumably different additives would modify the shape of the curve in different ways. The curve will also depend on the different individual dietary components. For example, in the work of Williams *et al.* (1991), the milk yield response occurred with hay as the forage but not with ammonia-treated straw. Much work remains to be done to delineate these relations for different diets.

The response to fungal feed additives, as with any feed additive such as protein supplements, must depend on animal requirements and management, as pointed out clearly by Chase (1989). Thus cows in early lactation responded better to YC than those in later stages (Harris and Lobo, 1988; Günther, 1989). Similarly, the response to AO is greatest in early as opposed to mid or late lactation (Wallentine *et al.*, 1986; Kellems *et al.*, 1987). These nutritional effects will complicate

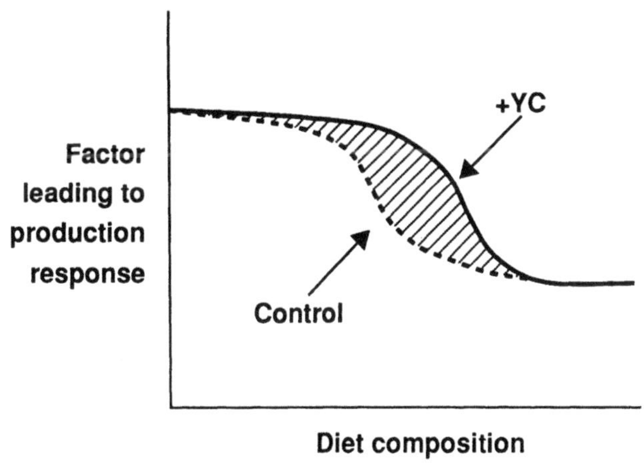

Figure 12.2 A hypothetical scheme outlining why the efficacy of fungal feed additives is dependent on dietary composition.

the investigation of dietary interaction, and it would help if the precise modes of action of YC and AO were known.

12.3.4 Effects on intake and digestion

Most studies indicate that fungal feed additives increase feed intake rather than alter feed conversion efficiency (Adams *et al.*, 1981; van Horn *et al.* 1984; Malcolm and Kiesling, 1986; Harris and Lobo, 1988; Edwards *et al.*, 1990; Gomez-Alarcon *et al.*, 1990; Sievert and Shaver, 1990; Williams *et al.*, 1991). Only occasionally is improved feed efficiency a possible benefit (Günther, 1989). A contrast can be drawn here with ionophores, which tend to have a minor, often depressive effect on intake but which improve feed conversion efficiency (Goodrich *et al.*, 1984). Williams and Newbold (1990) calculated that the improvement in the intake of dairy cows corresponded well with the observed effects on productivity. The main effects of fungal feed additives are therefore regarded as being *intake-driven*. Many factors are known to influence appetite, but the ones that have been considered for YC and AO in ruminants have been palatability, the rate of fibre digestion (thus directly affecting gut fill), the rate of digesta flow, and protein status.

Yeast extracts and *A. oryzae* fermentation products are widely used as flavour enhancers in human foods. Similar effects of YC and AO on the acceptability of feeds to ruminants cannot be ruled out. The products certainly have a pleasant odour, and Lyons (1987) and Rose (1987) suggested that the ability of yeast to produce glutamic acid could benefit the taste of feedstuffs supplemented with YC. While palatability improvements can certainly do no harm, there is now strong evidence that fungal feed additives have a more fundamental metabolic effect.

An improved feed intake would be expected to occur if fibre digestion in the rumen were increased. The latter is seen sometimes, but not always, when the measurement made is of total tract digestibility. Wiedmeier *et al.* (1987) found increases in DM, ADF and hemicellulose total tract digestibility with AO, YC and combined AO and YC in dry cows fed a mixed forage/concentrate diet. In three trials with cows, Gomez-Alarcon *et al.* (1990) observed that, with the exception of a diet containing high forage (63% alfalfa hay), AO caused increases in total tract DM, ADF and NDF digestibility. AO increased DM digestibility in a trial in which no significant intake response was seen (Gomez-Alarcon *et al.*, 1988a). In contrast, Arambel and Kent (1988), Arambel *et al.* (1987, 1990) and Oellermann *et al.* (1990) observed no changes in digestibility with heifers or cows fed AO, and Judkins and Stobart (1988) found no increase in digestibility in wethers fed AO. Harrison *et al.* (1988) and Williams and Newbold (1990) reported no effect with cows fed YC. A combined AO–YC product had no effect on the extent of digestion in heifers fed a 50% orchardgrass hay diet (Firkins *et al.*, 1990).

Total tract digestibility can, however, conceal profound changes in the site or rate of degradation of fibre in the tract. If fibre degradation in the rumen is stimulated, this might simply reduce the residue of material which is normally broken down in the hindgut, producing the same overall digestibility. Gomez-Alarcon (1988) found that AO and YC stimulated fibre breakdown in the rumen, effectively shifting more of the digestion to the rumen from the lower gut. Fondevila *et al.* (1990) observed that AO stimulated by 28% the rate of breakdown of straw suspended in nylon bags in the rumen, although the final extent of degradation was unchanged (Figure 12.3). Feeding YC to sheep gave similar effects on the digestion of hay (Chademana and Offer, 1990). No changes were found in OM, NDF or gross energy digestibility of three diets of differing forage : concentrate ratio, and the 48 h degradability of hay in the rumen was unchanged. However, 24 h OM degradation was increased by 11.6, 15.6 and 12.1% in low, medium and high forage diets respectively. Similar patterns of *in situ* breakdown were observed (Campos *et al.* 1990) with AO in non-lactating diary cows fed mainly corn stover and improved *in vitro* digestibility occurred at 24 but not 48 h with AO (Newbold *et al.*, 1991).

Another important factor that can affect intake is the outflow rate of digesta from the rumen, but the results with fungal feed additives have been mixed. Wiedmeier *et al.* (1987) found decreased liquid and particle outflow rates with AO and increases in the same rates with YC.

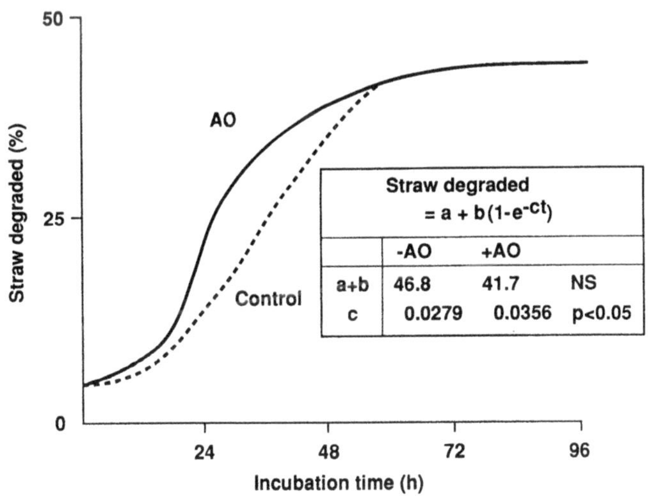

Figure 12.3 Degradation curve of straw incubated in nylon bags in the rumen of sheep receiving straw or straw + AO (Fondevila *et al.*, 1990). (a + b) is the maximum degradability and c is the intial rate of degradation.

Liquid outflow was stimulated by YC in growing steers under feedlot conditions (Adams *et al.*, 1981), but was not affected significantly in sheep receiving YC (Chademana and Offer, 1990). Sheep receiving straw showed no change in liquid outflow (Fondevila *et al.* 1990).

It can be concluded therefore that the enhanced intake which drives production responses to fungal feed additives is most likely due to an improved rate of breakdown of feedstuffs in the rumen. The stimulation of digestion, especially of fibre, need not affect the final ruminal degradability or total tract digestibility.

12.3.5 Influence on rumen fermentation

(a) Fermentation products and pH

Ruminant nutrition studies are often accompanied by estimates of some easily measured parameters in rumen fluid, including pH, volatile fatty acids (VFA) and ammonia concentrations. These often help to explain the effects of different dietary manipulations on host animal nutrition. With fungal feed additives, however, the trends that can be discerned, with the possible exception of rumen pH, tell us little about how YC and AO work.

YC and AO appear to be different in their effects on VFA concentrations in rumen fluid. The effects are always small and often insignificant (Chademana and Offer, 1990; Dawson *et al.*, 1990; Fondevila *et al.*, 1990; Oellermann *et al.*, 1990). The trend appears to be that both additives increase total VFA concentration. When effects on VFA ratios are observed, YC increases the proportion of propionate formed whereas AO increases acetate production (Adams *et al.*, 1981; Wiedmeier *et al.*, 1987; Harrison *et al.*, 1988; Firkins *et al.* 1990; Newbold *et al.*, 1990). On the other hand, Gomez-Alarcon *et al.*, (1990) observed a lower proportion of acetate with AO when fed to cows on a high concentrate diet and Edwards *et al.* (1990) found higher acetate concentrations in bulls receiving YC. Increased acetate production was observed in mixed continuous cultures *in vitro* with AO (Frumholtz *et al.*, 1989; Newbold *et al.*, 1991), but no change in VFA ratios occurred with YC (Dawson *et al.*, 1990). In short-term batch incubations *in vitro*, high concentrations (1 gl^{-1}) of AO or YC stimulated propionate production (Martin *et al.*, 1989; Martin and Nisbet, 1990).

Changes in VFA concentrations *per se* therefore seem to have little significance in how fungal feed additives exert their influence on rumen fermentation. Possibly of much greater significance are findings that YC stimulated the *rate* of *in vitro* VFA production from different substrates in rumen fluid taken from sheep receiving YC (Gray and Ryan, 1989; Ryan and Gray, 1989). Whether this effect is a genuine stimulation of VFA production per unit of microbial biomass, or is simply a consequence of a larger microbial population in the rumen of animals fed YC (see p. 331) remains to be established.

Methane production represents a substantial energy loss to the ruminant (Hungate, 1966). It is also intimately associated with the relative proportions of VFA that are produced (Demeyer and van Nevel, 1975) and the deamination of amino acids (Russell and Martin, 1984). In two *in vitro* studies, an increase in methane production was observed when YC was added to a batch system (Martin *et al.*, 1989; Martin and Nisbet, 1990). Surprisingly, increased hydrogen production was also observed (Martin and Nisbet, 1990). A decreased proportion of methane in the headspace gas was found when AO was added to a semi-continuous rumen fermenter (Rusitec, Frumholtz *et al.*, 1989), and methane production was decreased in calves when YC was included in the diet (Williams, 1988). Clearly more in *vivo* studies are required to establish how significant the effects of fungal feed additives are on methane production.

In studies where rumen ammonia concentration has been altered by fungal feed additives, it is usually decreased (Adams *et al.*, 1981; Chademana and Offer, 1990; Dawson, 1987; Harrison *et al.*, 1988; Newbold *et al.*, 1990; Oellermann *et al.*, 1990). However *in vivo* and *in vitro* increases have been seen with AO (Wiedmeier *et al.*, 1987; Gomez-Alarcon, 1988; Frumholtz *et al.*, 1989; Martin and Nisbet, 1990). Since the ammonia pool results from the nett effects of production from protein and urea on one hand and microbial demands for protein synthesis on the other, knowing how additives affect its size has only limited usefulness in telling us how the additives function. Arambel *et al.* (1987) suggested that AO might stimulate proteolysis as a consequence of its protease activity (Boing, 1983). An increased digestibility of crude protein is a response that has been observed with AO (Campos *et al.*, 1990; Wiedmeier *et al.*, 1987) and YC (Wiedmeier *et al.*, 1987). However, this may not always occur (Oellermann *et al.*, 1990), and direct measurement of enzyme activities in Rusitec showed that protease activity was decreased and deamination of amino acids was increased by AO (McKain *et al.*, 1991). A similar trend was seen with YC in the sheep rumen (Newbold, 1990).

Rumen pH is one of the most critical determinants of rumen function, particularly for the cellulolytic bacteria, which fail to grow at pH 6.0 and below (Stewart, 1977). Rumen pH falls as a result of increasing concentrate in the diet. This fall inhibits degradation of the fibrous components and causes, in part at least, negative associative effects between forages and concentrates. The degradability of the fibrous components of the diet is inhibited by adding concentrate above a certain proportion (Mould *et al.*, 1983). Fungal feed additives usually appear to increase rumen pH slightly (Wiedmeier *et al.*, 1987; Gomez-Alarcon *et al.*, 1990; Oellermann *et al.*, 1990), although this does not always happen (Arambel *et al.*, 1990; Chademana and Offer, 1990; Fondevila *et al.*, 1990; Gomez-Alarcon *et al.*, 1990) and in some experiments YC actually caused a fall in rumen pH (Harrison *et al.*, 1988; Edwards *et al.*, 1990). Increases in pH have also been recorded in *in vitro* fermentation systems (Frumholtz *et al.*, 1989; Dawson *et al.*, 1990).

Perhaps the most crucial aspect of how fungal feed additives affect rumen fermentation is often concealed within the experimentally derived VFA and ammonia concentrations and pH values that are reported. Variation in ammonia concentrations in rumen fluid from cows receiving yeast culture was less than controls, and microbial numbers were similarly more stable (Harrison *et al.* 1988). Harrison *et al.* concluded that ruminal fermentation was more stable in cows receiving yeast culture supplement. Experiments with steers fed barley with hay (Williams 1989, Williams *et al.* 1991) illustrate this point (Figure 12.4). Post-feeding peaks in lactate concentration and troughs in pH were markedly decreased in animals receiving YC.

The other small molecular weight compounds that have been shown to change with AO or YC addition to the diet are the minor VFA, which are formed largely by deamination of amino acids. Both YC and AO increased the percentage of branched-chain VFA in bovine rumen fluid (Wiedmeier *et al.*, 1987). Molar proportions of isovalerate were decreased and valerate was increased in cows receiving YC (Harrison *et al.*, 1988),

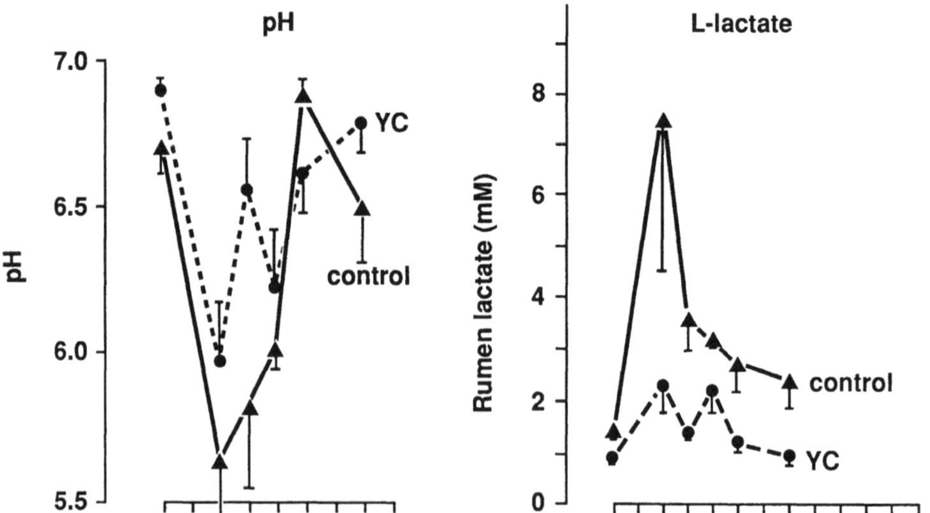

Figure 12.4 Influence of YC on rumen pH and lactate concentration in the rumen of steers after feeding (from Williams *et al.*, 1991)

and isovalerate concentration was similarly increased in cows receiving AO (Oellermann *et al.*, 1990). Similar *in vitro* observations have been made (Arambel *et al.*, 1987; Frumholtz *et al.*, 1989; Martin *et al.*, 1989). Dawson *et al.* (1990) found no difference in total iso-acids produced under *in vitro* or *in vivo* conditions with YC. These higher VFA stimulate the growth of the predominant rumen cellulolytic bacteria (Bryant, 1973)

(b) Microbial yield

Effects on microbial yield are variable. Wanderley *et al.* (1987) found that AO had no effect on yield, whereas Gomez-Alarcon *et al.* (1990) observed that microbial yield was increased by AO in two trials out of three done with cows. In the third trial, where AO had no effect, YC was included as another treatment, and it also failed to stimulate the yield. Wanderley *et al.* (1987) found that protein flow was increased, however, and Williams *et al.* (1990) measured protein flow in the duodenum of sheep, which tended to be increased by YC, leading to increased absorption of non-ammonia N. Urinary allantoin measurements in bulls implied that YC had improved microbial yield (Edwards *et al.*, 1990). In the study of Wanderley *et al.* (1987), it appeared that the increased protein flow was a consequence of both increased microbial protein and increased undegraded dietary protein in the digesta in the duodenum.

(c) Microbiology

Substantial increases in the total viable count (TVC) of anaerobic bacteria in the rumen when ruminants were fed fungal feed additives were first observed with AO (14%) and YC (30%) by Wiedmeier *et al.* (1987). Harrison *et al.* (1988) subsequently reported a 58% increase in TVC with YC, and Dawson *et al.* (1990) a nearly five-fold increase in TVC when steers were fed YC. Large *in vitro* increases have been also observed. Frumholtz *et al.* (1989) found a 79% increase in TVC with AO, a 90% increase was found in subsequent studies with AO (Newbold *et al.*, 1991), and an increase of more than ten-fold occurred in response to YC in continuous culture (Dawson *et al.*, 1990).

Clearly such large increases in TVC do not reflect changes in the total bacterial protein present, especially in view of small or no effects on microbial yield. Instead, they illustrate enormous changes in the viability of the bacterial cells that are present when fungal feed additives are used, especially when imperfect *in vitro* growth conditions are used. Thus YC and AO must in some way improve conditions for the growth of rumen bacteria.

The growth of cellulolytic bacteria is also stimulated by fungal feed additives. *In vivo* population sizes tend to increase proportionally by a little more than the increase in total population (Wiedmeier *et al.*, 1987;

Harrison *et al.*, 1988; Arambel *et al.*, 1990; Dawson *et al.*, 1990), but this does not always happen (Fondevila *et al.*, 1990). *In vitro* studies show that the effect on cellulolytic bacteria is sometimes considerably greater than that on the total population (Frumholtz *et al.*, 1989; Arambel *et al.*, 1990; Dawson *et al.*, 1990), although again this does not always hold (Newbold *et al.*, 1991).

Ciliate protozoa comprise up to half of the total microbial biomass in the rumen (Williams and Coleman, 1988) and are primarily responsible for the wasteful breakdown and resynthesis of bacterial protein that reduces microbial yield (Demeyer and van Nevel, 1979; Wallace and McPherson, 1987). They also contribute to cellulolysis (Coleman, 1985; Williams and Coleman, 1988). Yet, despite their evident importance, few protozoal counts appear to have been reported in ruminants fed fungal feed additives. Protozoal numbers were reduced by 45% in Rusitec when AO was added (Frumholtz *et al.*, 1989), but numbers tended to increase with AO in cows (Oellermann *et al.*, 1990).

The third major category of rumen microorganisms, namely the anaerobic fungi, which are highly cellulolytic (Orpin and Joblin, 1988), have likewise received little attention with regard to fungal feed additives. AO tended to increase fungal numbers in the rumen digesta of cattle receiving AO (Oellermann *et al.*, 1990). A second fungus, not identified, was present with *Aspergillus* attached to fibre particles in the duodenum of cattle receiving AO (Wanderley *et al.*, 1985). Anaerobic fungi were less numerous than aerobic fungi in the bovine rumen and were not increased by AO (Oellermann *et al.*, 1990). When AO was added directly to pure cultures of *Neocallimastix frontalis*, *N. patriciarum* and *Sphaeromonas communis*, it had no influence on gas production by these major species of rumen anaerobic fungi (C. J. Newbold, unpublished experiments). Thus most of the available evidence suggests that fungal feed additives have little effect on the natural population of anaerobic fungi growing in the rumen.

12.3.6 Other effects

A number of other effects have been attributed to fungal feed additives that may or may not be directly associated with their mode of action. In cattle subjected to high ambient temperatures, AO reduced rectal temperature and heat stress (Huber *et al.*, 1986; Huber, 1987). YC may also affect mineral metabolism, presumably due to the ion-binding properties of its cell wall (Rose, 1987). Improved zinc nutrition may explain some of the effects of YC (Williams, 1988) properties of its cell wall (Rose, 1987). Improved zinc nutrition may explain some of the effects of YC (Williams, 1988).

The findings that viable yeast survived passage through the tract to increase numbers in the duodenum and ileum of sheep (Newbold *et al.*, 1990) and *Aspergillus* spores were present in duodenal digesta of cows receiving AO (Wanderley *et al.*, 1985) could have important implications for a second site of action of fungal feed additives. Possible post-ruminal effects of fungal feed additives have been largely ignored until now. It is possible that the benefits noted with horses receiving YC, such as improved nutrient digestibility, which arise from stimulation of the caecal microflora (Pagan, 1990b), could also occur with ruminants.

12.3.7 Possible modes of action

Some of the claims made for the preparations, particularly in their mode of action, have been improbable. Furthermore, manufacturers are generally reluctant to describe how the products were selected. While this is understandable in a commercial sense, it has added fuel to the sceptics' arguments. We regard identifying the precise mode of action of present fungal feed additives, at the cellular and molecular level if possible, as being of utmost importance in establishing confidence in the products and in establishing a basis for future developments. Firstly, however, we must attempt to untangle the many reported effects into models that can be subjected to critical and experimental testing. It must also be established whether effects might be dose-related or dependent on growth of the yeast or *Aspergillus* in the rumen.

(a) Growth of yeasts and fungi in the rumen

Yeasts and moulds occur naturally in the microbial community of the rumen (Lund, 1974, 1980). Up to 1.3×10^5 yeasts ml^{-1} grew when dilutions of bovine rumen fluid were incubated at 25°C, but only $3.5 \times 10^3\ ml^{-1}$ grew at 39°C, suggesting that the yeasts present normally are essentially transient members of the community, entering with the fodder (Lund, 1974). Nine different species were identified, none of which was a *Saccharomyces* (Lund, 1974). Yeast numbers were similar in sheep (Newbold *et al.*, 1990). A natural yeast population was undetectable in some roughage-fed steers, and yeasts were also undetectable when the rumen fluid was used in *in vitro* continuous cultures (Dawson *et al.*, 1990). Thus yeasts, and particularly *Sac. cerevisiae*, are not normally significant members of the rumen microbial community. The temperature (39°C) and chemical composition of rumen fluid tended to be inhibitory to the *in vitro* growth of *Sac. cerevisiae* (Arambel and Tung, 1987).

Dawson (1987) obtained data from *in vitro* experiments which implied that *Sac. cerevisiae* might grow in the rumen. Subsequent experiments suggest that substantial growth of yeast is unlikely to occur, however. Yeast numbers increased from 2.5×10^5 to 4.7×10^5 ml^{-1} 4 h after feeding in cows receiving YC (Harrison *et al.*, 1988), and when YC was fed to sheep, yeast numbers in rumen fluid increased from 1.5×10^3 to 3.34×10^5 ml^{-1} after 1 h (Newbold *et al.*, 1990). When numbers in the sheep rumen were extrapolated back to zero time (Figure 12.5), the population of yeast corresponded to the number of viable cells added as YC with the feed. Numbers then declined at a rate of 0.17 h^{-1}, i.e. somewhat faster than would be expected simply from liquid dilution, indicating that any net growth of yeast in the rumen was insignificant and that cell death had occurred. Similar conclusions were made by Arambel and Tung (1989) using vivar chambers fitted with different sizes of membrane filters. *In vitro* continuous cultures (Dawson *et al.*, 1990) confirmed that viable yeasts do not increase in number in rumen fluid. A lack of viability should not be confused with a lack of metabolic activity, however. Ingledew and Jones (1982) found that *Sac. cerevisiae* was metabolically active in rumen fluid for up to 6 h.

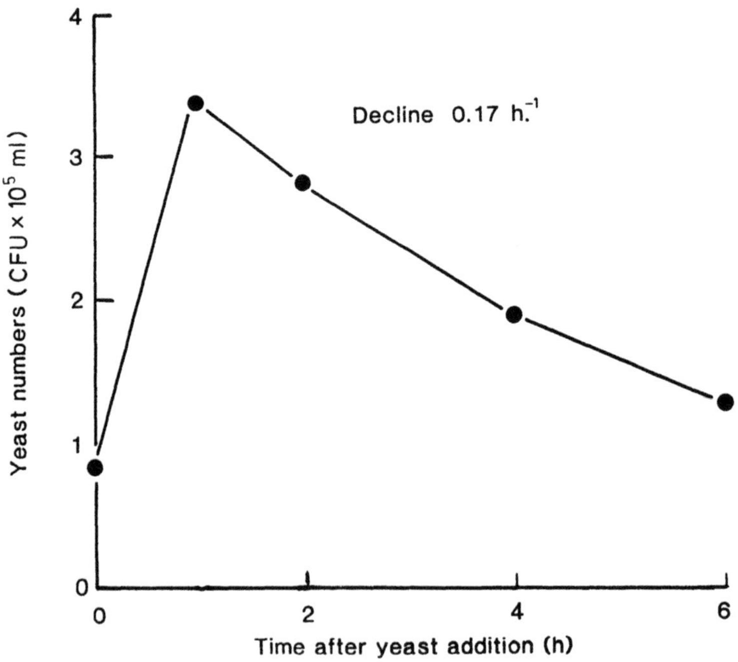

Figure 12.5 Viable counts of yeast in rumen fluid of sheep following the addition of 1.6×10^9 live yeast cells in YC. The rate of decline = 0.17 h^{-1} (Williams *et al.*, 1990)

Numbers of aerobic fungi in the rumen of straw-fed sheep were 4.2×10^5 ml^{-1} in control animals and 5.8×10^5 ml^{-1} in those receiving AO (Fondevila *et al.*, 1990). The aerobic fungal population was smaller in cows fed a mixed diet (1.7×10^3 ml^{-1}), and no significant increases occurred in response to increasing amounts of AO (Oellermann *et al.*, 1990). In view of the low viability of AO fermentation extracts as described earlier, perhaps these results are not surprising, and in view of its strictly aerobic mode of growth, we would not expect to observe *in vivo* multiplication of *Aspergillus*.

Yeast extract that did not contain live cells did not stimulate the *in vitro* growth of *Fibrobacter succinogenes* on cellulose in the same way as *Sac. cerevisiae* (Dawson, 1990). Similarly, when tested *in vitro* autoclaved YC (Dawson *et al.*, 1990) and autoclaved AO (Newbold *et al.*, 1991) were ineffective in stimulating bacterial numbers in mixed fermentations. Intriguingly, AO that had been sterilized by gamma-irradiation rather than autoclaving retained most of its stimulatory activity (Newbold *et al.*, 1991). Either a heat-sensitive nutrient is destroyed by autoclaving, or the AO has a metabolic activity that is destroyed by heat but not irradiation.

(b) Principal features of a sequential model

Rose (1987), Dawson (1987) and many others since have suggested how the fungal feed additives may work in terms of their many different individual effects on the animal and on rumen fermentation. It seems unlikely, however, that there are more than one or two critical primary sites at which YC or AO act to exert their ultimate nutritional benefit. Fungal feed additives must interact at the cellular or molecular level in the animal, probably in the rumen, to cause these primary effects, which then have secondary consequences and so on until the productivity benefit occurs. It is therefore vital to distinguish the important primary effects from the secondary ones if an understanding of fungal feed additives is to be obtained.

Offer (1990) attempted to bring together the various pieces of information into a single branched sequence, and this is a useful exercise. Figure 12.6 has simplified Offer's diagram and revised it to incorporate proposals made by Newbold (1990). A detailed analysis of each step can be done to assess its relative importance to the whole sequence, and therefore a better idea of the primary mode of action can be obtained.

Step 1, linking productivity with intake, seems to hold true in most studies where the two have been studied, as do steps 2a and 2b relating digestion and substrate supply to intake. It remains to be proved that the stimulation of cellulolysis and increased microbial protein flow are truly independent; faster fibre digestion leading to more rapid bacterial growth would decrease bacterial maintenance requirements leading to an improved yield (Hobson and Wallace, 1982).

An increased viable count of bacteria in the rumen is the most reproducible effect of fungal feed additives, and it seems reasonable that steps 3a and 3b should be consequences of improved bacterial viability. Thus we have to explain what stimulates rumen bacteria.

The imaginative work of Williams (1989) and Williams *et al.* (1991) suggested a logical sequence of effects similar to steps 4 to 6 in Figure

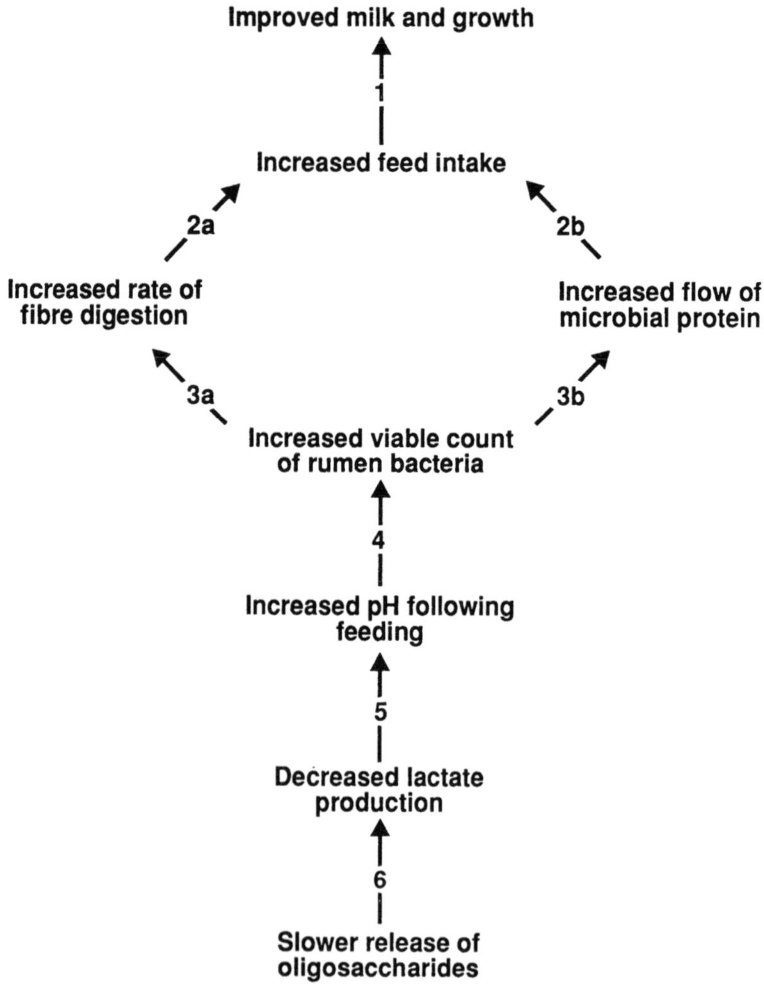

Figure 12.6 A sequential model for the mode of action of fungal feed addtives.

12.6. Stabilization of rumen pH (step 4) was suggested to be the reason for the improved microbial growth. This was particularly attractive as it would lead to a familiar explanation for the shape of Figure 12.2. Associative effects on cellulolysis cause the sigmoidal curve, and stabilizing pH would logically move the curve to the right, creating the window of efficacy. Despite earlier speculation about the properties of yeast cell walls (Rose, 1987), YC has no influence whatever on the buffering capacity of rumen fluid (Ryan, 1990). The stabilization of rumen pH must therefore have been a secondary effect of YC rather than the primary effect.

Lactic acid has a lower pK_a than the VFA, so the lower lactate concentrations observed (Williams, 1989; Williams *et al.* 1991; Newbold *et al.*, 1990) could conceivably have been responsible for the increased rumen pH. The strain of yeast used in some culture systems does not assimilate lactate, although other strains of *Sac. cerevisiae* do (NCYC, 1990), therefore the yeast did not ferment the lactate in these experiments. Alternatively, substances in the feed additives might have stimulated the removal of lactate by indigenous rumen bacteria, as shown in pure culture with the lactate-fermenting rumen bacterium, *Selenomonas ruminantium* (Nisbet and Martin, 1990a.b). The dicarboxylic acids fumarate and malate stimulated lactate transport and growth yield on lactate as did malate-containing soluble extracts of AO (Nisbet and Martin, 1990a) and YC (Nisbet and Martin, 1990b). The malate content of fungal feed additives was not high, however, considering the small amounts of product that are effective in the animal. Furthermore, autoclaving would not be expected to destroy

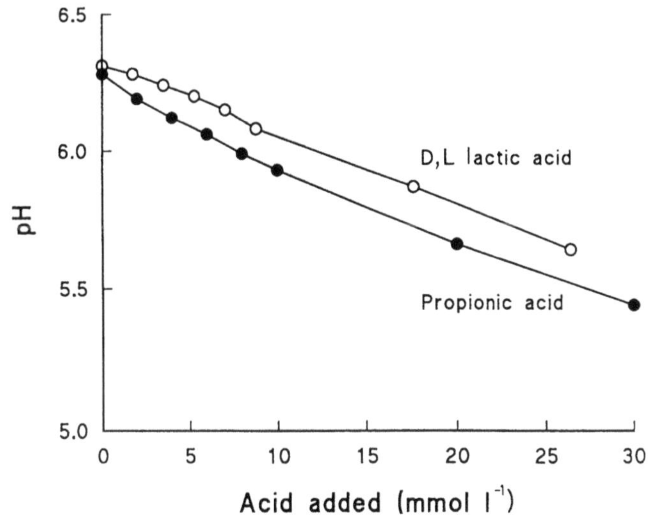

Figure 12.7 Influence of addition of D, Llactic acid and propionic acid to sheep rumen fluid.

malate, yet the products' activity was destroyed by autoclaving. It is possible that yeasts and fungi actually produce dicarboxylic acids in the rumen and thereby stimulate lactate uptake by rumen bacteria. The other possibility is that decreased lactate production could have been caused by a slower fermentation of dietary sugars – lower concentrations of soluble sugars were present in steers receiving YC (Williams *et al.*, 1991). Rapid bacterial growth in the rumen tends to cause increased lactate production (Hungate, 1966), and removal of sugars may have slowed growth sufficiently to decrease lactate production.

The main problems with the lactate–pH link are, firstly that the small quantities of lactate (up to 8 mM), observed (Williams, 1989; Williams *et al.* 1991 would not give a pH in rumen fluid which was significantly different from equivalent amounts of VFA (Figure 12.7), and, secondly, that the pH increases that are observed in most experiments are very small and insignificant. Furthermore, the *in vitro* stimulation of the rumen cellulolytic bacterium, *Fib. succinogenes*, by yeast (Figure 12.8; Dawson, 1990) is an extremely significant observation which almost certainly would not be explained by a change in culture pH, as only 10^4 yeast cells ml^{-1} were added. It can be concluded, therefore, that decreased rumen lactate concentrations must be a secondary consequence of fungal feed additives arising from, but not responsible for, improved bacterial growth.

Having eliminated some possible explanations, other suggestions remain to be evaluated in terms of the primary mechanism by which bacterial growth is stimulated.

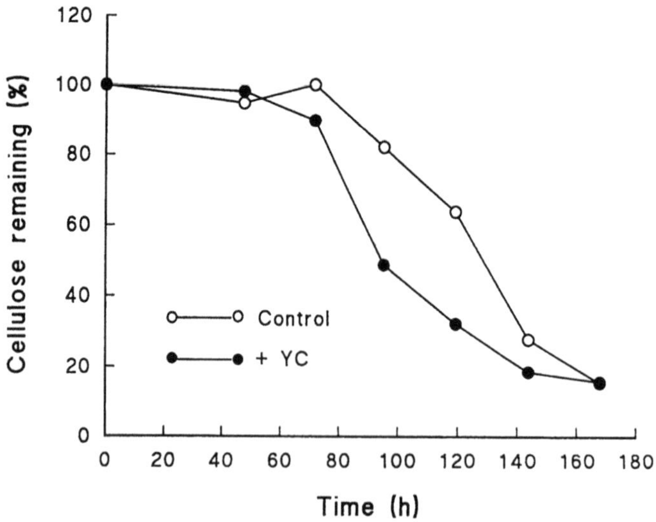

Figure 12.8 Influence of *Saccharomyces cerevisiae* on the growth of *Fibrobacter succinogenes* on cellulose (after Dawson, 1990)

1. The observed sugar accumulation by yeast might have a beneficial effect on cellulolytic bacteria by removing sugars that might otherwise depress cellulolytsis (Mould *et al.*, 1983). There seems to be no evidence to the contrary for YC and information is lacking for AO; this aspect is worthy of further investigation.
2. AO contains fibre-digesting enzymes, including cellulase, that might stimulate the growth of cellulolytic organisms and consequently the total population (Offer, 1990). However, these enzymes do not appear to enhance any of the fibre-degrading activities of rumen fluid (Newbold and Wallace, unpublished observations).
3. YC and AO may contain nutrients that stimulate rumen bacterial growth. Since autoclaved YC and AO do not work, the nutrient(s) must be heat-labile. Furthermore, the nutrient(s) would have to be absent from yeast extract, for reasons described above.
4. YC and AO may not contain the required nutrients, but might produce nutrients or cofactors under *in vivo* conditions. This would explain why autoclaved additives did not work, as enzymes would have been destroyed. The identity of this nutritional effect with higher VFA resulting from proteolytic and deaminative activities of the additives seems doubtful. The increases observed in their concentrations, while sometimes significant, were nevertheless small. Dicarboxylic acid production could occur, even independently of growth, and presumably yeasts and fungi that remain metabolically active would be able to produce other growth factors such as vitamins, despite being unable to grow. This possibility has not been eliminated by any experiments of which we are aware, and it remains an important area for future investigation.
5. Fungal feed additives may remove toxic factors in rumen fluid that inhibit the growth of rumen bacteria. The removal of toxic metal ions is possible for yeast, with its highly ionic cell wall (Rose, 1987). Scavenging of molecular oxygen (Rose, 1987) is another possibility. Molecular oxygen is much more toxic to *Fib. succinogenes* and other rumen bacteria than increased Eh (Marounek and Wallace, 1984), so traces of O_2 could be detrimental even without changing Eh. Presumably other compounds toxic to rumen bacteria are ingested with the feed or produced *in situ*, so detoxification is another mechanism that has not yet been eliminated.
6. Altered VFA and methane production in response to fungal feed additives could influence host animal productivity, as occurs with the ionophores which stimulate propionogenesis and decrease methanogenesis (Bergen and Bates, 1984). The effects of YC and AO seem to be opposite, however, as YC tends to stimulate propionate while AO increases acetate concentration. These effects are not particularly large and they are unlikely to give rise to increased

productivity. Surprisingly, *Sac. cerevisiae* stimulated acetate production relative to succinate (which forms propionate in the mixed ecosystem) in pure cultures of *Fib. succinogenes* (Dawson, 1990).

All of these explanations would have to deal with the dietary dependence of fungal feed additives. At present, our interpretation is that the primary effect probably occurs whatever the diet fed, but only becomes critical in some circumstances, such as at the transition of the associative effects curve, or when other stresses affect the rumen bacteria, particularly the cellulolytic organisms, detrimentally.

(c) An alternative model

The foregoing analysis of the importance of various effects and their likely position as primary or secondary effectors of bacterial stimulation leads us to propose a different model to that described in Figure 12.6. In this model (Figure 12.9), stimulation of bacterial growth is proposed to be the most critical property responsible for productivity improvements. Many of the arrows are actually reversed in comparison with Figure 12.6, and a more limited number of different possible

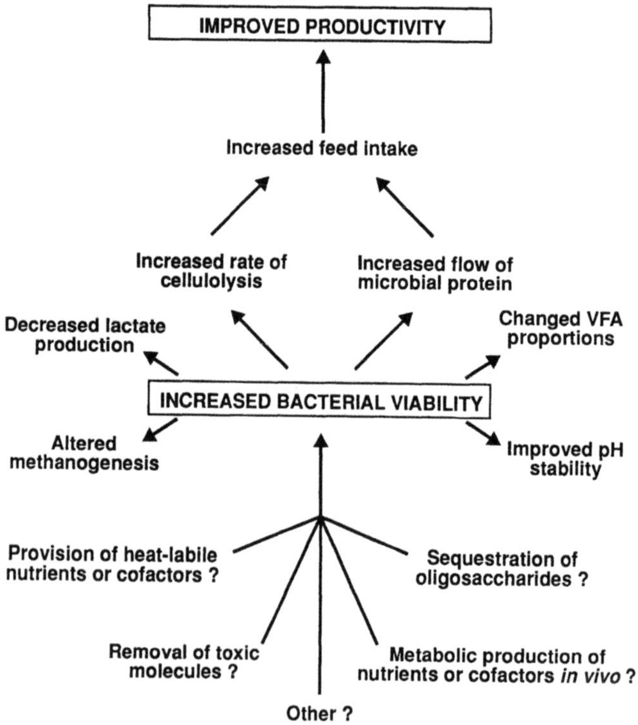

Figure 12.9 An alternative scheme for investigating the primary mode of action of fungal feed additives.

primary effects are offered. We hope it might be useful in the design of future experimentation aimed at defining the mode of action of fungal probiotics.

(d) Differences between yeasts and fungi

No firm pattern has emerged about whether YC and AO have different modes of action. AO tends to be more effective in higher forage diets, but the relation is not clear. The viability difference may or may not indicate different modes of action. Certainly the enzyme complements will be different. Until their precise modes of action have been elucidated, it would be prudent to assume that their modes of action are different. If so, the effects of AO and YC could be additive. Nutrient digestibility and cellulolytic bacterial numbers were higher when AO and YC were used in combination than when they were used alone (Wiedmeier *et al.*, 1987). Several preparations are now available that combine the two organisms.

12.3.8 Screening for fungal feed additives

The only sure test of the efficacy of any microbial additive is to determine the response of ruminants under production conditions, and this will be true until the mode of action of the present products is established unequivocally. Clearly the number of preparations which can be screened by full production trials is limited, and more rapid, preferably *in vitro* methods would be useful.

One of the difficulties of any *in vitro* system for screening is that fungal feed additives often appear to take time to work (Arambel *et al.*, 1987) – the work of Offer (1990) implies that the *in vivo* adaptation period might be several days – and the *in vitro* system would have to take account of adaptation. Continuous or semi-continuous culture system (Frumholtz *et al.*, 1989; Newbold *et al.*, 1990) or others (Dawson *et al.*, 1990) are fairly useful for this purpose. They are generally more stable than a comparable number of sheep, and an assessment of the ability of products to affect the fermentation can be done in two or three weeks. In other circumstances, adaptation may not be necessary for an *in vitro* effect to be obtained, and a simpler *in vitro* system may be useful: Gomez-Alarcon *et al.* (1988b) found that an increased *in vitro* digestibility occurred without any apparent need for adaptation.

What properties should be screened for? Gray and Ryan (1990) identified large short-term and smaller long-term differences in the rate of VFA produced in rumen fluid from sheep receiving YC. Thus the short-term (22 h) increases could be used as a rapid screen for new products. However, clearly an assay more closely related to the suspected mode of action would be preferable. Effects on DM disappearance may occur

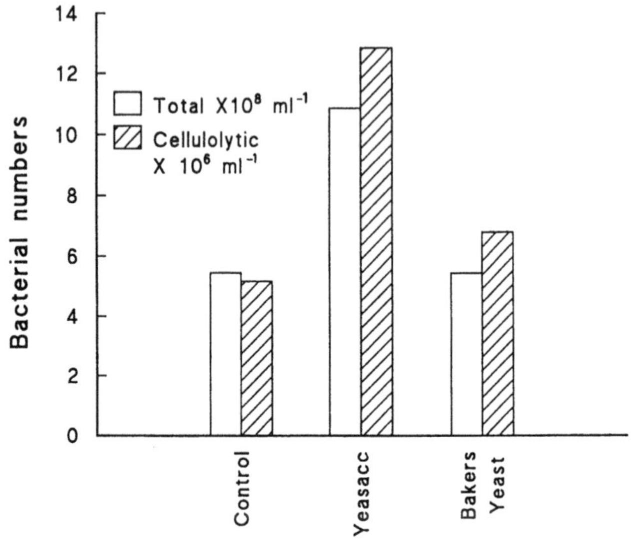

Figure 12.10 Influence of different yeast preparations on the stimulation of bacterial counts in Rusitec.

sometimes under *in vitro* conditions and they may be useful (Arambel *et al.*, 1987; Gomez-Alarcon *et al.*, 1988b). At present, we consider that *in vitro* screening for increased viable counts of bacteria in continuous or semi-continuous cultures is a suitable compromise between the full *in vivo* trial and other methods that would allow more preparations to be tested but would not allow any adaptation to occur. Few if any studies, whether *in vivo* or *in vitro*, have failed to pick up stimulation of bacterial numbers, either of total or cellulolytic bacteria. More extensive screening will require more rapid methods. Here the risk is greater if the wrong effect of the present products is given highest priority. However, cautious use of pure cultures seems to offer at least some prospect of success. Bacterial stimulation of the type observed by Dawson (1990) with *Fib. succinogenes* could be crucial, and it would be possible to screen many preparations. Alternatively, if lactate uptake does prove to be pivotal, a lactate uptake assay (Nisbet and Martin, 1990a,b) might be useful.

12.4 BACTERIAL PROBIOTICS FOR ADULT RUMINANTS

Relatively few experiments have been done in adult ruminants with the types of bacterial preparation that are used in young ruminants or non-ruminant animals. Jaquette *et al.* (1988) and Ware *et al.* (1988) reported significant increases (6.2 and 5.7% respectively) in milk production from cows receiving *L. acidophilus*.

The mode of action of a lactobacillus preparation in the rumen is difficult to imagine. Lactobacilli produce lactate, sometimes to the severe detriment of the animal in cases of lactic acidosis (Slyter, 1976). Seven species of rumen bacteria were unaffected by lactobacilli or enterococci that are used in bacterial probiotics to inhibit pathogens (Newman *et al.*, 1990). It is nonetheless possible that less common, possibly detrimental species are inhibited by lactobacilli, however, bacterial probiotics may have no advantage over fungal products in the adult animal. A mixed YC + *Lactobacillus* + *Enterococcus* preparation was little different to YC alone in its influence on rumen fermentation (Dawson *et al.*, 1990).

12.5 FUTURE DEVELOPMENTS

The next stage, as far as most workers are concerned, is to determine whether other strains or species of yeasts and fungi can be effective. Experiments with Rusitec have shown that different yeast strains can be effective, but to different extents (Figure 12.10). A new strain has been selected by its ability to stimulate cellulolysis by *Fib. succinogenes* more than the strain present in the existing product (Dawson, 1990). No information is available about other species or genera of yeast.

Different species in the genus *Aspergillus* may give effects comparable to *A. oryzae*. *A. niger* was at least as effective as *A. oryzae* in enhancing nutrient digestion in cows (Campos *et al.*, 1990). *In vitro* digestibility trials suggested that species of *Trichoderma* and *Penicillium* would be much less active than the two *Aspergillus* species (Tapia and Herrera-Saldana, 1989).

New strains might mean that less of the product will be required to be added to the feed, or perhaps might extend the dietary range where they are effective. The second generation of fungal probiotics will therefore presumably do much the same as the present products. The potential third generation is more exciting. Many benefits to rumen fermentation that were previously regarded as unrealistic are now being considered feasible as a by-product of recombinant DNA work with rumen bacteria (Teather, 1985; Forsberg *et al.*, 1986; Hespell, 1987 Russell and Wilson, 1988). It has been argued that even when rumen bacteria have been modified to express a new desired activity, natural selection will militate against the successful *in vivo* establishment of these organisms (Russell and Wilson, 1988). While there may be means of circumventing this problem (Wallace, 1989), it seems logical that the new activities should instead be introduced into yeasts or fungi, which already overcome the negative selection problem by being added conveniently to the rumen at each meal.

We look forward to the development of these new strains, which will hopefully mean that the rumen microbiologist, instead of following the nutritionist in an explanatory mode, as has been the role for so long, will instead lead advances in ruminant nutrition in years to come.

REFERENCES

AAFCO (1986) Official publication of the Association of American Feed Control Officials Incorporated.

Adams, D.C., Galyean, M.L., Kiesling, H.E., Wallace, J.D. and Finker, M.D. (1981) Influence of viable yeast culture, sodium bicarbonate and monensin on liquid dilution rate, rumen fermentation and feedlot performance of growing lambs and digestibility in lambs. *J. Anim. Sci.*, **53**, 780–9.

Aldrovandi, V., Ballarini, G., Caleffi, F. and Monetti, P.G. (1984) Use of *Lactobacillus acidophilus* in rearing veal calves. *Obiet. Document. Vet.*, **5**, 51–4.

Allison, B.C. and McCraw, R.L. (1989) Efficacy of Vita-ferm formula for stocker calves. *Anim. Sci. Newsletter A&T and NC State Uni.*, November pp. 12–18.

Arambel, M.J. and Tung, R.S (1987) Evaluation of *Saccharomyces cerevisiae* growth in the rumen ecosystem. *19th Biennial Conf. on Rumen Function, Chicago, Illinois.* p.29

Arambel, M.J., Wiedmeier, R.D. and Walters, J.L. (1987) Influence of donor animal adaptation to added yeast culture and/or *Aspergillus oryzae* fermentation extract on *in vitro* rumen fermentation. *Nutr. Rep. Int.*, **35**, 433–6.

Arambel, M.J. and Kent, B.A. (1988) Effect of yeast culture on milk production response and apparent nutrient digestibility in early lactating dairy cows. *J. Dairy Sci.*, **71** (Suppl. 1), 220.

Arambel, M.J., Kent, B.A. and Walters, J.L. (1990) Effect of graded levels of dried fermentation extract and cellulase enzyme on fermentation characteristics of nonlactating holstein heifers. *J. Dairy Sci.*, **73** (Suppl. 1), 236.

Bara, M. and Kmet, V. (1987) Effect of pectinase on rumen fermentation in sheep and lambs. *Arch. Anim. Nutr.*, **7/8**, 643–9.

Barrow, P.A., Brooker, B.E., Fuller, R. and Newport, M.J. (1980) The attachment of bacteria to the gastric epithelium of the pig and its importance in the microecology of the intestines. *J. Appl. Bacteriol.*, **48**, 147–54.

Bechman, T.J., Chambers, J.V. and Cunningham, M.D. (1977) Influence of *Lactobacillus acidophilus* on performance of young dairy calves. *J. Dairy Sci.*, **60** (Suppl. 1), 74.

Beeman, K. (1985) The effect of *Lactobacillus* spp. on convalescing calves. *Agri-practice*, **6**, 8–10.

Beharka, A.A., Nagaraja, T.G. and Morrill, J.L. (1990) Ruminal microbial and metabolic development in young calves fed calf starter supplemented with *Aspergillus oryzae* extract. *J. Dairy Sci.*, **73** (Suppl. 1), 220.

Bergen, W.G. and Bates, D.B. (1984) Ionophores: their effect on production efficiency and mode of action. *J. Anim. Sci.*, **58**, 1465–83.

Boing, J.T.P. (1983). Enzyme production. in *Industrial Microbiology*, 4th edn (G.Reed, Ed), pp. 685–689. Avi Publishing Co. Inc., Westport, Connecticut.

Bonaldi, G., Buratto, L., Darsie, G. *et al.* (1986). Use of *Saccharomyces cerevisiae* and *Lactobacillus acidophilus* in veal calves. *Obiet. Document. Vet.*, **7**, 49–53.

Bruce, B.B., Gilliland, S.E., Bush, L.J. and Staley, T.E. (1979) Influence of feeding cattle cells of *Lactobacillus acidophilus* on faecal flora of young dairy calves. *Oklahoma Anim. Sci. Res. Rep.*, 207.

Bryant, M.P. (1973) Nutritional requirements of the predominant rumen celluollytic bacteria *Fed. Proc.*, **32**, 1809–13.

Campos, M.R., Herrera-Saldana, R., Viniegas, G.G. and Diaz, C.M. (1990) The effect of *Aspergillus niger* and *Aspergillus oryzae* (Amaferm) as probiotics on in situ digestibility of a high fibre diet. *J. Dairy Sci.*, **73** (Suppl. 1), 133.

Carter, H.E. and Phillips, G.E. (1944) The nutritive value of yeast proteins. *Fed. Proc.*, **3**, 123–8.

Chademana, I. and Offer, N.W. (1990) The effect of dietary inclusion of yeast culture on digestion in the sheep. *Anim. Prod.*, **50**, 483–9.

Chase, L.E. (1989) Yeast in dairy cattle feeding programs. *Georgia Feed Conf.*, pp. 121–6.

Collins, E.A. and McCrea, C.T. (1978) Monensin sodium toxicity in cattle. *Vet. Record*, **103**, 386.

Coleman, G.S. (1985) The cellulase content of 15 species of entodiniomorphid protozoa. *J. Agric. Sci., Camb.*, **104**, 349–60.

Cushnie, G.H., Richardson, A.J. and Sharman, G.A.M. (1981) Procedures and equipment for the production and rearing of gnotobiotic lambs. *Lab. Anim.*, **15**, 199–204.

Dawson, K.A. (1987). Mode of action of the yeast culture, Yea-Sacc, in the rumen: a natural fermentation modifier, in *Biotechnology in the Feed Industry* (ed. T.P. Lyons), Alltech Technical Publications, Nicholasville, Kentucky, pp. 119–25.

Dawson, K.A. (1990) Designing the yeast culture of tomorrow: mode of action of yeast culture for ruminants and non-ruminants, in *Biotechnology in the Feed Industry* (ed. T.P. Lyons), Alltech Technical Publications, Nicholasville, Kentucky, pp. 59–78.

Dawson, K.A., Newman, K.E. and Boling, J.A. (1990) Effects of microbial supplements containing yeast and lactobacilli on roughage-fed ruminal microbial activities. *J. Anim. Sci.*, **68**, 3392–8.

Deaville, E.R. and Galbraith, H. (1990) Effect of dietary yeast and protein supplementation on growth and fibre characteristics of British angora goats. *Anim. Prod.*, **51**, 565.

Dehority, B.A. and Orpin, C.G. (1988) Development of, and natural fluctuations in, rumen microbial populations, in *The Rumen Microbial Ecosystem* (ed. P.N. Hobson), Elsevier, London, pp. 151–83.

Demeyer, D. and van Nevel, C.J. (1975) Methanogenesis, an integrated part of carbohydrate fermentation and its control, in *Digestion and Metabolism in the Ruminant* (eds I.W. McDonald and A.C.I. Warner), The University of New England Publishing Unit, Armidale, Australia, pp. 366–82.

Demeyer, D.I. and van Nevel, C.J. (1979) Effect of defaunation on the metabolism of rumen micro-organisms. *Brit. J. Nutr.*, **42**, 515–24.

Eckles, C.H. and Williams, V.M. (1925) Yeast as a supplementary feed for lactating cows. *J. Dairy Sci.*, **8**, 89–93.

Edwards, I.E., Mutsvangwa, T., Topps, J.H. and Paterson, G.F.M. (1990) The effect of supplemental yeast culture (Yeasacc) on patterns of rumen fermentation and growth performance of intensively fed bulls. *Anim. Prod.*, **51**, 579.

Ellinger, D.K., Muller, L.D. and Glantz, P.J. (1978) Influence of feeding fermented colostrum and *Lactobacillus acidophilus* on faecal flora and selected blood parameters of young dairy calves. *J. Dairy Sci.*, **61** (Suppl. 1), 126.

Fallon, R.J. and Harte, F.J. (1987) The effect of yeast culture inclusion in the concentrate diet on calf performance. *J. Dairy Sci.*, **70** (Suppl. 1), 119.

Firkins, J.L., Weiss, W.P., Eastridge, M.L. and Hull, B.L. (1990) Effects of feeding fungal culture extract and animal-vegetable fat on degradation of hemicellulose and on ruminal bacterial growth in heifers. *J. Dairy Sci.*, **73**, 1812–22.

Fondevila, M., Newbold, C.J., Hotten, P.M. and Orskov, E.R. (1990) A note on the effect of *Aspergillus oryzae* fermentation extract on the rumen fermentation of sheep fed straw. *Anim. Prod.*, **51**, 422–5.

Fonty, G., Gouet, P., Jouany, J.P. and Senaud, J. (1987) Establishment of the microflora and anaerobic fungi in the rumen of lambs. *J. Gen. Microbiol.*, **133**, 1835–43.

Forsberg, C.W., Crosby, B. and Thomas, D.Y. (1986) Potential for manipulation of the rumen fermentation through the use of recombinant DNA techniques. *J. Anim. Sci.*, **63**, 310–25.

Fox, S.M. (1988) Probiotics: intestinal inoculants for production animals. *Veterinary Medicine*, August 1988, pp. 806–30.

Frumholtz, P.P., Newbold, C.J. and Wallace, R.J. (1989) Influence of *Aspergillus oryzae* fermentation extract on the fermentation of a basal ration in the rumen simulation technique (Rusitec). *J. Agric. Sci., Camb.*, **113**, 169–72.

Fuller, R. (1989) Probiotics in man and animals. *J. Appl. Bacteriol.*, **66**, 365–78.

Gilliland, S.E., Bruce, B.B., Bush. L.J. and Staley, T.E. (1980) Comparison of two strains of *Lactobacillus acidophilus* as dietary adjuncts for young calves. *J. Dairy. Sci.*, **63**, 964–72.

Gomez-Alarcon, R.A.G. (1988) Effects of *Aspergillus oryzae* on milk production, feed utilization and rumen fermentation in lactating dairy cows. Ph.D. Dissertation, Univ. of Arizona, Tucson.

Gomez-Alarcon, R., Huber, J.T., Higginbotham, G.E. and Wanderley, R. (1986) Effect of feeding a culture of *A. oryzae* on rumen and total tract digestion in dairy cows. *J. Dairy Sci.*, **69** (Suppl. 1), 291.

Gomez-Alarcon, R., Dudas, C. and Huber, J.T. (1987) Effect of *Aspergillus* (Amaferm) and yeast on feed utilization by Holstein cows. *J. Dairy Sci.*, **70** (Suppl. 1), 344.

Gomez-Alarcon, R., Oliveirs, J. and Huber, J.T. (1988a) Effect of *Aspergillus oryzae* on *in vitro* and *in situ* rumen fermentation. *J. Anim. Sci.*, **66** (Suppl. 1), 638.

Gomez-Alarcon, R., Wiersma, F., Ammon, D. *et al.* (1988b) Effect of feeding Amaferm (*Aspergillus oryzae* extract) to cows in early lactation on milk yields and related parameters. *J. Dairy Sci.*, **71** (Suppl. 1), 302.

Gomez-Alarcon, R., Dudas, C. and Huber, J.T. (1990) Influence of cultures of *Aspergillus oryzae* on rumen and total tract digestibility of dietary components. *J. Dairy Sci.*, **73**, 703–10.

Goodrich, R.D., Garrett, J.E., Gast, D.R. *et al.* (1984) Influence of monensin on the performance of cattle. *J. Anim. Sci.*, **58**, 1484–98.

Gray, W.R. and Ryan, J.P. (1989) Effect of yeast culture on volatile fatty acid levels in ovine rumen fluid incubated with oats, barley and hay. *Biochem. Soc. Trans.*, **17**, 390–2.

Gray, W.R. and Ryan, J.P. (1990) Two distinct modes of action, namely *ad initio* and *ad finem* of the yeast culture Yeasacc on ruminal fermentation in sheep. *Biochem. Soc. Trans.*, **18**, 349–50.

Guard, C.L. (1986) Pathophysiology of diarrhoea due to *Escherichia coli* in neonatal calves, in *Aspects of Digestive Physiology in Ruminants* (eds A. Dobson and M.J. Dobson), Comstock Publishing Associates, Ithaca, pp. 203–16.

Günther, K.D. (1989) Yeast culture's success under German dairy conditions, in *Biotechnology in the Feed Industry* (ed. T.P. Lyons), Alltech Technical Publications, Nicholasville, Kentucky, pp. 39–46.

Harris, B., van Horn, H.H., Manookian, K.E., *et al.* (1983) Sugar cane silage, sodium hydroxide and steam pressure treated sugar cane bagasse, corn silage, cottonseed hulls, sodium bicarbonate and *Aspergillus oryzae* product in complete rations for lactating cows. *J. Dairy Sci.*, **66**, 1474–85.

Harris, B. and Lobo, R. (1988) Feeding yeast culture to lactating dairy cows. *J. Dairy Sci.*, **77** (Suppl. 1), 276.

Harrison, G.A., Hemken, R.W., Dawson, K.A. *et al.* (1988) Influence of addition of yeast culture supplement to diets of lactating cows on ruminal fermentation and microbial populations. *J. Dairy Sci.*, **71**, 2967–75.

Hatch, R.C., Thomas, R.O. and Thayne, W.V. (1973) Effect of adding *Bacillus acidophilus* to milk fed baby calves. *J. Dairy Sci.*, **56**, 682.

Hefel, H. (1980) Cited by Fox, S.M. (1988), Probiotics: intestinal inoculants for production animals. *Vet. Med.*, August, 806–30.

Hespell, R.B. (1987) Biotechnology and modifications of the rumen microbial ecosystem. *Proc. Nutr. Soc.*, **46**, 407–13.

Hobson, P.N. and Wallace, R.J. (1982) Microbial ecology and activities in the rumen: part II. *CRC Crit. Rev. Microbiol.*, **9**, 253–320.

Holdeman, L.V., Cato, E.P. and Moore, W.E.C. (1977) *Anaerobe Laboratory Manual*, 4th edn, Virginia Polytechnic Institute and State University, Virginia.

Hollister, A.G., Cheeke, P.R., Robinson, K.L. and Patton, N.M. (1990) Effects of dietary probiotics and acidifiers on performance of weanling rabbits. *J. Appl. Rabbit Res.*, **13**, 6–9.

Huber, J.T. (1987) Fungal additives for lactating dairy cows. *22nd Pacific Northwest Animal Nutrition Conf.*

Huber, J.T., Higginbotham, G.E. and Ware, D. (1985) Influence of feeding vitaferm, containing an enzyme-producing culture from *Aspergillus oryzae*, on performance of lactating cows. *J. Dairy Sci.*, **68** (Suppl. 1), 30.

Huber, J.T., Higginbotham, G.E. and Gomez-Alarcon, R. (1986) Influence of feeding an *A. oryzae* culture during hot weather on performance of lactating dairy cows. *J. Dairy Sci.*, **69** (Suppl. 1), 290.

Hughes, J. (1988) The effect of a high strength yeast culture in the diet of early weaned calves. *Anim. Prod.*, **46**, 526.

Hungate, R.E. (1966) *The Rumen and Its Microbes*, Academic Press, New York.

Ingledew, W.M. and Jones, G.A. (1982) The fate of live brewers yeast slurry in bovine rumen fluid. *J. Inst. Brewing*, **88**, 18–20.

James, R.E., McGilliard, M.L. and Hartman, D.A. (1984) Calf mortality in Virginia dairy herd improvement herds. *J. Dairy Sci.*, **67**, 908–18.

Jaquette, R.D., Dennis, R.J. and Coalson, J.A. (1988) Effect of feeding viable *Lactobacillus acidophilus* (BT1386) on performance of lactating dairy cows. *J. Dairy Sci.*, **71** (Suppl. 1), 219.

Jonsson, E. (1985) Lactobacilli as probiotics to pigs and calves. A microbiological approach. *Raprt Sveriges Lantbruksuniverst Institutionen for Husdjurens Utrording och Vard*, No 148.

Jonsson, E. and Olsson, I. (1985) The effect on performance, health and faecal microflora of feeding *Lactobacillus* strains to neonatal calves. *Swed. J. Agric. Res.*, **15**, 71–6.

Jordan, R.M. and Johnston, L. (1990) Yeast culture supplemented lamb diets. *J. Anim. Sci.*, **68** (Suppl. 1), 493.

Judkins, M.B. and Stobart, R.H. (1988) Influence of two levels of enzyme preparation on ruminal fermentation, particulate and fluid passage and cell wall digestion in wether lambs consuming either a 10% or 25% grain diet. *J. Anim. Sci.*, **66**, 1010–15.

Karney, T.L., Johnson, M.C. and Ray, B. (1986) Changes in the lactobacilli and coliform populations in the intestinal tract of calves from birth to weaning. *J. Anim. Sci.*, **63** (Suppl. 1), 446–7.

Kay, R.M. and Poole, P. (1988) *Lactobacillus acidophilus* as a probiotic in the rearing of purchased calves. *Anim. Prod.*, **46**, 525.

Kellems, R.O., Johnston, N.P., Wallentine, M.V. *et al.*, (1987) Effect of feeding *Aspergillus oryzae* on performance of cows during early lactation. *J. Dairy Sci.*, **70** (Suppl. 1), 345.

Kmet, V., Jonecova, Z., Bomba, A. and Nemcova, R. (1988) Stimulation of the development of rumen microflora in calves with microbial preparations. *Zivocisna Vyoba*, **33**, 23–6.

Lee, R.W. and Botts, R.L. (1988) Evaluation of single oral dosing and continuous feeding of *Streptococcus faecium* M74 (Syntabac) on performance of incoming feedlot cattle. *J. Anim. Sci.*, **66** (Suppl. 1), 460.

Lund, A. (1974) Yeasts and moulds in the bovine rumen. *J. Gen. Microbiol.*, **81**, 453–62.

Lund, A. (1980) Yeasts in the rumen contents of musk oxen. *J. Gen. Microbiol.*, **121**, 273–6.

Lyons, T.P. (1987) Yeast culture. A natural feed additive for all species. *The Feed Compounder*, August 1987, pp. 20–3.

Lyons, T.P. (1990). Biotechnology: risk or revolution – A review of Alltech's position. in *Biotechnology in the Feed Industry* (ed. T.P. Lyons), pp. 1–35. Alltech Technical Publications, Nicholasville, Kentucky.

Maeng, W.J., Kim, C.W. and Shin, H.T. (1987) Effect of a lactic acid bacteria concentrate (*Streptococcus faecium* Cernelle 68) on growth rate and scouring prevention in dairy calves. *Korean J. Dairy Sci.*, **9**, 204–10.

Malcolm, K.J. and Kiesling, H.E. (1986) Influence of live yeast culture and whole cottonseed on milk production and butterfat production of lactating dairy cows and rumen fermentation of steers. *J. Anim. Sci.*, **63** (Suppl. 1) 473.

Marounek, M. and Wallace, R.J. (1984) Influence of culture E_h on the growth and metabolism of the rumen bacteria *Selenomonas ruminantium*, *Bacteroides amylophilus*, *Bacteroides succinogenes* and *Streptococcus bovis* in batch culture. *J. Gen. Microbiol.*, **130**, 223–9.

Martin, S.A., Nisbet, D.J. and Dean, R.G. (1989) Influence of a commercial yeast supplement on the *in vitro* ruminal fermentation. *Nutr. Rep. Int.*, **40**, 395–403.

Martin, S.A. and Nisbet, D.J. (1990) Effects of *Aspergillus oryzae* fermentation extract on fermentation of amino acids, bermudagrass and starch by mixed ruminal microorganisms *in vitro*. *J. Anim. Sci.*, **68**, 2142–9.

Massip, A. and Pondant, A. (1975) Facteurs associés à la morbidité chez les veaux résultats d'une enquete realisée en ferme. *Ann. Med. Vet.*, **119**, 491–534.

McKain, N., Newbold, C.J. and Wallace, R.J. (1991) Combined effects of *Aspergillus oryzae* fermentation extract (Amaferm) and monensin on fermentation in the rumen simulation technique (Rusitec). *Anim. Prod.* **52**, 593. (Abstract)

Mitchell, I.G. and Kenworthy, R. (1976) Investigations on a metabolite from *Lactobacillus bulgaricus* which neutralizes the effect of enterotoxin from *Escherichia coli* pathogenic for pigs. *J. Appl. Bacteriol.*, **41**, 163–74.

Mould, F.L., Orskov, E.R. and Mann, S.O. (1983) Associative effects of mixed feeds. I. Effects of type and level of supplementation and the influence of the rumen fluid pH on cellulolysis *in vivo* and dry matter digestion of various roughages. *Anim. Feed Sci. Technol.*, **10**, 15–30.

Muralidhara, K.S., Sheggeby, G.G., Elliker, P.R. *et al.* (1977) Effect of feeding lactobacilli on the coliform and lactobacillus flora of intestinal tissue and faeces from piglets. *J. Food. Protect.* **40**, 288–95.

NCYC (1990) *National Collection of Yeast Cultures, Catalogue of Cultures 1990*, AFRC Institute of Food Research, Norwich.

Newbold, C.J. (1990) Probiotics as feed additives in ruminant diets. *51st Minnesota Nutrition Conf.*, pp. 102–18.

Newbold, C.J., Williams, P.E.V., McKain, N. *et al.* (1990) The effects of yeast culture on yeast numbers and fermentation in the rumen of sheep. *Proc. Nutr. Soc.*, **49**, 47A.

Newbold, C.J., Brock, R. and Wallace, R.J. (1991) Influence of autoclaved or irradiated *Aspergillus oryzae* fermentation extract on the fermentation in the rumen simulation technique (Rusitec). *J. Agric. Sci., Camb.*, **116**, 159–62.

Newman, K.E., Dawson, K.A. and Morehead, M.C. (1990) Antagonistic activities of bacterial isolates from probiotic feed supplements upon pathogenic and rumen bacteria. *J. Anim. Sci.*, **68** (Suppl. 1), 505.

Nisbet, D.J. and Martin, S.A. (1990a) Effect of dicarboxylic acids and *Aspergileus oryzae* fermentation extract on lactate uptake by the ruminal bacterium *Selenomonas ruminantium*. *Appl. Environ. Microbiol.*, **56**, 3515–18.

Nisbet, D.J. and Martin, S.A. (1990b) Effect of Yea-sacc[1026] on lactate utilisation by the ruminal bacterium *Selenomonas ruminantium*, in *Biotechnology in the Feed Industry* (ed. T.P. Lyons), Alltech Technical Publications, Nicholasville, Kentucky, pp. 463–7.

Oellermann, S.O., Arambel, M.J., Kent, B.A. and Walters, J.L. (1990) Effect of graded amounts of *Aspergillus oryzae* fermentation extract on ruminal characteristics and nutrient digestibility in cattle. *J. Dairy Sci.*, **73**, 2412–16.

Offer, N.W. (1990) Maximising fiber digestion in the rumen: the role of yeast culture, in *Biotechnology in the Feed Industry* (ed. T.P. Lyons), Alltech Technical Publications, Nicholasville, Kentucky, pp. 79–96.

Orpin, C.G. and Joblin, K.N. (1988) The rumen anaerobic fungi, in *The Rumen and its Microbes* (ed. P.N. Hobson), Elsevier, London, pp. 129–50.

Orskov, E.R. (1982) *Protein Nutrition in Ruminants*, Academic Press, London.

Ozawa, K., Yabu-uchi, K., Yamanaka, K., *et al.* (1983) Effect of *Streptoccus faecalis* BIO-4R on intestinal flora of weanling piglets and calves. *Appl. Environ. Microbiol.*, **45**, 1513–18.

Pagan, J.D. (1990a) Manipulation of hindgut fermentation in horses, in *Biotechnology in the Feed Industry* (ed. T.P. Lyons), Alltech Technical Publications, Nicholasville, Kentucky, pp. 97–112.

Pagan, J.D. (1990b) Effect of yeast culture supplementation on nutrient digestibility in mature horses. *J. Anim. Sci.*, **68** (Suppl. 1), 371.

Phillips, W.A. and von Tungeln, D.L. (1985) The effect of yeast culture on the poststress performance of feeder calves. *Nutr. Rep. Int.*, **32**, 287–94.

Pond, K.R. and Goode, L. (1985) *The Evaluation of Probios for Wean-Stressed Lambs*, Pioneer Hi-Bred International, Des Moines, Iowa.

Reiter, B. (1978) Antimicrobial systems in milk. *J. Dairy Res.*, **45**, 131–47.

Reiter, B., Marshall, V.M. and Phillips, S.M. (1980) The antibiotic activity of the lactoperoxidase-thiocyanate-hydrogen peroxide system in the calf abomasum. *Res. Vet. Sci.*, **28**, 116–22.

Rose, A.H. (1987) Yeast, a microorganism for all species: a theoretical look at its mode of action, in *Biotechnology in the Feed Industry* (ed. T.P. Lyons), Alltech Technical Publications, Nicholasville, Kentucky, pp. 113–18.

Roy, J.H.B. (1980) *The Calf*, 4th edn., Butterworth, London.

Russell, J.B. and Martin, S.A. (1984) Effects of various methane inhibitors on the fermentation of amino acids by mixed rumen microorganisms *in vitro*. *J. Anim. Sci.*, **59**, 1329–38.

Russell, J.B. and Strobel, H.J. (1989) Mini-review: the effect of ionophores on ruminal fermentation. *Appl. Environ. Microbiol.*, **55**, 1–6.

Russell, J.B. and Wilson, D.B. (1988) Potential opportunities and problems for genetically altered rumen microorganisms. *J. Nutr.*, **118**, 271–9.

Ryan, J.P. (1990) The suggestion that the yeast cell *Saccharomyces cerevisiae* may absorb sufficient hydrogen ions to increase ruminal pH is untenable. *Biochem. Soc. Trans.*, **18**, 250–351.

Ryan, J.P. and Gray, W.R. (1989) Effect of high strength yeast culture *ab initio* utilizes a residual source of volatile fatty acids in strained ruminal fluid from hay fed sheep. *Biochem. Soc. Trans.*, **17**, 392–3.

Savage, D.C. (1977) Microbial ecology of the gastrointestinal tract. *Ann. Rev. Microbiol.*, **31**, 107–33.

Sievert, S.J. and Shaver, R.D. (1990) Effects of nonfiber carbohydrate and *Aspergillus oryzae* fermentation extract on intake, milk production and digestion in lactating dairy cows. *J. Dairy Sci.*, **73** (Suppl. 1), 23.

Slyter, L.L. (1976) Influence of acidosis on rumen function. *J. Anim. Sci.*, **43**, 910–29.

Smith, H.W. (1965) The development of the flora of the alimentary tract in young animals. *J. Pathol. Bacteriol*, **90**, 495–513.

Smith, H.W. (1971) The bacteriology of the alimentary tract of domestic animals suffering from *E. coli* infection. *Ann. NY Acad. Sci.*, **176**, 110–25.

Stewart, C.S. (1977) Factors affecting the cellulolytic activity of rumen contents. *Appl. Environ. Microbiol.*, **33**, 497–502.

Stuart, J.C. (1978) An outbreak of monensin poisoning in adult turkeys. *Vet. Records*, **102**, 303–4.

Svozil, B., Danek, P., Kumprecht, I. and Zobac, P. (1987) The efficiency of different contents of the bacterium *Streptococcus faecium* M-74 in the nutrition of calves. *Zivocisna Vyroba*, **32**, 265–71.

Tapia, M.N. and Herrera-Saldama, R. (1989) The effect of four fungal compounds as probiotics on *n vitro* dry matter disappearence of different feedstuffs. *J. Anim. Sci.*, **67** (Suppl. 1), 521 (Abstract)

Teather, R.M. (1985) Application of gene manipulation to rumen microflora. *Can. J. Anim. Sci.*, **65**, 563–74.

Terrell, S.S., Vaughn, J.B., Bertrand, W.E. *et al.*(1984). On trial: an enzyme producing supplement for dogs. *Vet. Med. Small Anim. Clin.*, **79**, 1367–77.

Theodorou, M.K., Beever, D.E., Haines, M.J. and Brooks, A. (1990) The effect of a fungal probiotic on intake and performance of early weaned calves. *Anim. Prod.*, **53**, 577.

Thivend, P., Toullec, R. and Guillotea, P. (1979) Digestive adaptation in the preruminant, in *Digestive Physiology and Metabolism in Ruminants* (eds Y. Ruckebusch and P. Thivend), MTP Press, London, pp. 561–85.

Tournut, J. (1989) Applications of probiotics to animal husbandry. *Rev. Sci. Tech. Off. Int. Epiz.*, **8**, 551–6.

Umberger, S.H., Notter, D.R., Webb, K.E. and McClure, W.H. (1989) Evaluation of a *Lactobacillus* inoculant on feedlot lamb performance. *Anim. Sci. Res. Rep., Virginia Agric. Exp. Stat., Virginia Poly.*, **8**, 40–5.

Underdahl, N.R., Torres-Medina, A. and Doster, A.R.(1982) Effect of *Streptococcus faecium* C63 in control of *Escherichia coli* induced diarrhoea in gnotobiotic pigs. *Amer. J. Vet. Res.*, **43**, 2227–32.

United States Department of Agriculture (1988) *Agricultural Statistics 1988*, US Government Printing Office, Washington.

van Horn, H.H., Olayiwole, M.B., Wilcox, C.J. *et al.* (1976) Effects of housing, feeding management, and ration formulation on calf growth and feed intake. *J. Dairy Sci.*, **59**, 924–30.

van Horn, H.H., Harris, B., Taylor, M.J. *et al.* (1984) By-product feeds for lactating dairy cows: effects of cottonseed hulls, sunflower hulls, corrugated paper, peanut hulls, sugercane bagasse and whole cottonseed with additives of fat, sodium bicarbonate and *Aspergillus oryzae* product on milk production. *J. Dairy Sci.*, **67**, 2922–38.

Wadstrom, T. (1984) *Streptococcus faecium* M74 in control of diarrhoea induced by a human enterotoxigenic *Escherichia coli* strain in an infant rabbit model. *Zent. Bakteriol. Mikriobiol. Hygiene.* **A257**, 357–63.

Wallace, R.J. (1989) Identification of slowly metabolized sugars and sugar derivatives that could be used to establish new or modified microbial species in the rumen. *Curr. Microbiol.*, **19**, 271–4.

Wallace, R.J. and McPherson, C.A. (1987) Factors affecting the rate of breakdown of bacterial protein in rumen fluid. *Brit. J. Nutr.* **58**, 313–23.

Wallentine, M.V., Johnston, N.P., Andrus, D. *et al.* (1986) The effect of feeding an *Aspergillus oryzae* culture-vitamin mix on the performance of lactating dairy cows during periods of heat stress. *J. Dairy Sci.*, **69** (Suppl. 1), 294.

Wanderley, R.C., Huber, J.T., Theurer, C.B. and Poore, M. (1985) Ruminal digestion of protein and fibre in duodenally cannulated cows treated with Vitaferm. *J. Dairy Sci.*, **68** (Suppl. 1), 31.

Wanderley, R., Al-Dehneh, A., Theurer, C.B. *et al.* (1987) Fibre disappearance and nitrogen flow from the rumen of cows fed differing amounts of grain with or without fungal addition. *J. Dairy Sci.*, **70** (Suppl. 1), 180.

Ware, D.R., Read, P.L. and Manfredi, E.T. (1988) Lactation performance of two large dairy herds fed *Lactobacillus acidophilus* strain BT 1386 in a switchback experiment. *J. Dairy Sci.*, **71** (Suppl. 1), 219.

Warner, R.G., Bernholdt, A.F., Grippen, C.H. and Loosli, J.K. (1953) The influence of diet on the development of the ruminant stomach. *J. Dairy Sci.*, **36**, 599.

Warner, R.G., Grippen, C.H., Flatt, W.P. and Loosli, J.K. (1955) Further studies on the influence of diet on the development of the ruminant stomach. *J. Dairy Sci.*, **38**, 605.

Wells, B. and Mason, T.R. (1976) Yeast culture and probiotic for feedlot lambs. *J. Anim. Sci.*, **43** (Suppl. 1), 337.

Whitlock, R.H., White, N.A., Rowland, G.N. and Plue, R. (1979) Monensin toxicosis in horses: clinical manifestations. *Proc. Amer. Assoc. Equine Practitioners*, **23**, 473–86.

Wiedmeier, R.D. (1989) Optimizing the utilization of low quality forages through supplementation and chemical treatment. *Utah Beef Cattle Field Day*, **9**, 17–21.

Wiedmeier, R.D., Arambel, M.J. and Walters, J.L. (1987) Effect of yeast culture and *Aspergillus oryzae* fermentation extract on ruminal characteristics and nutrient digestibility. *J. Dairy Sci.*, **70**, 2063–8.

Williams, A.G. and Coleman, G.S. (1988) The rumen protozoa, in *The Rumen and its Microbes* (ed. P.N. Hobson), Elsevier, London, pp. 77–128.

Williams, P.E.V. (1988) Understanding the biochemical mode of action of yeast culture, in *Biotechnology in the Feed Industry* (ed. T.P. Lyons), Alltech Technical Publications, Nicholasville, Kentucky, pp. 79–100.

Williams, P.E.V. (1989) The mode of action of yeast culture in ruminant diets: a review of the effect on rumen fermentation patterns. In *Biotechnology in the Feed Industry* (ed. T.P. Lyons), Alltech Technical Publications, Nicholasville, Kentucky, pp. 65–84.

Williams, P.E.V. and Newbold, C.J. (1990) Rumen probiosis: The effects of novel microorganisms on rumen fermentation and ruminant productivity, in *Recent Advances in Animal Nutrition 1990* (eds W. Haresign and D.J.A. Cole), Butterworth, London, pp. 211–27.

Williams, P.E.V., Walker, A. and MacRae, J.C. (1990) Rumen probiosis: The effects of addition of yeast culture (viable yeast (*Saccharomyces cerevisiae*) plus growth medium) on duodenal protein flow in wether sheep. *Proc. Nutr. Soc.*, **49**, 128A.

Williams, P.E.V., Tait, C.A.G., Innes, G.M. and Newbold, C.J. (1991) Effects of the inclusion of yeast culture (*Saccharomyces cerevisiae* plus growth medium) in the diet of dairy cows on milk yield and forage degradation and fermentation patterns in the rumen of sheep and steers. *J. Anim. Sci.* **69**, 3016–26.

Wolter, R., Henry, N., Jacquot, L., *et al.* (1987) Probiotics in animal nutrition. Experimental study of their efficiency in rats and beef calves. *Rec. Med. Vet.*, **163**, 1131–8.

Wren, B. (1987) Probiotics – fact or fiction. *Large Anim. Vet.*, Nov/Dec., 28–30.

Youanes, Y.D. and Herdt, T.H. (1987) Changes in small intestine morphology and flora associated with decreased energy digestibility in calves with naturally occurring diarrhoea. *Amer. J. Vet. Res.*, **48**, 719–25.

Ziolecka, A., Osinska, Z. and Ziolecki, A. (1984) The effect of stabilized rumen extract on growth and development of calves. 1. Liveweight gain and efficiency of feed utilization. *Z. Tierphysiol. Tierrernahrg. Futtermit.*, **51**, 13–20.

Ziolecki, A. and Briggs, C.A.E. (1961) The microflora of the rumen of the young calf. II. Source, nature and development. *J. Appl. Bacteriol.*, **24**, 148–63.

Ziolecki, A., Kwiatkowska, E. and Laskowska, H. (1984) The effect of stabilized rumen extract on growth and development of calves. 2. Digestive activity in the rumen and development of microflora in the rumen and faeces. *Z. Tierphysiol. Tierrernahrg. Futtermit.*, **51**, 20–31.

Chapter Thirteen

Probiotics for humans

BARRY R. GOLDIN AND SHERWOOD L. GORBACH

13.1 INTRODUCTION

This chapter will review the current status of probiotics in the area of human health and nutrition, including some reference of the use of probiotics in foods. The development of new strains of bacteria that can either improve current medical use or be used for medical problems in which probiotics have not been previously considered applicable or effective will be discussed. In this regard a review of both selective procedures and bioengineering techniques will be discussed.

The chapter will be divided into several broad categories. The first section will review the colonization and composition of the intestinal microflora. This aspect is being reviewed to put into context the environment in which probiotics operate and the difficulties and limitations of probiotic usage in this complex environment. The second section will review the list type of probiotics available and review the previous and current use of probiotics in human health and nutrition. The effectiveness of bacterial supplements in a medical context will also be evaluated.

The third section will discuss the properties a probiotic requires for survival, growth and physiological impact in the human gastrointestinal tract. Based on these considerations the fourth section will discuss the future development of probiotics with specific characteristics that can be useful in the context of human medical problems, nutrition and maintenance of good health. A review of selective criteria and bioengineering prospects will be discussed in the context of producing organisms that could serve specific purposes.

13.2 COLONIZATION OF THE GASTROINTESTINAL TRACT

Colonization of the gastrointestinal tract of newborn infants occurs within a few days of birth (Haenel, 1970). The type of delivery, dietary constituents and gestational age influence the colonization pattern. The initial period of bacterial colonization in the colon takes place over approximately a two-week period. During this period the bacterial colonization is similar for formula- and breast-fed infants. *Escherichia coli* and *Streptococcus* are almost always the first organism to appear, at concentrations between 10^8 and 10^{10} organisms per gram of faeces (Grutte *et al.*, 1965). Several anaerobic organisms, namely *Bifidobacterium, Clostridium* and *Bacteroides*, often then take up residence in the gastrointestinal tract. In breast-fed infants a major decrease in the bacterial populations of *E. coli* and *Streptococcus* then occurs, as well as partial or complete disappearance of *Clostridium* and *Bacteroides* resulting in predominance of *Bifidobacterium*. In formula-fed infants these reductions or disappearances do not take place, resulting in a more complex flora (Haenel and Benedig, 1975; Ellis-Pegler *et al.*, 1975; Rotimi and Duerden, 1981; Stark and Lee, 1982). The relatively simple flora of the breast-fed baby remains until dietary supplementation occurs. Upon introduction of other foods to the diet of the breast-fed infant there is a return of *E. coli, Streptococcus* and *Clostridium* to the faeces. The differences between breast-fed and formula-fed infants disappear. There is then a transition period which continues into the second year of life in which the intestinal flora evolves to resemble that of the adult.

13.2.1 Composition and distribution of microflora

There are more then 400 bacterial types that have been identified in human faeces (Finegold *et al.*, 1974; Moore and Holderman, 1974). Their findings indicate the intestinal microflora constitute a complex ecosystem. The most prevalent anaerobic bacteria are *Bacteroides, Clostridium, Bifidobacterium, Eubacterium, Peptococcus, Fusobacterium* and *Peptostreptococcus*. The bacteria of the human gastrointestinal tract are predominately strict anaerobes and outnumber facultative bacteria by a factor of 10^2 to 10^4.

The stomach of healthy individuals is populated by very low numbers of bacteria. The saliva from the oral cavity is a source of gastric bacteria, however, the highly acidic gastric juice destroys most of these organisms (Drasar and Hill, 1974). Gram-positive facultative bacteria such as lactobacilli and streptococci are most commonly isolated from gastric juice. The number of microorganisms isolated in the stomach is between 10^1 and 10^2 ml^{-1} of gastric juice.

The upper small intestine in healthy humans is also sparsely populated with organisms similar to those found in the stomach. However, the counts rise to 10^2 to 10^4 organism ml^{-1} of contents. The ileum and caecum are heavily populated with microorganisms and the composition is much more complex. The most common organisms found in the lower small intestine are those that have been listed above as most prevalent in the gastrointestinal tract (Drasar *et al.*, 1969). The total bacterial counts vary between 10^4 and 10^8 ml^{-1} of ileal contents. Within the colon the bacterial concentration rises dramatically and is between 10^{11} and 10^{12} ml^{-1} of faecal material. The composition of the colonic flora, as would be expected, comprises the bacteria listed above as most prevalent in the gastrointestinal tract.

13.2.2 Metabolic activities of microflora

The metabolic capacity of the colonic bacteria is extremely diverse (Scheline, 1973, 1980). Any compound taken orally or any substance entering the intestine *via* the biliary tract, *via* the blood, or by secretion directly into the lumen is a potential substrate for bacterial transformation. This concept will be discussed in more detail later in this chapter.

13.3 CURRENT USE OF PROBIOTICS

Table 13.1 presents a list of organisms that are orally ingested by humans either in fermented foods or as food additives or as bacterial preparations. This is a relatively comprehensive list derived from the authors' search of the literature, but is by no means complete. It is not to be implied that most or all of these organisms are beneficial and in some cases the organisms primarily function to preserve food. The list is included to provide a reference for the type of organisms that have been used for various purposes by the human population.

13.4 NUTRITIONAL BENEFITS OF PROBIOTICS

The nutritional benefits of probiotics have been most extensively investigated in regard to the fermentation of food products with lactobacilli. The general parameters that have been studied are the effect of this fermentation on the quantity, availability, digestibility and assimilability of nutrients. Alm (1982a) reported that, as a result of bacterial proteolysis, yoghurt has higher levels of free amino acids as compared to milk. During the process of yoghurt production an increase in protein digestibility has been noted (Breslaw and Kleyn, 1973). Animal studies have revealed that a number of different fermented products, such as, acidophilus milk, *Lactobacillus lactis* fermented milk, kefir, yoghurt and cultured buttermilk caused increased growth and feed efficiency in several different animal models when compared to unfermented dairy products (Hargrove and Alford, 1978; McDonough *et al.*, 1982; Friend and Shahani, 1984; Broussalian and Westhoff, 1983).

Table 13.1 Organisms used as probiotics in the food and agricultural industry.

Organism	Comment
Saccharomyces boulardii	Non-pathogenic yeast used for treatment of diarrhoea
Lactobacillus acidophilus	As a supplement in dairy products and used for fermentations; numerous health claims
L. plantarum	In dairy products, pickled vegetables and silage
Lactobacillus GG	In yoghurt and whey drink; numerous health claims
L. casei subsp. *rhamnosus*	In dairy products and silage
L. brevis	In dairy products and silage
L. delbrueckii subsp. *bulgaricus*	For the production of yoghurt; health claims have been made
Streptococcus salivariusr subsp. thermophilus	For the production of yoghurt
Bifidobacterium bifidum	Component of new dairy products and in preparation for new born; health claims
Bif. infants	Similar to *Bif. bifidum*
Enterococcus *faecium*	Being introduced in certain health products; health claims
Lactococcus lactis subsp. *lactis* and *cremoris*	Used in production of buttermilk and certain cheeses

When experiments were performed in which fractions of yoghurt were recombined with analogous control milk fractions, a heat-labile substance having a molecular weight of greater than 20,000 appeared to be involved in the increased food efficiency properties observed when yoghurt is fed to animals (Hargrove and Alford, 1980).

Evidence has been presented indicating the bioavailability of calcium, zinc, iron, manganese, copper and phosphorus is increased in yoghurt when compared to milk (McDonough *et al.*, 1983). Those findings are supported by the observation that the growth rate differential between rats fed yoghurt and milk can be reduced by supplementation of the milk with minerals, especially iron, zinc, manganese and copper (McDonough *et al.*, 1983).

Lactobacilli require B vitamins for growth; however, these organisms are also capable of synthesizing certain vitamins. Studies have demonstrated an increase in riboflavin and niacin in yoghurt (Alm, 1982a; Deeth and Tamine, 1981), B_6 pantothenic in cheddar cheese, and B_{12} in cottage cheese (Deeth and Tamine, 1981) and folic acid in a variety of products including yoghurt, bifidus milk and kefir (Shahani and Chandan, 1979). Cultured dairy products contain higher levels of B vitamins, compared to similar products prepared by direct acidification (Deeth and Tamine, 1981). Riboflavin and thiamin have also been shown to increase during the preparation of *Lactobacillus* fermented products (Rajalakshmi and Vanaji, 1967).

These studies when taken together indicate that fermented dairy products such as yoghurt contain higher levels of specific vitamins and minerals and protein and carbohydrate in a more digestible form. The changes noted resulted in better growth of laboratory animals fed yoghurt when compared with unfermented dairy products.

13.5 THERAPEUTIC BENEFITS OF PROBIOTICS

Fermented dairy products have been reported to be effective in the treatment of a number of disorders, including diarrhoea, constipation, colitis, pathogenic recolonization of the intestinal tract, flatulence, gastric acidity, gastroenteritis, hypercholesterolaemia, hepatic encephoclopathy and carcinogenesis. The data that have been generated supporting these claims will be discussed in this section.

13.5.1 Carcinogenesis

Fermented dairy products or starter cultures used by the dairy industry have been shown to either inhibit chemically induced colon tumours or transplantable tumour lines in rodents.

Evidence for the anti-tumour effect of *L. acidophilus* comes from studies conducted in our laboratory. Oral supplementation of the diet with viable *L. acidophilus* of human origin, which is bile resistant, caused a significant decline in three different faecal bacterial enzymes (Goldin and Gorbach, 1977, 1984a). The decline in faecal enzyme activity was noted in humans and rats. The bacterial enzymes that were affected included β-glucuronidase, azoreductase and nitroreductase. These enzymes catalyse the conversion of procarcinogens to proximal carcinogens in the large bowel.

These studies were extended in an animal model of colon cancer induced by the chemical carcinogen, 1,2-dimethylhydrazine (DMH). The activation of DMH to a proximate carcinogen occurs in the large intestine, and the bacterial enzyme β-glucuronidase is involved in this process. Suppression of this enzyme might reduce DMH activation and subsequent tumour formation. In our experiments DMH-treated animals were given *L. acidophilus* in powdered form and compared with controls (Goldin and Gorbach, 1980). At 20 weeks, 40% of the *L. acidophilus* treated animals had colon tumours, compared to 77% of the controls ($p<0.02$). At 36 weeks, however, 73% of the *L. acidophilus* treated animals and 83% of the controls had colon tumours. These studies show that the addition of this strain of *Lactobacillus* to the diet can delay colon tumour formation by prolonging the induction of the latent period.

In another study from our laboratory (Goldin and Gorbach, 1984b), we have shown that oral *L. acidophilus* supplementation to the diet in rats lowers the amount of carcinogenic amines excreted in the faeces after feeding procarcinogen precursors to these animals.

This corroborates our earlier findings that lactobacilli suppress the metabolic activity of the colonic microflora and in this manner may reduce the formation of carcinogens in the large intestine.

Reddy *et al.* (1983) and Friend *et al.* (1982) investigated the inhibitory effect of yoghurt on the proliferation of Ehrlich ascites tumour cells in male Swiss mice. They observed that feeding yoghurt resulted in 28–35% reduction in the number of tumour cells when compared to control groups fed milk. DNA synthesis in the tumour line of animals receiving yoghurt was only 75% of that found in animals fed a commercial diet. Shahani *et al.* (1983) reported that feeding milk and colostrum fermented with *L. acidophilus* resulted in 16–41% reduction in tumour proliferation. Bogdanov *et al.* (1978) observed that *L. bulgaricus* possessed a potent antitumour activity. They isolated three glycopeptides which had biological activity against sarcoma-180 and solid Ehrlich ascites tumour.

The evidence for antitumour activity or inhibition of carcinogenesis is not extensive, however, under certain defined conditions specific strains of bacteria and fermented dairy products may have some protective role.

13.5.2 Treatment of intestinal infections

For many years there have been claims that various probiotics or fermented dairy products can be used to prevent or treat a variety of gastrointestinal disorders. The results of experiments in this area are not consistent. In this section some of the positive studies will be cited.

Gotz *et al.* (1979) observed that when a commercial powder preparation containing lactobacillus was given prophylactically, a reduction in the rate of diarrhoea from 14% to 0% was noted in patients receiving ampicillin for treatment of a variety of disorders. The reduction in the rate of diarrhoea was statistically significant, Hitchins *et al.* (1985) treated rats that had been infected intragastrically with *Salmonella enteritidis* with either yoghurt or milk. The yoghurt-fed rats had a significant reduction in the rate of mortality and body weight gain associated with salmonellosis. Zychowicz *et al.* (1974, 1975) treated 31 Polish children suffering from salmonella- or shigella-induced diarrhoea with acidophilus milk. Forty-three percent of the children with a salmonella infection and 67% of the children with shigella became asymptomatic shortly after receiving the acidophilus milk. Continued administration of acidophilus milk resulted in elimination of diarrhoeal symptoms in all of the children. Niv *et al.* (1963) treated infantile diarrhoea with yoghurt. Forty-five children hospitalized for diarrhoea were randomized with respect to aetiologies and severity of diarrhoea and age into two groups. One group was treated with neomycin/kaolin/pectin and the other groups received only yoghurt. The group receiving yoghurt had 50% shorter duration of disease compared to the other group. Clements *et al.* (1983) found that only one of two lots of a *Lactobacillus* preparation reduced the volume and duration of diarrhoea in a patient population that had been given neomycin. Another batch had no effect on the neomycin-induced diarrhoea. This type of result is not surprising given the variability in the viability of many probiotic commercial preparations. Alm (1983) reported that the carrier state for salmonella was shortened in infected individuals who drank 500 ml of acidophilus milk.

In summation there are studies indicating that probiotics may be useful in treating gastrointestinal disorders.

13.5.3 Hepatic encephalopathy

Hepatic encephalopathy is a neurologic disorder associated with liver failure and elevated blood ammonia levels. Ammonia is normally produced in the intestine from urea by the action of bacterial urease. The ammonia is absorbed and detoxified in the liver. In patients suffering from liver failure, the detoxification mechanism is impaired and the ammonia levels rise in the circulating blood. Treatment with *L. acidophilus* and/or neomycin results in a lowering of faecal urease and a decrease in ammonia levels of patients with hepatic encephalopathy (Macbeth *et al.*, 1965; Read *et al.*, 1966). These metabolic events are associated with improvement in the clinical status of the patient.

13.5.4 Treatment of hypercholesterolaemia

Elevated plasma cholesterol has been positively associated with a higher rate of coronary heart disease. There are currently a number of drugs available to lower plasma cholesterol, however, non-pharmacological agents that could accomplish this reduction would be preferable. In this context there have been a number of studies using probiotics in different forms to determine if they have cholesterol-lowering properties.

In studies with laboratory rats consumption of milk fermented with *Streptococcus salivarius* subsp. *thermophilus* and mixed with animal feed caused lower blood cholesterol values when compared with rats fed skim milk supplemented feed (Rao *et al.*, 1981); this effect could not be ascribed to simple redistribution of cholesterol between the plasma, liver and other body pools (Pulsani and Rao, 1983). In other animal studies, consumption of a diet of *L. acidophilus* fermented milk supplemented with a commercial feed lowered serum cholesterol levels in weanling rats (Grunewald, 1982). Gilliland *et al.* (1985) studied pigs which were fed a high cholesterol diet in order to induce elevated serum cholesterol. When these animals were simultaneously fed *L. acidophilus* the rise in serum cholesterol was inhibited. *In vitro* studies with this *L. acidophilus* strain showed that it assimilated cholesterol from the culture medium, leading the authors to speculate that the organism bound cholesterol in the intestinal lumen, thereby reducing its absorption into the bloodstream. Thakur and Jha (1981) reported that yoghurt and milk reduced dietary cholesterol-induced hypercholesterolaemia in rabbits; yoghurt had a greater effect than milk in these studies.

Hepner *et al.* (1979) showed that dietary supplementation with yoghurt for one week, decreased serum cholesterol in human volunteers. Fermented milk has been credited with counteracting hypercholesterolaemia among the Masai tribe, who have a high consumption of this fermented product (Mann and Spoering, 1974).

Although there are positive animal and human studies attributing plasma cholesterol-lowering properties to probiotics, there are no convincing double-blind human intervention trials to support these preliminary studies.

13.5.5 The ability of probiotics to increase the tolerance for lactose in lactase-deficient individuals

Lactose, a milk carbohydrate, causes intestinal distress when consumed by people with a deficiency in the intestinal mucosal enzyme β-galactosidase (lactase). Such individuals must restrict their intake of milk and dairy products. During fermentation, lactobacilli produce lactase which hydrolyses milk lactose to glucose and galactose. Up to 50% of lactose in acidophilus milk and yoghurt is utilized by this enzyme during fermentation. The production of lactase increases steadily during yoghurt fermentation. Lactose also can be released from bacterial cultures during digestion, indicating that bacterial lactase is present in the intestine of persons consuming yoghurt (Kilara and Shahani, 1976).

There have been several additional studies that further supported the findings cited above. Alm (1982a) monitored the rise in serum glucose (derived from lactose) in control subjects and in lactose-intolerant subjects who were given a 500 ml dose of milk or yoghurt. The lactose-intolerant subjects had a much lower rise in serum glucose compared with controls, indicating reduced lactose utilization. When yoghurt was given to lactose-intolerant subjects, their serum glucose levels approached the level of controls given milk, showing increased utilization of lactose from yoghurt. Moreover, lactose-intolerant subjects reported abdominal distress when drinking milk, but they did not report any such problems when given yoghurt.

Kim and Gilliland (1983) noted that administration of fermented acidophilus milk markedly decreased the breath level of hydrogen in lactose-intolerant subjects when compared with the high hydrogen breath levels when taking unfermented milk. Kolars *et al.* (1984) used the breath hydrogen measurement to demonstrate that lactose intolerant subjects given 18 g of lactose in yoghurt had only about one-third as much hydrogen excretion as those given a similar amount of lactose in milk or water. Analysis of duodenal aspirates indicated significant lactase activity in the intestine 1 h after ingestion of yoghurt.

These results indicate markedly improved intestinal absorption of lactose in yoghurt compared with unfermented milk. In conjunction with the improved absorption of lactose, less diarrhoea, flatulence and abdominal distension was noted in individuals eating yoghurt when compared with those ingesting a similar lactose load from milk or water solutions. The evidence for a positive role for probiotics in permitting lactose-intolerant individuals to consume dietary products containing lactose is among the most convincing of all the health claims for probiotics.

Savaiano *et al.* (1984) demonstrated that yoghurt was far superior to cultured milk (buttermilk) or pasteurized yoghurt in enhancing the digestion of lactose. Pasteurization of yoghurt reduced lactase levels 10-fold and reduced bacterial cell counts 100-fold. Interestingly, these investigators found that eight of nine subjects fed cultured milk experienced gastrointestinal distress, whereas all subjects fed pasteurized yoghurt were symptom-free, even though the amount of malabsorbed lactose was similar. McDonough *et al.* (1987) using breath-hydrogen measurements studied 14 lactose-intolerant subjects. The subjects were fed yoghurt, heated yoghurt, yoghurt plus lactose, heated yoghurt plus lactose, sweet acidophilus milk and sonicated sweet acidophilus milk. The study showed that the greatest increase in breath hydrogen occurred when subjects drank milk or sweet acidophilus milk (18.7 and 18.3 ppm, respectively). Intermediate increases in breath hydrogen were noted for heated yoghurt (15.7 ppm), yoghurt plus lactose, i.e. lactose added to bring the concentration to the pre-fermented level (14.1 ppm), and sonicated sweet acidophilus milk (12.4 ppm). The lowest incremental increase in breath hydrogen was noted for heat-treated yoghurt plus lactase (8.1 ppm) and for yoghurt (5.9 ppm). These investigators concluded that both the reduction of lactose during fermentation and the presence of indigenous bacterial lactase are responsible for the increased ability to tolerate lactose in yoghurt; de Wit *et al.* (1988) recently confirmed these findings in studies that showed a 4- to 7-fold greater increase in breath hydrogen in subjects drinking milk or eating heated yoghurt compared to individuals eating regular yoghurt.

13.6 MORE RECENT DEVELOPMENTS IN THE AREA OF PROBIOTICS AND HEALTH

13.6.1 *Saccharomyces boulardii*

Surawicz *et al.* (1989a) have reported that a non-pathogenic yeast, *Saccharomyces boulardii*, is effective in the treatment of antibiotic-induced diarrhoea (AAD) in hospitalized patients. They found in a prospective hospital based study that 22% of patients receiving a placebo developed diarrhoea after antibiotic treatment compared to 9.5% of patients receiving *Sac. boulardii*. Treatment with *Sac. boulardii* also reduced the incidence of recurring diarrhoea caused by *Clostridium difficile* toxin (Surawicz *et al.* 1989b; Kimmey *et al.*, 1990). This condition is also initiated by antibiotic use.

13.6.2 *Lactobacillus* GG

In our laboratory we have isolated a strain of *Lactobacillus* designated as GG, which colonizes the human intestinal tract and adheres more tightly to human intestinal and buccal cells than other strains of *Lactobacillus* or *Streptococcus* (Goldin and Gorbach, 1987). When individuals consume the GG strain either in a lyophilized form or as a fermented milk product, it is consistently in faeces in concentrations detected $>10^6$ g^{-1} (Gorbach, 1990). The GG strain elaborates an anti-microbial substance which has a broad spectrum of activity against a range of bacteria, including *C. difficile*.

Based on these observations *Lactobacillus* GG has been administered to date to 32 patients (Gorbach *et al.*, 1987; Gorbach, 1990; Bennett *et al.* 1991) suffering from antibiotic-induced *C. difficile* toxin caused recurring diarrhoea. All 32 patients improved symptomatically after receiving *Lactobacillus* GG. Twenty-seven (84% of total) patients were cured by a single course of treatment based on a minimum follow-up of two months. The other five patients relapsed within 10 days following *Lactobacillus* GG treatment and were retreated. Three of these patients were cured and the other two were lost to follow-up. These results indicate the effectiveness of a probiotic in a defined medical setting.

There have been other studies indicating that *Lactobacillus* GG can be effective in treating diarrhoea. Siitonen *et al.*(1990) reported a study in which subjects were given erythromycin with either pasteurized yoghurt or a *Lactobacillus* GG fermented yoghurt. The subjects receiving the *Lactobacillus* GG yoghurt had a lower incidence of diarrhoea. In another study investigators found (Oksanen *et al.*, 1990) that administration of a lyophilized preparation of *Lactobacillus* GG to Finnish travellers to Turkey resulted in a 39% protection rate from diarrhoea for one group of travellers to a specific resort area in Turkey.

13.7 PROPERTIES REQUIRED FOR PROBIOTICS TO BE EFFECTIVE IN NUTRITIONAL AND THERAPEUTIC SETTINGS

A probiotic can be used exogenously or endogenously to enhance nutritional status and/or the health of the host. In the case of exogenous use microorganisms are most commonly used to ferment various foods and by this process can preserve and make nutrients more bioavailable. In addition, microorganisms can metabolize sugars, such as lactose in yoghurt, making this food more acceptable for consumption by individuals suffering from lactose intolerance. However, the most interesting properties that probiotics acting exogenously can have is the production of substances that may be antibiotics, anticarcinogens or have other pharmaceutical properties. The properties required for exogenously derived benefits from probiotics are the ability to grow in the food or the media in which the organism is placed, and the specific metabolic properties which result in the potential beneficial effects stated above. The selection of organisms that can be helpful therapeutically and nutritionally would be based on specific properties that are desired. This can be achieved by either classical biological selection techniques or genetic engineering (see later section).

Probiotics that are ingested by the host and exert their favourable properties by virtue of residing in the gastrointestinal tract have to have certain properties in order to exert an effect. A discussion of these properties is presented below.

13.7.1 Survivability

Microorganisms introduced orally have to, at least, transiently survive in the stomach and small intestine. Although this appears to be a rather minimal requirement, many bacteria including the yoghurt-producing bacteria *L. delbrueckii* subsp. *bulgaricus* and *S. salivarius* subsp. *thermophilus* often do not survive to reach the lower small intestine (Robins-Browne and Levine, 1981). The reason for this appears to be the low pH of the stomach. In fasting individuals the pH of the stomach is between 1.0 and 2.0 and most microorganisms, including lactobacilli, can only survive from 30 seconds to several minutes under these conditions (Conway *et al.*, 1987). Therefore, in order for a probiotic to be effective, even the selection of strains that can survive in acid at pH 3.0 for sometime would have to be introduced in a buffered system such as milk, yoghurt or food.

13.7.2 Bile resistance

The small intestine and colon contain relatively high concentrations of bile acids which can inhibit growth or kill many bacteria. Therefore, it is essential that probiotics, to be effective, should be able to grow in 0.15–0.30% oxgall-agar.

13.7.3 Adhesion

It is not clear if adhesion to the intestinal epithelium is essential for the persistence of a probiotic in the human intestinal tract. However, adhesion seems to be a property that enhances long-term survival. The ability of microorganisms to adhere to epithelial cells is to a large extent species specific, although this may be relative. Strains of toxigenic *E. coli*, for example, have been shown to selectively bind to pig intestinal cells and other toxigenic strains bind only to human intestinal cells (Deneke *et al.* 1983). Kleeman and Klaenhammer (1982) have studied the binding capacity of 32 human *L. acidophilus* isolates to human foetal intestinal cells. When calcium was added to the binding assay all the strains adhered. In the absence of calcium only four strains showed adhesion properties. Therefore, it appears that non-specific and specific binding mechanisms are both involved. Conway and Kjelleberg

(1989) have demonstrated that binding of *L. fermentum* to rat stomach squamous epithelium is mediated by a specific protein with a MW of 12 000 to 13 000. This may be the case for specific binding of many different microorganisms and would explain the species specificity.

13.7.4 Antimicrobial production

As indicated previously, the intestinal microflora is a complex ecosystem. Introducing new organisms into this highly competitive environment is difficult. Thus organisms that can produce a product or products that will inhibit the growth or kill existing organisms in the intestinal milieu have a distinct advantage. There have been numerous reports (Axelsson *et al*, 1987; Silva *et al*, 1987; Fernandes *et al*, 1987) of organisms currently being used as probiotics producing such substances. Table 13.2 lists some of these antimicrobial-producing organisms.

Table 13.2. – Antimicrobial substances.

Probiotic	*Compound*
Lactobacillus GG	Wide spectrum antibiotic
L. acidophilus	Acidolin
	Acidophilin
	Bacteriocin
	Lactocidin
L. bulgaricus	Bulgarican
L. plantarum	Lactolin
L. brevis	Lactobacillin
	Lactobrevin
L. reuteri	Reuterin

13.8 FUTURE DEVELOPMENT OF PROBIOTICS FOR HUMAN USE

13.8.1 Survivability and adhesion

In the previous section the requirement for probiotic survival and adhesion to the epithelium in the gastrointestinal tract was discussed. However, in the past, quantitative evaluation of these parameters has not been employed in selecting organism that may be beneficial to humans. Screening of organisms for their ability to survive in the human gastrointestinal tract is not difficult. The selection of human bacterial isolates will enhance the possibility of finding organisms that will survive. The isolates can then be tested by administering orally between 10^9 and 10^{11} viable organism in a single dose with an appropriate buffering agent and the bacterial counts of the specific organism are then measured in the faeces over a several week period. This technique is most successful if the natural flora does not contain the organism being tested or only in small numbers. The first question of transient survival can be established in 48–96 hours. The evaluation of the ability of the organism to permanently establish in the gastrointestinal tract, by proliferation, can be established by continuous appearance in the faeces over several weeks to several months. The faecea counts should exceed 10^6 g^{-1} of faeces. The application of this screen for selecting probiotics should be encouraged in the future.

There are several tests for determining if a prospective probiotic can bind to intestinal epithelium. Radiolabelling the microorganism with an amino acid and then counting for adhering radioactivity in either ileal cells recovered from ileostroma effluent or from buccal cells obtained by gently scraping the inside of the cheek are effective methods. Good adhesion properties should enhance the possibility of long-term survival of the organism in the intestinal tract by countering the peristaltic action of the intestine.

13.8.2 Antimicrobial production

The growth media filtrates and sonicates from the bacterial cells of prospective probiotics should be tested for bactericidal and bacteriostatic activity in well-plates against a wide variety of pathogens (Silva *et al.*, 1987). The ability of probiotics to establish in the gastrointestinal tract will be enhanced by their ability to eliminate competitors.

13.8.3 Immunological enhancement

In recent years there have been several reports indicating that lactobacilli used in dairy products can enhance the immune response of the host (Perdigón *et al.*, 1989; de Simone *et al.*, 1989a, b; Goulet *et al.*, 1989). Organisms that have been identified as having this property are *Bifidobacterium longum, L. acidophilus, L. casei* subsp. *rhamnosum* and *L. helviticus*. In the future, prospective probiotics, in the appropriate settings (anticancer, infection resistance, etc.), should be tested for enhancement of the immunological response. The measurements that should be considered are lymphocyte proliferation, interleukin 1, 2 and 6, tumour necrosis factor, prostaglandin E production and serum total protein, albumin, globulin and gamma interferon, as well as a delayed skin hypersensitivity test.

13.9 FUTURE APPLICATIONS OF PROBIOTICS

13.9.1 Production of nutrients

Sands and Hankin (1974, 1976) have selected for high lysine-producing mutants of *L. plantarum*, *L. acidophilus* and *L. delbrueckii* subsp. *bulgaricus*. The organisms were grown in the presence of increasing concentrations of the lysine analogue S-2-aminoethyl-cysteine. Organisms were selected for their ability to grow in presence of the analogue which competes with lysine and is therefore normally a lysine inhibitor. Confirmation of higher lysine production was verified by using as a marker *Leuconostoc mesenteroides* which requires lysine for growth. The high lysine-producing mutant of *L. plantarum* has been used to increase the lysine content of silage and the *L. acidophilus* and *L. delbrueckii* subsp. *bulgaricus* mutants have been used to ferment soybean milk to make a high lysine-containing yoghurt. These investigators have suggested the same technique can be used with mutant yeast employed in brewing beer. A lysine-excreting yeast has been described that will grow on hydrocarbons (Leavitt and Ryan, 1974).

This technique for selecting lysine-producing mutants can be expanded to give mutants that produce other essential amino acids and vitamins. The probiotics can then be introduced in foods by fermentation or by supplementation. This would enrich the food in these amino acids or vitamins. In addition, if a probiotic is designed to have a long-term survival capacity in the gastrointestinal tract, an *in situ* production mechanism can be established there. This would be particularly effective for vitamins, such as, B_{12}, K and E, which would be required by the host in relatively small amounts. These techniques would be very valuable in underdeveloped nations whose population often suffers from severe amino acid and vitamin deficiencies.

13.9.2 Cholesterol lowering

As mentioned earlier, Gilliland *el al.* (1985) have shown that dietary elevation of plasma cholesterol levels in pigs can be prevented by introduction of a *L. acidophilus* strain that is bile resistant and assimilates cholesterol. The strain used was a pig isolate. This technique can be used for humans by selecting human isolates of *L. acidophilus* or *Bif. bifidum* or other probiotics that have long-term survival in the gastrointestinal tract and at the same time efficiently assimilate cholesterol. The effect on plasma cholesterol concentration would be dependent on the extent of assimilation, as well as other factors difficult to assess without actually assessing the effect in human studies.

13.9.3 Anticancer probiotics

Studies in our laboratory have indicated that diet and antibiotics (Goldin and Gorbach, 1981, 1984a) can lower the generation of carcinogens in the colon and reduce chemically induced tumours. These effects appear to be mediated through the intestinal microflora. Additional studies have shown that the introduction of *L. acidophilus* into the diet lowered the incidence of chemically induced colon tumours in rats (Goldin and Gorbach, 1980). A possible mechanism for these anticancer effects relies on inhibiting intestinal bacterial enzymes that convert procarcinogens to more proximal carcinogens. This technique can be expanded in the future by testing probiotics for their ability to inhibit the growth of organisms normally found in the flora that have high activities of enzymes such as β-glucuronidase, nitroreductase, azoreductase and β-glycosidase or the capability for nitrososation. The ability of probiotics to deactivate faecal mutagens (Bruce *et al.*, 1977) can also be a marker used to introduce organisms that lower cancer risk.

13.9.4 Enhancement of immunological properties

As discussed previously, there are now numerous immunological parameters that can be tested in humans non-invasively. This affords the opportunity to conduct human studies on the immunological enhancing properties of probiotics that can survive over extended periods of time in the intestinal tract. A human trial has been initiated in our laboratory with *Lactobacillus* GG. Probiotics can be developed – either by selection or genetic engineering – that have specific antigenic sites which could then be placed in the host to evoke the appropriate immunological response to protect against various bacterial intestinal infections. There are a great many possibilities to explore and it is an area for future research.

13.9.5 Production of hormones and other agents

The possibility of genetically engineering strains of bacteria that can produce substances such as insulin, androgens, oestrogens, growth hormone or cholesterol-lowering compounds, just to mention, a few is intriguing. The ability to produce *in situ* over a long period of time drugs or hormones that are constantly required by individuals suffering from various diseases (i.e. diabetes and hyperchlosteraemia) is of particular interest. There are problems to this approach, however; e.g. control of production and contamination of normal individuals with the organism. The first problem may be solved by establishing the maximum achievable production level of the organism in the gut and thereby setting an upper limit on dose. The contamination problem may be more difficult to solve, although antibiotic sensitivity can be introduced into the strains, so that the organism could be rapidly eliminated if a normal individual is infected with a specifically designed probiotic. This idea may have too many regulatory problems associated with it; however, it is still something that may have potential use in human disease regulation.

13.10 TECHNIQUES FOR PROBIOTIC MODIFICATION

13.10.1 Natural selection and mutagenesis

The selection of desirable probiotics can in many cases be achieved by selecting organisms under pressure, such as by depleting a required nutrient or introducing an inhibitor of a required substance and taking advantage of the spontaneous forward mutation rate to select the appropriate organisms. The introduction of mutagens, such as N-methyl nitroso guanidine (MNNG), etc., can enhance the efficiency of this process. Simple screening of organisms for properties, such as cholesterol assimilation, or epithelial adhesion, although tedious, can also be employed to achieve the desired result.

13.10.2 Genetic engineering

The use of genetic engineering to produce new strains of *Streptococcus* and *Lactobacillus* is now being performed (de Vos, 1987; Chassy, 1987). The technique involves plasmid transfer using different gene transfer systems, such as transduction, conjugation, fusion, transaction and transformation (de Vos, 1987). Plasmid-free strains have been developed as recipients. Most of these studies have used antibiotic resistance as a marker for transfer of genetic information. An example of the success of this technique has been the ability to convert a lactose-deficient *S. sanguis* to lac+ using a conjunctive lactose plasmid containing all three specific enzymes required for the lac+ property from a lac+ *S. lactose*.

The possibilities in this area are just being realized and the ability to produce organisms that have specific proteases, such as serine protease, which is required for starting caseine degradation or a number of other proteases, is now actively being pursued. The extension of this work to have probiotics produce antibiotics, hormones, etc., is being actively pursued by pharmaceutical companies with noted success in exogenous production of insulin and many other biologically important compounds. The challenge in the future will be to produce organisms that can thrive in the host and at the same time produce nutrients and other substances that promote the health of the human host.

REFERENCES

Alm, L. (1982a) Effect of fermentation on lactose, glucose and galactose content in milk and suitability of fermented milk products for lactose individuals. *J. Dairy Sci.*, **65**, 346–52.

Alm, L. (1982b) Effect of fermentation on B-vitamin content of milk in Sweden. *J. Dairy Sci.*, **65**, 353–9.

Alm, L. (1983) The effect of *Lactobacillus acidophilus* administration upon the survival of *Salmonella* in randomly selected human carriers. Prog. *Food Nutr. Sci.*, **7**, 13–17.

Axelsson, L., Chung, T.C., Dobrogosz, W.J. and Lindgren, L.E. (1987) Discovery of a new antimicrobial substance produced by *Lactobacillus reuteri*. *FEMS Microbial Rev.* **46** (3), 65.

Bennett, R.G., Gorbach, S.L., Goldin, B.R. *et al.* (1991), Treatment of relapsing *Clostridium difficile* diarrhea with lyophilyzed *Lactobacillus* GG. Submitted for publication.

Bogdanov, I.G., Velichkov, V.T. and Gurvich, A.I. (1978) Antitumor action of glycopeptide from the cell wall of *Lactobacillus bulgaricus*. *Bull. Exp. Biol. Med.*, **84**, 1750–3.

Breslaw, E.S. and Kleyn, D.H. (1973) *In vitro* digestibility of protein in yogurt at various stages of processing. *J. Food Sci.*, **38**, 1016–21.

Broussalian, J. and Westhoff, D. (1983) Influence of lactose concentration on milk and yogurt growth of rats. *J. Dairy Sci.*, **66**, 438–43.

Bruce, E.R., Varhese, A.J., Furrer, R. and Land, P.C. (1977) A mutagen in the feces of normal humans, in *Origins of Human Cancer*, Cold Spring Harbor Lab, New York.

Chassy, B.M. (1987) Prospects for the genetic manipulation of lactobacilli. *FEMS Microbiol. Rev.*, **46** (3), 197–312.

Clements, M.L., Levine, M.M. and Ristaino, P.A. (1983) Exogenous lactobacili fed to man: Their fate and ability to prevent diarrheal disease. Prog. *Food Nutr.*, **7**, 29–37.

Conway, P.L., Gorbach, S.L. and Goldin, B.R. (1987) Survival of lactic acid bacteria in the human stomach and adhesion to intestinal cells. *J. Dairy Sci.*, **70**, 1–12.

Conway, P.L. and Kjelleberg, S. (1989) Protein-mediated adhesion of *Lactobacillus fermentum* strain 737 to mouse stomach squamous epithelium. *J. Gen. Microbiol.*, **135**, 1175–86.

Deeth, H.C. and Tamine, A.Y. (1981) Yogurt: nutritive and therapeutic aspects. *J. Food Protect.*, **44**, 78–86.

Deneke, C.F, McGowan, K., Thorne, G.M. and Gorbach, S.L. (1983) Attachment of entertoxigenic *Eschericha coli* to human intestinal cells. *Infect. Immun.*, **39**, 1102–6.

De Simone, C., Salvatori, B.B., Jirillo, E. *et al.* (1989a) Modulation of immune activities in humans and animals by dietary lactic acid bacteria, in *Yogurt: Nutritional and Health Properties*, National Yogurt Assoc., McLean, Virginia.

De Simone, C., Salvatori, B.B., Jirillo, E. *et al.* (1989b) Yogurt and the immune response, in *Les Laits Fermentes*, John Libbey, London.

De Vos, W.M. (1987) Gene cloning and expression in lactic streptococci. *FEMS Microbiol. Rev.* **46** (3), 281–95.

De Wit, O., Pochart, P. and Desjeux, J.F. (1988) Breath hydrogen concentration and plasma glucose, insulin and free fatty acid levels after lactose, milk, fresh or heated yogurt ingestion by healthy young adults with or without lactose malabsorption. *Nutrition*, **4**, 131–5.

Drasar, B.S. and Hill, M.J. (1974) *Human Intestinal Flora*, Academic Press, New York.

Drasar, B.S., Shiner, M. and McLeod, G.H. (1969) Studies on the intestinal flora. I. The bacterial flora of the gastrointestinal tract in health and achlorhydric persons. *Gastroenterology*, **56**, 71–9.

Ellis-Pegler, R.B., Crabtree, C. and Lambert, H.P. (1975) The faecal flora of children in the United Kingdom. *J. Hyg.* (Camb), **75**, 135–42.

Fernandes, C.F., Shahani, K.M. and Amer, M.A. (1987) Therapeutic role of dietary lactobacilli and lactobacillic fermented dairy products. *FEMS Microbiol. Rev.*, **46** (3), 343–56.

Finegold, S.M., Atterbery, H.R. and Sutter, V.L. (1974) Effect of diet on human fecal flora: Comparison of Japanese and American diets. *Amer. J. Clin. Nutr.*, **27**, 1546–9.

Friend, B.A., Farmer, R.E. and Shahani, K.M. (1982) Effect of feeding and intraperitoneal implantation of yogurt culture cells on Ehrlich ascites tumor. *Milchwiss.*, **37**, 708–10.

Friend, B.A. and Shahani, K.M. (1984) Nutritional and therapeutic aspects of lactobacilli. *J. Appl. Nutr.*, **36**, 125–53.

Gilliland, S.E., Nelson, C.R. and Maxwell, C. (1985) Assimilation of cholesterol by *Lactobacillus acidophilus*. *Appl. Environ. Microbiol.* **49**, 377–81.

Goldin, B.R. and Gorbach, S.L. (1977) Alterations in fecal microflora enzymes related to diet, age, lactobacillus supplements, and dimethylhydrazine. *Cancer*, **40**, 2421–6.

Goldin, B.R. and Gorbach, S.L. (1980) Effect of *Lactobacillus acidophilus* dietary supplementation on 1,2-dimethylhydrazine dihydrochloride-induced intestinal cancer in rats. *J. Natl Cancer Inst.*, **64**, 263–5.

Goldin, B.R. and Gorbach, S.L. (1981) The effect of antibiotics on rat intestinal tumors induced by dimethylhydrazine. *J. Natl Cancer Inst.*, **67**, 877–80.

Goldin, B.R. and Gorbach, S.L. (1984a) Effect of milk and *Lactobacillus* feeding on human intestinal bacterial enzyme activity. *Amer J. Clin. Nutr.*, **39**, 756–61.

Goldin, B.R. and Gorbach, S.L. (1984b) Alterations of the intestinal microflora by diet, oral antibiotics and *Lactobacillus*. Decreased production of free amines from aromatic nitro compounds, azo dyes and glucuronides. *J. Natl Cancer Inst.*, **73**, 689–95.

Goldin, B.R. and Gorbach, S.L. (1987) *Lactobacillus* GG: A new strain with properties favorable for survival, adhesion and antimicrobial activity in the gastrointestinal tract. *FEMS Microbiol. Rev.*, **46** (3), 72.

Gorbach, S.L., (1990) Lactic acid bacteria and human health. *Ann. Med.*, **22**, 37–41

Gorbach, S.L., Chang, T.W. and Goldin, B.R. (1987) Successful treatment of relapsing *Clostridium difficile* colitis with *Lactobacillus* GG. *Lancet*, **ii**, 1519.

Gotz, V., Romankiewicz, J.A., Moss, J. and Murray, H.W. (1979) Prophylaxis against ampicillin-associated diarrhea with a lactobacillus preparation. *Amer. J. Hosp. Pharmacol.*, **36**, 754.

Goulet, J., Saucier, L. and Maineau, S. (1989) Stimulation of the non-specific immune response of mice by fermented milks, in *Yogurt: Nutritional and Health Properties*, National Yogurt Assoc., McLean, Virginia.

Grunewald, K.K. (1982) Serum cholesterol levels in rats fed skim milk fermented by *Lactobacillus acidophilus*. *J. Food Sci.*, **47**, 2078–9.

Grutte, F.K., Horn, R. and Haenel, H. (1965) Ernahrung und biochemisch – mikrookologische Vorgange in Enddarm von Sauglingen. *Z. Kinderh.*, **93**, 28–39.

Haenel, H. (1970) Human normal and abnormal gastrointestinal flora. *Amer. J. Clin. Nutr.*, **23**, 1433–9.

Haenel, H. and Benedig, J. (1975) Intestinal flora in health and disease. *Prog. Food Nutr. Sci.*, **1**, 21–64

Hargrove, R.E. and Alford, J.A. (1978) Growth rate and feed efficiency of rats fed yogurt and other fermented milks. *J. Dairy Sci.*, **61**, 11–19.

Hargrove, R.E. and Alford, J.A. (1980) Growth response of weanling rats to heated, aged, fractionated, and chemically treated yogurts. *J. Dairy Sci.*, **63**, 1065–72.

Hepner, G., Fried, R., St Jean, S. *et al.* (1979) Hypocholestermic effect of yogurt and milk. *Amer. J. Clin. Nutr.*, **32**, 19–24.

Hitchins, A.D., Wells, P., McDonough, F.E. and Wong, N.A. (1985) Amelioration of the adverse effect of a gastrointestinal challenge with *Salmonella enteritidis* on weanling rats by a yogurt diet. *Amer. J. Clin. Nutr.*, **41**, 91–100.

Kilara, A. and Shahani, K.M. (1976) Lactose activity of cultured and acidified dairy products. *J. Dairy Sci.*, **61**, 2031–5.

Kim, H.S. and Gilliland, S.E. (1983) *Lactobacillus acidophilus* as a dietary adjunct for milk to aid lactose digestion in humans. *J. Dairy Sci.*, **66**, 959–66.

Kimmey, M.B., Elmer, G.W., Surawicz, C.M. and McFarland, L.V. (1990) Prevention of further recurrences of *Clostridium difficile* colitis with *Saccharomyces boulardii*. *Digest Dis. Sci.*, **35**, 897–901.

Kleeman, E.G. and Klaenhammer, T.R. (1982) Adherence of *Lactobacillus* species to human intestinal cells. *J. Dairy Sci.*, **65**, 2063–9.

Kolars, J.C., Levitt, M.D., Aouji, M. and Savaino, D.A. (1984) Yogurt – an autodigesting source of lactose. *New Engl. J. Med.*, **310**, 1–3.

Leavitt, R.L. and Ryan, F.X. (1974) L-Lysine decarboxylase as a specific inhibitor of growth of a species *Pichia*. *J. Gen. Microbiol.*, **80**, 311–13.

Macbeth, W.A.A.G., Kass, E.H. and McDermott, W.V. Jr (1965) Treatment of hepatic encephalopathy by alteration of intestinal flora with *Lactobacillus acidophilus*. *Lancet*, **i**, 399–403.

Mann, G.V. and Spoerig, A. (1974) Studies of a surfactant and cholesteraemia in masai. *Amer. J. Clin. Nutr.*, **27**, 464–9.

McDonough, F.E., Hitchins, A.D. and Wong, N.P. (1982) Effects of yogurt and freeze dried yogurt on growth stimulation of rats. *J. Food Sci.*, **47**, 1463–5.

McDonough, F.E. Wells, P., Wong, N.P. *et al.* (1983) Role of vitamins and minerals in growth stimulation of rats fed yogurt. *Fed. Proc.*, **42**, 556.

McDonough F.E., Hitchins, A.D., Wong, N.P. *et al.* (1987) Modification of sweet acidophilus milk to improve utlization by lactose-intolerant persons. *Amer. J. Clin. Nutr.*, **45**, 570–4.

Moore, W.E.C. and Holdeman, L.V. (1974) Human fecal flora: The normal flora of 20 Japanese-Hawaiians. *Appl. Microbiol.*, **27**, 961–79.

Niv, M., Levy, W. and Greenstein, N.M. (1963) Yogurt in the treatment of infantile diarrhea. *Clin. Pediat. (Phila)*, **2**, 407–11.

Oksanen, P.J., Salminen, S., Saxelin, M. *et al.* (1990) Prevention of travellers diarrhea by *Lactobacillus GG*. *Ann. Med.*, **22**, 53–6.

Perdigón, S., Alvarez, S., Noder de Macias, M.E. and Medici, M. (1989) Effect of lactic acid bacteria orally administered and of yoghurt on the immune system, in *Les Laits Fermentes*, John Libby, London.

Pulsani, SR. and Rao, D.R. (1983) Whole body liver and plasma cholesterol levels in rats fed thermophilnus, bulgaricus and acidophilus milks. *J, Foud Sci.*, **48**, 280–1

Rajalakshmi, R. and Vanaji, K. (1967) Chemical and biological evaluation of the effects of fermentation on the nutritive value of foods prepared from rice and grain. *Brit. J. Nutr.*, **21**, 467–73.

Rao, D.M., Chawan, C.B., and Pulusani, S.R. (1981) Influence of milk and thermophilic milk plasma cholesterol levels and hepatic cholesterogenesis in rats. *J. Food Sci.*, **46**, 1339–41.

Read, A.E., McCarthy, C.F., Heaton, K.W., and Laidlow, J. (1966) *Lactobacillus acidophilus* (ENPAC) in treatment of hepatic encephalopathy. *Brit. Med J.*, **1**, 1267–9.

Reddy, G.V., Friend, B.A, Shahani, K.M, and Farmer, R.E. (1983) Antitumor activity of yogurt components *J. Food Protect.*, **46** 8–11.

Robins-Browne, R.M. and Levine, M. (1981) The fate of ingested lactobacilli in the proximal small intestine. *Amer J. Clin. Nutr.*, **34**, 514–19.

Rotimi, V.O. and Duerden, B.I. (1981). The development of the bacterial flora in normal neonates. *J. Med. Microbiol.*, **34** 51–62.

Sands, D.C. and Hankin, L. (1974) Selecting lysine-excreting mutants of lactobacilli for used in food and feed enrichment. *Appl. Microbiol.*, **28**, 523–4.

Sands, D.C. and Hankin, L. (1976) Fortification of foods by fermentation with lysine-excreting mutants of lactobacilli. *J. Agric. Food Chem.*, **24**, 1104–6.

Savaiano, D.A., Abouelanouar, A., Smith, D.E. and Levitt, M.O.. (1984) Lactose malabsorption from yogurt, pasteurized yogurt, sweet acidophilus milk, and cultured milk in lactose-deficient individuals. *Amer J. Clin. Nutr.*, **40**, 1219–23.

Scheline, R.R. (1973) Metabolism of foreign compounds by gastrointestinal microorganisms. *Pharmocol. Rev.*, **25**, 451–523.

Scheline, R.R.(1980) Drug metabolism by the gastrointestinal microflora. *Monogr. Pharmacol. Physiol.*, **5**, 551–80.

Siitonen, S., Vapoatalo, H., Salminen, S. *et al.* (1990) Effect of *Lactobacillus* GG yoghurt in prevention of antibiotic-associated diarrhoea. *Ann. Med.*, **22**, 57–9.

Silva, M., Jacobus, N.V., Deneke, C. and Gorbach, S.L. (1987) Antimicrobial substance from a human *Lactobacillus* strain. *Antimicrob. Agents Chemother.*, **31**, 1231–3.

Shahani, K.M., and Chandan, R.C. (1979) Nutritional and healthful aspects of cultured and culture-containing dairy foods. *J. Dairy Sci.*, **62**, 1685–94.

Shahani, K.M., Friend, B.A. and Bailey, P.J. (1983), Antitumor activity of fermented colostrum and milk. *J. Food Protect.*, **46**, 385–6.

Stark, P.L. and Lee, A. (1982) The microbial ecology of the large bowel of breast-fed and formula-fed infants during the first year of life. *J. Med. Microbial.*, **15**, 189–203.

Surawicz, C.M., Elmer, G.W., Speelman, P. *et al.* (1989a) Prevention of antibiotic-associated diarrhea by *Saccharomyces boulardii*: A prospective study. *Gastroenterolgy*, **96**, 981–8.

Surawicz, C.M., McFarland, L.V., Elmer, G. and Chinn, J. (1989b) Treatment of recurrent *Clostridium difficile* colitis with vancomycin and *Saccharomyces boulardii*. *Amer. J. Gastroenterol.*, **84**, 1285–7.

Zychowicz, C., Surazynmska, A., Sietwierska, B. and Ciephinska, T. (1974) Effect of *Lactobacillus acidophilus* cultures (acidophilus milk) on the carrier state of shigella and salmonella organisms in children. *Pediatria Polska*, **49**, 997–1003.

Zychowicz, C., Kowalczyk, S. and Ciephinska, T. (1975) Results of administration of *Lactobacillus acidophilus* cultures (acidophilus milk) in an endemic focus of dysentery. *Pediatria Polska*, **50**, 429–35.

Chapter Fourteen

Problems and prospects

ROY FULLER

14.1 INTRODUCTION

It is a big step from demonstrating that the gut microflora protects against disease to the identification of the specific organisms responsible for the effect. It is obvious from the foregoing chapters that the one constant feature of the probiotic effect is its variability. Although under defined experimental conditions probiotics give significant responses, in the field the results are often unpredictable. However, the inconsistency is not as great as it at first appears. Probiotics are not a single entity and many factors can influence their performance. A previous chapter has discussed the properties used to select the currently available probiotic preparations (see Chapter 9). In this chapter we shall consider the reasons for the apparent inconsistency of the response to probiotics selected in this way and suggest ways in which the future development can work towards and ultimately ensure a product that is dependable.

14.2 FACTORS AFFECTING THE PROBIOTIC RESPONSE

The interaction of a microbial feed supplement with the host animal and its gut microflora resulting in the expression of the probiotic effect is by its nature a very complex phenomenon. Some of the factors which influence the outcome are discussed below.

14.2.1 Composition of host animal gut flora

It seems likely that probiotics act by affecting the composition of the host's intestinal microflora in some way (see Chapter 2). Thus a prerequisite for a positive response is the presence of a microbe which has an adverse effect, such as depression of growth rate. It therefore follows that if the growth-depressing organism is absent, then no growth promotion will be obtained. This type of variation occurs with other growth promotants such as antibiotics. Similarly, no probiotic response will be obtained if the probiotic organism has already been acquired naturally. The requirement for probiotic supplementation has been created by the unnatural lifestyles imposed on animals and humans. The extreme example is the chicken where the egg is removed from the influence of the mother hen and hatched in a clean incubator where it cannot acquire its protective flora because it has no contact with intestinal microorganisms from adult chickens. The same limited access to intestinal organisms operates with neonatal mammals but here the isolation is usually less complete. However, in the situation where a human baby is delivered by caesarian section into an incubator the mother's influence is minimal and changes in the gut microflora are detectable (Hall *et al.*, 1990). Currently available probiotics are often dismissed as ineffective because they are too simple and do not contain any of the strictly anaerobic bacteria which in experimental studies have been shown to be important in the development of the protective effect. However, under practical conditions the animal may only be lacking one or a few strains of the total collection which go to make up the protective flora. In the chicken it has been shown (Mead and Impey, 1987) that 48 strains are required to produce a protective effect against colonization of the gut by salmonellae. This collection contains unidentified strains of Gram-positive anaerobic cocci, Gram-positive anaerobic rods and budding bacteria as well as representatives from the genera *Escherichia*, *Streptococcus*, *Bacillus*, *Bacteroides*, *Fusobacterium*, *Lactobacillus*, *Eubacterium*, *Propionibacterium*, *Clostridium* and *Bifidobacterium*.

Although the lactobacilli alone will not increase resistance to gut colonization by salmonellae, if they are omitted from the protective flora the amount of protection is reduced. So one can envisage a situation where the animal has all but a few of the essential strains needed to produce protection. In such a situation the administration of lactic acid bacteria could improve the protection and give a positive result. Whether or not the probiotic contains the 'right' organisms depends on which ones are missing from the gut flora and this may be the basis of some of the variation observed.

14.2.2 Dosing regime

Different trials using the same probiotic may use different dosing regimes; one may be continuous and another may use a single dose. Although the minimum effective dose is not known, it has been found that the probiotic effect disappears after the cessation of dosing. Trials with rats and humans show that the probiotic effect is lost after dosing stops (Cole and Fuller, 1984; Goldin and Gorbach, 1984) and in pigs and chickens the probiotic organisms cannot be recovered from the gut 7 days after dosing (Jonsson, 1986; Votava *et al.*, 1987).

Permanent colonization of the gut by ingested organisms does not seem to be a likely outcome. Even indigenous microorganisms do not always remain permanently in the gut. Although the flora in gross terms remains constant, it can change in detail. Thus although the count of *Escherichia coli* may be the same throughout the life of an animal, the dominant serotype will change periodically (Sears and Brownlee, 1952). Similarly, in baby pigs the dominant lactobacilli as determined by plasmid profiling will change during the first 14 days of life (Tannock *et al.*, 1990). This dynamic interchange of strains within species indicates that although an introduced strain may colonize the gut, it will only remain there until it is displaced by another naturally acquired organism which is better adapted to occupy the niche.

14.2.3 Age and type of animal

The gut microflora, physiology and immune status of an animal change as the animal gets older, and it is not the same animal during the neonatal period as it is during post-weaning and in the adult state. Pollman *et al.* (1980) found that starter pigs responded better to probiotic supplementation than did growing–finishing pigs. Similarly the response of ruminants to fungal probiotics is different at different stages of lactation.

Because the flora is still in a state of flux during the neonatal period, it may be a general rule that it is easier to influence the flora during this period rather than in later life when a relatively stable flora has been established. It would seem advisable that supplementation with probiotics should start as soon after birth as possible. Changes in diet can also occur during this period. Brockett and Tannock (1981) found that dietary changes affected the appearance of the epithelium-associated lactobacillus flora in the mouse stomach. They were of the opinion that the relative amounts of palmitic and oleic acid in the diet were important in this respect. It is well known that a change occurs in the gut flora of human babies when they are weaned either on to formula milk or a solid diet (Yuhara *et al.*, 1983). Milk has been shown to contain factors which stimulate the growth of bifidobacteria (Neut *et al.*, 1984) and could have an influence on the response to probiotics containing these organisms. The suppressing effect that *Lactobacillus acidophilus* supplementation had on enzyme activity was diet dependent. No effect was found in rats fed a grain diet but a positive response was obtained with rats fed a meat diet (Goldin and Gorbach, 1977).

14.2.4 Quality assurance

The viability of a probiotic product is of paramount importance but unfortunately it cannot always be assumed to be as high as the figure quoted on the label. Gilliland and Speck (1977) examined 15 probiotic supplements which claimed to contain high counts of viable *L. acidophilus*. In the event only seven were found to contain lactobacilli and of those seven only three contained *L. acidophilus*. Since viability is not always checked before experimental trials are conducted it is possible that some of the negative results were due to the absence of a sufficient number of viable cells in the preparation being tested. In four of these preparations lactobacilli other than *L. acidophilus* were present; they were classified as *L. bulgaricus*, *L. brevis*, *L. fermentum*, *L. lactis*, *L. casei* subsp. *rhamnosus*, *L. delbrueckii* and *L. plantarum*. Unless in cases like this, the species present in the preparation is checked the result can be misleading and would certainly give a false impression when two probiotics alleged to contain *L. acidophilus* were being compared. Taxonomy is not a strong point with the probiotic producers and no microbial identification should be accepted without checking.

The other complicating factor that should be borne in mind is strain variation, which can occur within the same species. It is quite possible that two probiotics based on the same species of bacterium will give different results. Barrow *et al.* (1980) found that in their isolates from pigs there were large differences between strains of the same species in their ability to attach to pig epithelial cells. They also found large strain differences in growth rate. Differences in metabolism between strains of the same species also exist. It is therefore essential when making a comparison between two preparations with the same species to know whether the cultures used to produce the probiotic preparation originated from the same strain.

14.2.5 Type of preparation

Probiotics come in a variety of different forms and one type of preparation may be more suitable than another for a particular animal. If colonization of the gut is necessary for the probiotic response it is important to remember that bacteria show host specificity with respect to this and that what may be a good probiotic for a pig will not necessarily be effective in a chicken. The colonization host specificity is reflected in the adhesion to epithelial cells. Barrow *et al.* (1980) tested lactobacilli and streptococci isolated from a wide range of different sources and found that, with two exceptions, only those isolated from pigs adhered to pig epithelial cells.

In the past probiotic strains have been selected in a rather empirical way but recent awareness of host specificity has led some producers to carry out *in vitro* tests which detect those strains which are most likely to colonize the gut of the animal being fed. For example, tests for epithelial adhesion, growth rate, bile tolerance and resistance to low pH may be of use in selecting strains. However, it should be remembered that *in vitro* results are not always transposable to the *in vivo* situation. Probably the most suitable *in vitro* method for assessing the activity of probiotic organisms is the continuous flow culture technique. It has proved its usefulness in demonstrating changes in the microbial ecology of the rumen and its potential for the study of the flora of the monogastric should be explored more fully.

The importance of colonization may be obviated by feeding the probiotic continuously whereby a high count of the probiotic organism can be maintained in the gut without its colonization and growth. Some probiotic organisms like *Aspergillus oryzae*, which are unlikely to grow in the rumen but have an effect on its metabolic activity, must be operating in this way.

But even if probiotics are designed to be fed continuously it is important to maximize the survival in the gut and tests such as those listed above may still be relevant to the selection of the most effective strains.

14.2.6 Production methods

The behaviour of an organism in the intestinal tract can be affected by the way in which it is grown and harvested. For example, the expression of epithelial adhesion can be affected by the type of carbohydrate energy source provided for growth (Fuller, 1975) or the presence of milk in the habitat in which the adhesion is occurring (Conway *et al.*, 1987). Growth of *Enterococcus faecium* in milk enhances its protective effect against *E. coli* diarrhoea in pigs (Ushe and Nagy, 1985). The suspension of probiotic organisms in milk may enhance their ability to adhere and in consequence improve colonization of the gut. The stage of the growth cycle at which the probiotic organisms are collected may also influence the adhesion to epithelial cells. Cook *et al.*(1988) found that *L. casei* showed better adherence to uroepithelial cells when in the stationary phase than when tested in the early log phase of growth. The presence of calcium in the growth medium can affect the survival of the organism after freezing or freeze-drying, which is often an essential feature of the probiotic production process.

Some of these factors may be operating during the production of yoghurt or other fermented milks. For example, yoghurt can be produced either by fermenting milk at 42°C for 3–4 hours or by fermenting at 30°C overnight. Both processes end when the fermentation has reached a predetermined pH of about 4.2. Although both products are called yoghurt and are fermented with the same starter (*L. delbrueckii* subsp. *bulgaricus* and *Streptococcus salivarius* subsp. *thermophilus*) they may, by virtue of their different growth conditions, end up as two different biochemical products and therefore have different effects when fed to animals.

14.3 FUTURE DEVELOPMENTS

Further development of probiotic products is to some extent dependent upon the availability of an effective and reliable preparation which gives consistent positive results. It is only in this way that any development work done can be confidently related to the probiotic effect. When such a strain is available, development work may proceed along the following lines.

14.3.1 Mode of action

At present we know very little of the way in which probiotics work or the site in the gut in which they express their activity. Much of the research effort has been directed towards assessing the effect of probiotics on the composition of the microflora. The viable counts made are usually based on broad groups of organisms (e.g. lactobacilli, streptococci, total anaerobes). This type of analysis may fail to show up differences within these groups. For example, Garvie *et al.* (1984) showed that when rats were fed yoghurt there were changes at the species level. However, this type of study is time consuming and may be difficult to interpret in terms of the significance to the host. Results of more direct impact on the host are often obtained by measuring the effect of supplementation on selected metabolic parameters (see Chapter 3).

Our knowledge of gut microecology suggests that the mechanism of the probiotic effect may be in one or more of the following areas:

1. *Competition for nutrients.* This would seem to be an important factor in microbial competition in the gut (see Chapter 6).
2. *Competition for adhesion receptors on the gut epithelium* (see Chapter 11). This is an important colonization factor which can usefully be employed in selection of probiotic strains, but the effect may be too specific to explain the suppression of one organism by another.
3. *Production of antibacterial substances* (see Chapter 5). More effort should be made to demonstrate whether these substances which are

so active *in vitro* are present in the gut and if so whether they have a role in controlling the composition of the gut microflora.

4. *Stimulation of immunity* (see Chapter 7). This work, which is in its early stages, should be intensified and the relation of translocation (see Chapter 4) to effective immune stimulation should be examined.

At present the only reliable means of identifying an active organism is to do an animal field trial to see if it produces the desired result. This is a very time-consuming and expensive exercise. However, when we know how probiotics operate we shall be able to look for key biochemical features in the laboratory and select potential candidates for field trials. In the present state of our knowledge all we can do is use characters like epithelial adhesion and resistance to acid to maximize the colonization potential of the strains being used but with no real confidence that this is necessarily related to the manifestation of the probiotic effect.

14.3.2 Genetic manipulation

The techniques of genetic engineering would on the face of it allow us to have unlimited access to new strains (see Chapter 8). However, we should remember that although genetic material can be transferred from cell to cell, not all of the genetic information will be expressed, and if expressed may not remain stable under the conditions of large-scale fermentation or subsequently when growing in the gut. Moreover, there are strict guidelines which limit the release of genetically modified organisms into the environment. While these factors may limit the influence of genetic manipulation on the development of new probiotics, the potential of this technique is enormous and cannot be ignored.

When we have detailed information on the mode of action of probiotics we may (depending on its nature) be able to introduce the probiotic effect into an organism which permanently colonizes the intestinal tract. The ability to transfer colonization factors between lactobacilii is already possible. A strain of *L. acidophilus* isolated from a pig gained the ability to adhere to mouse stomach epithelium after transformation with genomic DNA from a strain of *L. fermentum* isolated from a mouse (McCarthy *et al.*, 1988).

We may also be able to incorporate protective antigens from pathogenic bacteria into harmless intestinal commensals such as lactic acid bacteria and capitalize on their ability to stimulate the immune system (see Chapter 7).

The technique also has potential in the field of nutrition. If cellulytic enzymes could be incorporated into lactobacilli it would enable animals like poultry, which have no capacity for fibre digestion, to utilize the plant cell wall material present in the diet.

Depending on the mechanism involved in the expression of the probiotic effect it may be possible to switch on the activity of the material in the appropriate part of the gut by triggering its release in response to some feature characteristic of the environment in that region of the intestinal tract. Genetic engineering could also be used to increase resistance to acid so that probiotics could survive passage through the stomach.

Resistance to heat would also be an advantage enabling producers to include probiotics in feed without risking subsequent damage by the heat engendered during the pelleting process. However, without spore production the amount of induced heat resistance would be limited.

The use of probiotics in relief of lactase deficiency is one area where already it may be possible to use genetic engineering to advantage because in this particular case the mechanism (i.e. provision of lactase) is known.

14.3.3 Non-viable probiotics

When the biochemical basis of the probiotic activity is known, it may be possible to produce the effect by feeding the substance responsible for the activity produced by the viable supplement. The production of such a supplement would be much easier because there would be no requirement to sustain the viability of the product. It would, of course, be more expensive because an additional step, extraction, would be necessary. The yield might be subject to improvement by genetic manipulation without the attendant problem of environmental release of genetically altered viable microorganisms.

Although they are, strictly speaking, not probiotics, microbial growth stimulants can be considered here. The colonization and metabolism of probiotic organisms may be stimulated by chemical supplements. The factors which stimulate the growth of bifidobacteria are a particular example (see Chapter 3). If a synergistic effect between growth factors and the probiotic effect can be established, this may be a way of improving the response of animals to probiotic organisms.

14.3.4 Improved quality assurance

The claims made for probiotic preparations must be accurate and stricter regulations are required to ensure this. The viability should be sustained throughout the stated shelf-life of the product. However, routine assessment of this may be difficult when a preparation contains more than one species of a particular genus.

Attention should also be paid to the correct identification of the organisms used. If a specific name is given it should be the correct one so that cross-product comparisons can be made. At the moment the standard of microbial taxonomy leaves a lot to be desired as can be judged by the fact that names like *Streptococcus faecium*, *Streptococcus thermophilus* and *Lactobacillus bulgaricus* are still being used long after they have become invalid.

14.4 SUMMARY

The use of probiotics in animal husbandry is now an acceptable practice and is on the increase. Unfortunately many of the products have been developed in a rather empirical way. What this book has attempted to do is to provide a scientific background to the probiotic concept and show the way in which currently available probiotics have been developed (see Chapters 10–13). Hopefully it will stimulate the necessary research that will provide the basis for a more rational approach to the selection of probiotic strains in the future. The live nature of probiotics creates unique features and problems compared with antibiotics and other drugs. Future research and development may enable us to identify the biochemical features responsible for the probiotic effects and give rise to a second generation of probiotics which are non-viable.

REFERENCES

Barrow, P.A., Brooker, B.E., Fuller, R. and Newport, M.J. (1980) The attachment of bacteria to the gastric epithelium of the pig and its importance in the micro-ecology of the intestine. *J. Appl. Bacteriol.*, **48**, 147–54.

Brockett, M. and Tannock, G.W. (1981) Dietary components influence tissue associated lactobacilli in the mouse stomach. *Can. J. Microbiol.*, **27**, 452–5.

Cole, C.B. and Fuller, R. (1984) A note on the effect of host specific fermented milk on the coliform population of the neonatal rat gut. *J. Appl. Bacteriol.*, **56**, 495–8.

Conway, P.L., Gorbach, S.L. and Goldin, B.R. (1987) Survival of lactic acid bacteria in the human stomach and adhesion to intestinal cells. *J. Dairy Sci.*, **70**, 1–12.

Cook, R.L., Harris, R.J. and Reid, G. (1988) Effect of culture media and growth phase on the morphology of lactobacilli and on their ability to adhere to epithelial cells. *Curr. Microbiol.*, **17**, 159–66.

Fuller, R. (1975) Nature of the determinant responsible for the adhesion of lactobacilli to chicken crop epithelial cells. *J. Gen. Microbiol.*, **87**, 245–50.

Garvie, E.I., Cole, C.B., Fuller, R. and Hewitt, D. (1984) The effect of yoghurt on some components of the gut microflora and the metabolism of lactose in the rat. *J. Appl.Bacteriol.*, **56**, 237–45.

Gilliland, S.E. and Speck, M.L. (1977) Enumeration and identity of lactobacilli in dietary products. *J. Food Protect.*, **40**, 760–2.
Goldin, B.R. and Gorbach S.L. (1977) Alterations in fecal microflora enzymes related to diet, age, lactobacillus supplements and dimethylhydrazine. *Cancer*, **40**, 2421–6.
Goldin, B.R and Gorbach, S.L. (1984) The effect of milk and lactobacillus feeding on human intestinal bacterial enzyme activity. *Amer. J. Clin. Nutr.*, **39**, 756–61.
Hall, M.A., Cole, C.B., Smith, S.L., *et al.* (1990) Factors influencing the presence of faecal lactobacilli in early infancy. *Arch. Dis. Child.*, **65**, 185–8.
Jonsson, E. (1986) Persistence of *Lactobacillus* strain in the gut of suckling piglets and its influence on performance and health. *Swed. J. Agric. Res.*, **6**, 43–7.
McCarthy, D.M., Lin, J.H.C., Rincker, L.A. and Savage, D.C. (1988) Genetic transformation in *Lactobacillus* sp. strain 100–33 of the capacity to colonize the nonsecreting gastric epithelium in mice. *Appl. Environ. Microbiol.*, **54**, 416–22.
Mead, G.C. and Impey, C.S (1987) The present status of the Nurmi concept for reducing carriage of food poisoning salmonellae and other pathogens in poultry, in *Elimination of Pathogenic Organisms from Meat and Poultry* (ed. F.J.M. Smulders), Elsevier, Amsterdam, pp.57–77.
Neut, C., Assy Seka, N., Romond, C. *et al.* (1984) Extraction à partir du lait maternal de facteurs de croissance spécifiques à *Bifidobacterium bifidum*. *Rev. Inst. Pasteur Lyon.*, **17**, 1–11.
Pollman, D.S., Danielson, D.M. and Peo, E.R. (1980) Effects of microbial feed additives on performance of starter and growing-finishing pigs. *J. Anim. Sci.*, **51**, 577–81.
Sears, H.J. and Brownlee, I. (1952) Further observations in the persistence of individual strains of *Escherichia coli* in the intestinal tract of man. *J. Bacteriol.*, **63**, 47–57.
Tannock, G.W., Fuller, R. and Pederson, K. (1990) *Lactobacillus* succession in the piglet digestive tract demonstrated by plasmid profiling. *Appl. Environ. Microbiol*, **56**, 1310–16.
Ushe, T.C. and Nagy, B. (1985) Inhibition of small intestinal colonization of enterotoxigenic *Escherichia coli* by *Streptococcus faecium* M74 in pigs. *Zbl. Bakt. Hyg., IAbt. Orig. B.*, **181**, 374–82.
Votava, J., Kumprecht, I. and Borovan, L. (1987) Radioactive labelling of bacteria in probiotic preparations Lactiferm and Microferm and their single use in broilers. *Sbor. ved. praci VUVZ Pohorelice*, **20**, 257–76
Yuhara, T., Isojima, S., Tsuchiya, F. and Mitsuoka, T. (1983) On the intestinal flora of bottle-fed infants. *Bifidobacteria Microflora*, **2**, 33–9.

Index

www.ingramcontent.com/pod-product-compliance
Ingram Content Group UK Ltd.
Pitfield, Milton Keynes, MK11 3LW, UK
UKHW020230290926
15470UKWH00002B/462
* 9 7 8 0 4 1 2 4 0 8 5 0 2 *